Progress in Sensory Physiology

Editors: H. Autrum, D. Ottoson, E.R. Perl, R.F. Schmidt, H. Shimazu, W.D. Willis
Editor-in-Chief: D. Ottoson

Volume 2
1981. 103 figures. V, 187 pages
ISBN 3-540-10923-4

Contents: *R. Necker:* Thermoreception and Temperature Regulation in Homeothermic Vertebrates. – *G.A. Manley:* A Review of the Auditory Physiology of the Reptiles. – *C.A. Smith:* Recent Advances in Structural Correlates of Auditory Receptors.

Volume 3
W.D. Willis
Control of Nociceptive Transmission in the Spinal Cord
1982. 51 figures. VI, 159 pages
ISBN 3-540-11510-2

Contents: Introduction: Centrifugal Control of Sensory Pathways. – Behavioral Evidence for Descending Control of Nociceptive Transmission. – Pharmacology of Analgesia Due to Descending Control Systems. – Descending Control of the Flexion Reflex. – Descending Control of Spinal Cord Nociceptive Neurons. – Correlations Between the Descending Control of Spinal Cord Nociceptive Pathways and the Operation of the Analgesia Systems.

Volume 4
1983. 41 figures. V, 118 pages
ISBN 3-540-12498-5

Contents: *N. Mei:* Sensory Structures in the Viscera. – *G.R. Martin:* Schematic Eye Models in Vertebrates. – *A. Gallego:* Organization of the Outer Plexiform Layer of the Tetrapoda Retina.

Volume 5
1985. 83 figures. V, 174 pages
ISBN 3-540-15339-X

Contents: *R.C. Hardie:* Functional Organization of the Fly Retina. – *H. Machemer, J.W. Deitmer:* Mechanoreception in Ciliates. – *M. Lindauer, H. Martin:* The Biological Significance of the Earth's Magnetic Field. – *H. Bleckmann:* Perception of Water Surface Waves: How Surface Waves are Used for Prey Identification Communications.

Progress in

Sensory Physiology 6

Editors: H. Autrum D. Ottoson
E.R. Perl R.F. Schmidt H. Shimazu W.D. Willis

Editor-in-Chief: D. Ottoson

With Contributions by
T. Sato K.B. Døving J.A. Coles S. Mense

Springer-Verlag
Berlin Heidelberg New York Tokyo

Editor-in-Chief

Professor Dr. David Ottoson
Karolinska Institutet, Fysiologiska Institutionen II
Solnavägen 1, 10401 Stockholm 60, Sweden

Editors

Professor Dr. Hansjochem Autrum
Zoologisches Institut der Universität München
Luisenstrasse 14, 8000 München 2, Germany

Professor Dr. Edward R. Perl
Department of Physiology
University of North Carolina at Chapel Hill
Chapel Hill, NC 27514, USA

Professor Dr. Robert F. Schmidt
Physiologisches Institut der Universität
Röntgenring 9, 8700 Würzburg, Germany

Professor Dr. Hiroshi Shimazu
Department of Neurophysiology
University of Tokyo, Institute of Brain Research
7.3.1. Hongo, Bunkyo Ku, Tokyo, Japan

Professor Dr. William D. Willis
The Marine Biomedical Institute
University of Texas Medical Branch
Galveston, TX 77550, USA

With 86 Figures and 10 Tables

ISBN-13:978-3-642-70413-0 e-ISBN-13:978-3-642-70411-6
DOI: 10.1007/978-3-642-70411-6

Typesetting: K+V Fotosatz GmbH, Beerfelden

2121/3140-543210

Contents

Receptor Potential in Rat Taste Cells

T. Sato

Department of Physiology, Nagasaki University School of Dentistry,
7-1 Sakamoto-machi, Nagasaki 852, Japan

1 Introduction

The physiological significance of gustatory sensation has at least three aspects: (1) the maintenance of life by ingesting nutritive substances and rejecting harmful substances; (2) the control of secretion of saliva, gastric juice, and pancreatic juice containing the digestive enzymes and (3) the pleasures of foods and drinks in humans.

Mammalian taste cells monitor the foods and drinks coming into the oral cavity and discriminate their chemical properties. Although most of the foods and drinks contain many different kinds of molecules, in the case of the physiological experiments in the laboratory we usually restrict the number of pure chemicals representing the so-called four basic taste qualities in order to understand the fundamental mechanisms of taste sensation.
There are two different types of sensory cells, the primary and the secondary. The primary sensory cell, like an olfactory cell, has an afferent axon which extends directly from the cell body and reaches the central nervous system. On the other hand, the secondary sensory cell, like a taste cell, does not have a directly extended afferent axon, but makes synaptic contacts with the axon of the first-order sensory neuron. The resting membrane potential of both sensory cells changes in response to an adequate stimulus applied to the cells, which usually initiates a slow depolarization. This is called the receptor potential or generator potential. The receptor potential is the precursor of the generation of impulses in the gustatory nerve fiber through the chemical synapse between the taste cells and the fiber terminals. In this review, I describe the various properties of receptor potentials evoked by the four basic taste stimuli in rat taste cells as investigated in our experiments with an intracellular recording technique.

2 Taste Bud Cells

The taste cells on the mammalian tongue surface are situated in the taste buds of the lingual papillae. There are four types of papillae on the mammalian tongue, and the taste buds are present on the fungiform, foliate, and circumvallate papillae. The fungiform papillae in adult rats, which were used in our experiments, are cylindrical protrusions of the tongue mucosa, about 0.4 – 0.5 mm in height and 0.2 – 0.4 mm in diameter. A single taste bud lies at the top of the fungiform papilla in the rat. The bud measures 50 – 80 μm in the long axis and 40 – 60 μm in the short axis (Farbman 1965). The ultrastructure of mammalian taste buds at the electron microscope level has been studied extensively with rabbit foliate papillae by Murray and his collaborators (Murray 1973). They classified the taste bud cells into four distinct types: type I (dark) cell, type II (light) cell, type III cell, and type IV (basal) cell (Murray 1969, Murray et al. 1969). The tips of all the cell types, except those of the basal cells, reach the taste pore. The existence of the four types of taste bud cells has been confirmed in the fungiform, foliate, and circumvallate papillae in various species of mammals (Murray and Murray 1970, Murray 1971, Murray et al. 1972). It was found that the total number of cells in single taste buds of rabbit foliate papillae varies from 30 to 80. The shares of the cell types were calculated to be 55% – 75% for type I cells, 15% – 30% for type II cells, 5% – 15% for type III cells and 1% – 5% for type IV cells (Murray 1971, 1973).
Murray et al. (1969) found the classical picture of afferent synaptic contacts between the type III cells and the nerve fibers in the rabbit foliate papilla taste buds, and they considered that the type III cells may be gustatory receptors. In the region of contact between a type III cell and an afferent nerve ending, there

exists an increased density of the synaptic membranes and an aggregation of empty synaptic vesicles and dark-cored vesicles in the cytoplasm adjacent to the nerve ending. These characteristics have also been found in the taste buds of fungiform and circumvallate papillae (Murray et al. 1972). Subsequently, more detailed investigations on the taste buds of rat (Fig. 1) and mouse have been carried out by Takeda and her collaborators (Takeda and Hoshino 1975, Takeda 1976, 1977, 1981, Takeda and Kitao 1980, Takedo et al. 1982), and they con-

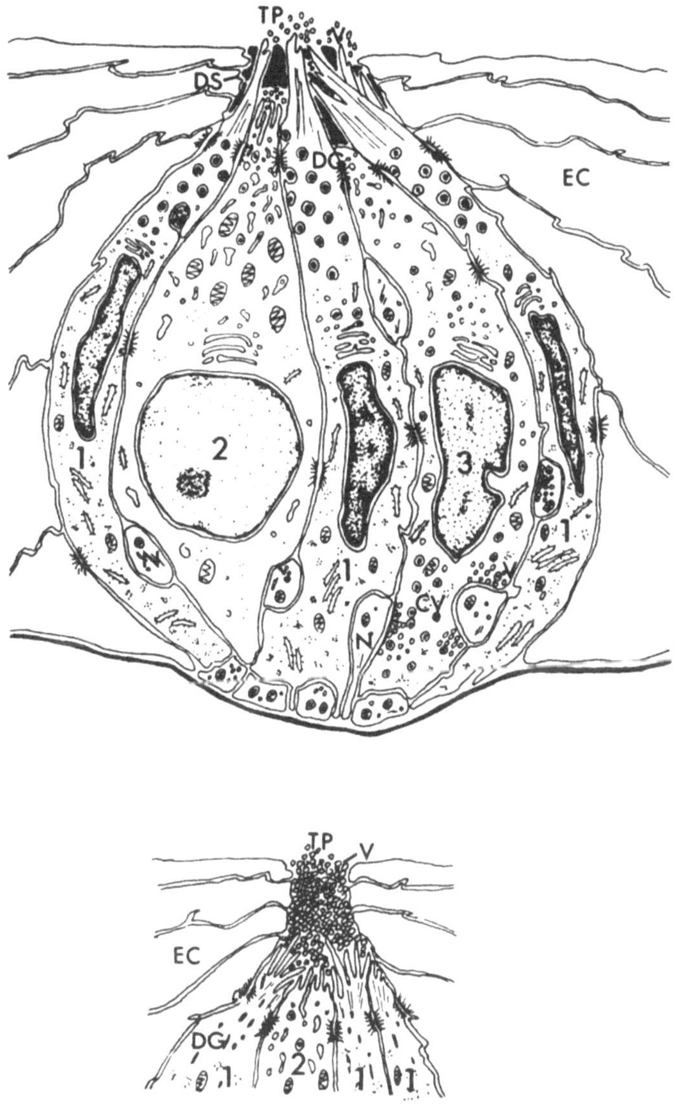

Fig. 1. Longitudinal sections of taste buds from the circumvallate papilla (*top*) and the fungiform papilla (*bottom*) in the rat. *EC*, surrounding epithelial cells; *TP*, taste pore; *DS*, dense substance; *V*, pore vesicle; *DG*, dense granules; *CV*, cored vesicles; *N*, nerves; *1*, type I cells; *2*, type II cells; *3*, type III cells (taste cells). (Takeda and Hoshino 1975)

firmed that the type III cells are gustatory cells. Takeda (1979), using autoradiography, found that type I, II, and III cells are all separately differentiated from the type IV (basal) cells. Furthermore, by means of fluorescence histochemistry and electron microscopy, Takeda and Kitao (1980) and Takeda et al. (1982) found that administration of monoamine precursors leads to an accumulation of small dense-cored vesicles at the presynaptic membrane of the afferent synapse and an increase in the number and density of large dense-cored vesicles only in type III cells. It is concluded that only the type III cells are capable of taking up and storing monoamines, which may act as neurotransmitters from the taste cell to the afferent nerve ending.

3 Resting Potential

3.1 Identification of Taste Cell

The resting potentials of taste cells of many animals have been measured with an intracellular microelectrode, e.g. frog (Sato 1969, 1972a, Esakov and Byzov 1971, Akaike et al. 1973, 1976, Sato and Beidler 1975), mudpuppy (West and Bernard 1978), catfish (Teeter and Kare 1974), rat (Kimura and Beidler 1961, Tateda and Beidler 1964, Ozeki 1970, 1971, Ozeki and Sato 1972, Sato 1976a, Sato and Beidler 1979, 1982, 1983a, b), hamster (Kimura and Beidler 1961), and mouse (Tonosaki and Funakoshi 1983). The mean resting potentials of the taste cells in these animals range from −18 to −40 mV (Sato 1980). The resting potential of taste cells in the mammalian tongue has been measured only in fungiform papillae.

When a glass capillary microelectrode is advanced into a taste bud cell from the surface of the fungiform papilla, it probably penetrates the taste cell body. The penetration of a taste cell within a taste bud or a taste disk by a microelectrode is signaled by a sudden appearance of both a negative resting potential and an effective membrane resistance, which can be measured with the aid of an electrical bridge (Ozeki 1971, Ozeki and Sato 1972, Sato 1972a, b, Sato and Beidler 1975, Akaike et al. 1976, West and Bernard 1978). The microelectrode is assumed to have penetrated a taste cell according to the following criteria: (a) a change in the resting membrane potential in either direction upon application of chemical stimuli; (b) a change in the cell membrane resistance during taste stimulation; (c) a return of the slow potential and altered resistance to the initial control level after rinsing the tongue; and (d) a disappearance of the resting potential and the membrane resistance after withdrawal of the microelectrode from the cell.

The identification and penetration of a mammalian taste cell by the microelectrode seem to be more difficult than those of a frog taste cell. To record from mammalian taste cells, the microelectrode must be advanced through the taste pore a few microns in diameter; otherwise, the microelectrode will be broken easily. On the other hand, a frog taste disk 100–400 μm in diameter almost covers the summit of the fungiform papilla and comprises mainly the cell body of supporting cells in the upper layer and that of taste cells in the lower layer (Graziadei and DeHan 1971), so that a microelectrode penetrates the taste cell through the supporting cell. The shift in the negative resting potential upon penetration occurs in two or three steps, which is part of the evidence of taste cell penetration in frog. However, in the case of the rat taste bud, cytoarchitectural studies show that the taste bud cells do not form horizontal layers composed of different cell types, but that the cells beneath a taste pore occupy the taste bud like a cluster of bananas. Therefore, the penetration by a microelectrode of the taste bud through the pore usually does not result in the appearance of a resting potential by two or three steps. Ozeki and Sato (1972) investigated extensively the electrophysiology of rat taste bud cells in the fungiform papillae. The electrophysiological responses of the taste bud cells to the application of taste stimuli suggested the presence of two kinds of cells. One cell type responds to chemical stimuli with a decrease in the membrane potential (a depolarization), whereas the other cell type does not respond to stimuli. Ozeki and Sato (1972) considered that the chemically depolarized taste bud cells may be gustatory cells, and that the chemically nondepolarized cells may be other kinds of cells such as supporting cells.

Since the type III cells in the mammalian taste bud can be identified as gustatory cells only by a careful study with an electron microscope, this cell type was long regarded as type II (Murray 1973). A single taste bud or disk cell can be stained iontophoretically with a dye-filled microelectrode that was used first to make an intracellular recording of the cell's receptor potential (Ozeki and Ōura 1972, Ozeki 1973, Sato and Beidler 1975, West and Bernard 1978); but it is very difficult at present to determine whether the stained cell on which the receptor potential was recorded is truly a gustatory (type III) cell.

According to the electrophysiological criteria used by Ozeki and Sato (1972), we can divide rat taste bud cells into two types on the basis of intracellular responses to the four basic taste stimuli: One is a taste cell which responds to taste stimuli and the other is a nontaste cell which does not respond to taste stimuli.

3.2 *Values of Resting Potential*

The resting potential of rat taste cells in fungiform papillae was first measured by Kimura and Beidler (1956, 1961) and subsequently by Tateda and Beidler (1964). The resting potential is −30 to −50 mV when the tongue surface is adapted to water. Ozeki and Sato (1972) obtained in 120 rat taste cells a mean resting potential of −40 mV, with a range of −19 to −85 mV, when the tongue surface was adapted to 41.4 m*M* NaCl, which is equivalent to the NaCl concentration of the rat's resting saliva (Hije 1969). It is known that the magnitude of the resting potential in a taste cell is dependent on the adapting solution (Akaike et al. 1976). Adapting the tongue to a lower concentration of NaCl results in a higher magnitude of resting potential. Thus, resting potential values of taste cells can be compared only under the condition that the same adapting solution is employed. Sato and Beidler (1982) measured carefully the resting potential of rat taste cells adapted to two different solutions, distilled water and 41.4 m*M* NaCl. The mean resting potential was -36 ± 1 mV (SE, $n = 78$) with a range of 20 to −60 mV in 41.4 m*M* NaCl solution, and -50 ± 3 mV (SE, $n = 22$) with a range of −20 to −70 mV in distilled water. In frog taste cells it has been suggested that the resting potential may be related to the diffusion potential of K^+ and Na^+ ions across the membrane of the taste cell (Sato and Sugimoto 1978). It is considered that the ionic mechanism for initiation of resting potential in rat taste cells is similar to that in frog taste cells.

4 Physiological Properties of Receptor Potential

4.1 Shape

In frog taste cells, Sato and Beidler (1975) found that even when a relatively high concentration of a stimulus is applied to the tongue, two shapes of receptor potentials, depolarizing and hyperpolarizing, appear depending on the kind of taste cell. The curve forms of receptor potentials of rat taste cells elicited by the four basic taste stimuli under either 41.4 m*M* NaCl or water adaptation fall largely into three types: depolarization alone (type A); depolarization preceded by an initial phasic hyperpolarization (type B); and hyperpolarization alone (type C) (Fig. 2). More strictly, the type A response sometimes includes a depolarization followed by an after-hyperpolarization, the type B response includes a depolarization preceded by a large initial hyperpolarization, and the type C response includes a hyperpolarization followed by an after-depolarization. The distribution of the three shapes of receptor potentials elicited by the four basic taste

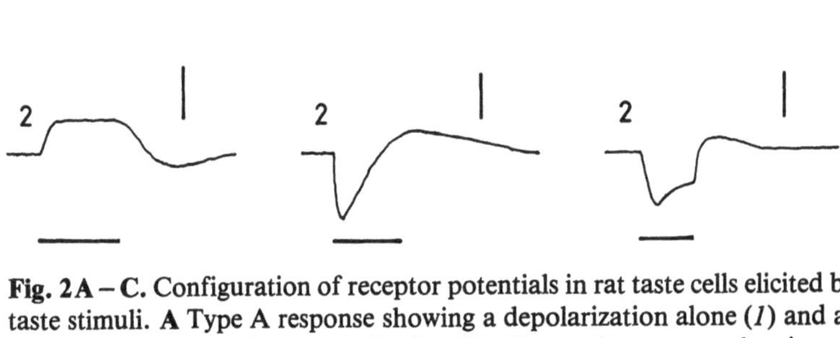

Fig. 2A – C. Configuration of receptor potentials in rat taste cells elicited by the four basic taste stimuli. **A** Type A response showing a depolarization alone (*1*) and a depolarization followed by an after-hyperpolarization (*2*); **B** type B response showing a depolarization preceded by an initial large hyperpolarization (*1*) and a depolaization preceded by an initial large hyperpolarization (*2*); **C** type C response showing a hyperpolarization alone (*1*) and a hyperpolarization followed by an after-depolarization (*2*). Taste stimuli: **A** (*1*) 0.5 *M* NaCl, (*2*) 0.5*M* NaCl; **B** (*1*) 0.01 *M* HCl, (*2*) 0.5 *M* sucrose; **C** (*1*) 0.02 *M* Q-HCl, (*2*) 0.01 *M* HCl. *Horizontal bars* under each record denote duration of taste stimulus application. *Vertical bars* denote 10 mV. The adapting and rinsing solution was a 41.4 m*M* NaCl solution. All records were obtained from different taste cells. (Sato and Beidler 1982)

Table 1. Distribution of three types of rat taste cell responses to four basic taste stimuli

Taste stimuli	Taste cell response							
	Under 41.4 m*M* NaCl adaptation				Under water adaptation			
	Type A (%)	Type B (%)	Type C (%)	No. of cells	Type A (%)	Type B (%)	Type C (%)	No. of cells
0.5 *M* NaCl	97	0	3	38	100	0	0	18
0.02 *M* Q-HCl	28	28	44	36	72	6	22	18
0.01 *M* HCl	27	51	22	37	66	6	28	18
0.5 *M* sucrose	38	21	41	34	63	6	31	16

Totals of the three types of responses elicited by each stimulus under each adapting solution are taken as 100%. (Sato and Beidler 1982)

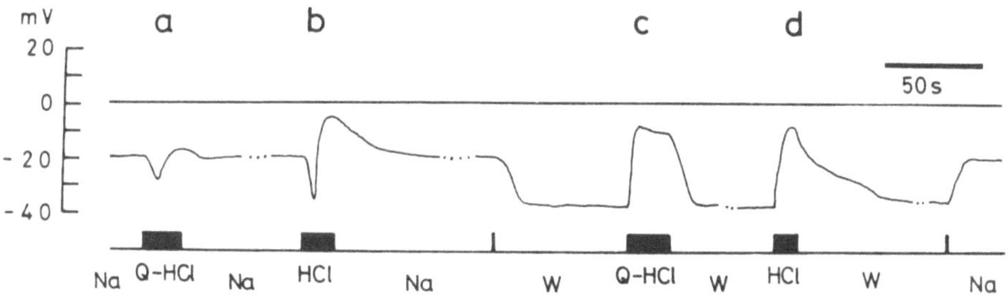

Fig. 3. Effect of adapting solution on the shape and amplitude of rat taste cell responses to stimulation with 0.02 *M* Q-HCl (*a* and *c*) and 0.01 *M* HCl (*b* and *d*). *Heavy bars* show duration of stimulus application. Adapting and rinsing solution: 41.4 m*M* NaCl (*Na*) and distilled water (*W*). (Sato and Beidler 1982)

stimuli is shown in Table 1. Nearly all responses to 0.5 *M* NaCl are of type A under either 41.4 m*M* NaCl or water adaptation. The percentages of type A responses to 0.02 *M* quinine-HCl (Q-HCl), 0.01 *M* HCl, and 0.5 *M* sucrose under water adaptation are much higher than of type A responses to the same stimuli under 41.4 m*M* NaCl adaptation. On the other hand, the percentages of type B responses to the three taste stimuli under water adaptation are much lower than those of type B responses under 41.4 m*M* NaCl adaptation. The shapes and magnitudes of receptor potentials in a rat taste cell obtained under adaptation to the two different solutions are shown in Fig. 3. Under 41.4 m*M* NaCl adaptation, the resting potential in this cell is −20 mV, and the response to 0.02 *M* Q-HCl or 0.01 *M* HCl is of type B (records a and b). As soon as the 41.4 m*M* NaCl adapting solution is replaced with water, the resting potential level reaches −35 mV, and the response shape elicited by 0.02 *M* Q-HCl or 0.01 *M* HCl changes into type A (records c and d). The amplitudes of the depolarizing potentials obtained in response to Q-HCl and HCl under 41.4 m*M* NaCl adaptation are smaller than those under adaptation to water.

The following experimental results mentioned in later sections strongly support the assumption that the intracellular recordings in our experiments were made from gustatory receptor cells: (1) similarity in the relationships between stimulus concentration and response magnitude in rat taste cells (Fig. 4) and nerves; (2) similarity in the ratios of response magnitude evoked by the four basic taste stimuli in rat taste cells and nerves (Table 9); and (3) after change of the adapting solution, a parallel change in either the dose-response curves or the response rates in rat taste cells and nerves (Figs. 4, 17, and 18; Tables 8 and 9).

4.2 Relation with Stimulus Concentration

Sato (1980) mentioned that the relationships between taste stimulus concentrations and receptor potential magnitudes in a taste cell are dependent on the taste stimuli, the adapting and rinsing solutions, and the solvents in which taste substances are dissolved. With rat taste cells which are adapted to water and rinsed with it after stimulation, the magnitude of depolarization in response to a taste stimulus in water increases as the concentration is raised above the threshold concentration, but no change in the membrane potential is evoked by a taste stimulus below the threshold (Kimura and Beidler 1961, Tateda and Beidler 1964). Ozeki and Sato (1972) examined the relation between stimulus concentration and membrane potential in rat taste cells adapted to 41.4 m*M* NaCl; they found that stimulus concentrations above threshold give rise to a depolarization, but those below threshold give rise to a hyperpolarization in some cells. Ozeki and Sato (1972) showed that the sensitivities and thresholds of individual rat taste cells to a stimulus differ from one cell to another. The thresholds of rat taste cells for NaCl, sucrose, Q-HCl, and HCl lie at 0.05 – 0.1 *M*, 0.01 – 0.05 *M*, 0.001 – 0.005 *M*, and 0.0005 – 0.001 *M*, respectively. Sato and Beidler (1982) also examined the relationships between taste stimulus concentrations and receptor potential amplitudes with rat taste cells adapted to either 41.4 m*M* NaCl or water. Three examples of relatively sensitive cells exposed to the taste stimuli NaCl (A), Q-HCl (B), and HCl (C) are shown in Fig. 4. The taste cell responses under water adaptation show depolarizations over a wide range of stimulus concentrations, unlike the responses under 41.4 m*M* NaCl adaptation. The slope of the dose-response curve under water adaptation is less than under 41.4 m*M* NaCl adaptation. The threshold concentration for initiation of depolarizing receptor potentials is 1.0 – 1.3 log units smaller under water adaptation than under 41.4 m*M* NaCl adaptation.

In rat tongue adapted to water, the threshold concentrations that initiate responses of the whole chorda tympani nerve lies at 0.0003 – 0.0005 *M* for NaCl, 0.00001 – 0.00003 *M* for Q-HCl, 0.0001 – 0.0003 *M* for HCl, and 0.01 – 0.03 *M* for sucrose (Pfaffmann et al. 1967, Yamada 1980). There are no experimental data concerning the thresholds of the chorda tympani responses of rat tongue adapted to 41.4 m*M* NaCl. As shown in Fig. 4, thresholds for initiating depolarizing responses of water-adapted rat taste cells to NaCl, Q-HCl and HCl are 0.001 *M*, 0.00001 *M*, and 0.00003 *M*, respectively. These values are very similar to those of the chorda tympani responses obtained by Pfaffmann et al. (1967) and Yamada (1980).

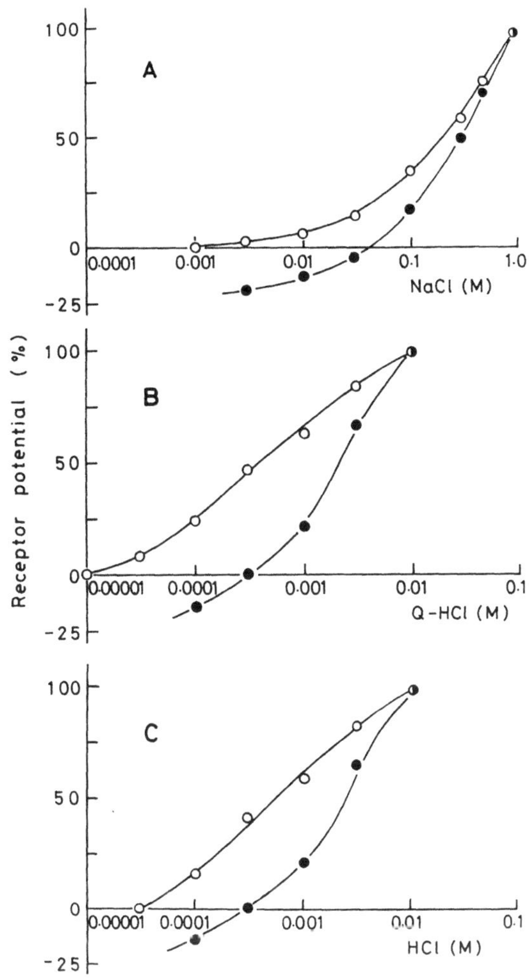

Fig. 4A – C. Relations between taste stimulus concentration and receptor potential amplitude in rat taste cells under adaptation to 41.4 m*M* NaCl solution (●) and distilled water (○). The magnitudes of the receptor potentials are expressed as percent of the maximum response obtained with each adapting solution. Taste stimuli: **A** NaCl, **B** Q-HCl and **C** HCl. Each graph is from a different taste cell. (Sato and Beidler 1982)

Since the relationships between stimulus concentrations and taste all depolarizations evoked by the four basic taste stimuli are almost the same as the relationships between stimulus concentrations and gustatory neural responses (Pfaffmann et al. 1967, Yamada 1980), the depolarizations in the rat taste cells are regarded as an excitatory receptor potential. On the other hand, as mentioned in later sections, the ratio of gustatory neural responses initiated by the four basic taste stimuli coincides only with the ratio of taste cell responses composed of both depolarizations and hyperpolarizations (Table 9); thereforc, the hyperpolarizations in the rat taste cells evoked by the four basic taste stimuli are regarded as an inhibitory receptor potential.

4.3 Time Course

Since the work of Kimura and Beidler was published in 1961, it has been accepted that the receptor potential appearing in a taste cell in response to a chemical shows a depolarization sustained during a period of the stimulation. The rise times of the depolarizing receptor potentials in a taste cell and of the gustatory neural responses in frogs are known to change as a function of the flow rate of the taste stimulus (Sato 1976b). As the flow rate is increased, the rise times of both the taste cell receptor potential and the gustatory neural response decrease exponentially.

Kimura and Beidler (1961) very slowly applied various taste stimuli, including the four basic taste qualities, to rat taste cells and obtained rise times of 5 – 15 s for the depolarizing responses; the exact flow rate of the taste solutions was not stated in that report. With a flow rate of 1.2 ml/min, Ozeki and Sato (1972) recorded rise times of 5 – 10 s for depolarizations of rat taste cells in response to 0.5 *M* NaCl, 0.5 *M* sucrose, and 0.02 *M* Q-HCl, but a long rise time of 40 – 50 s in response to 0.01 *M* HCl. Sato and Beidler (1982) examined in addition the rise and fall times of rat taste cell responses, depolarizing and hyperpolarizing, by applying taste stimuli at the rate of 1.4 ml/min. Under 41.4 m*M* NaCl adaptation, the mean times to the peak after the onset of a depolarizing or hyperpolarizing receptor potential in response to the four basic taste stimuli are shown in Fig. 5.

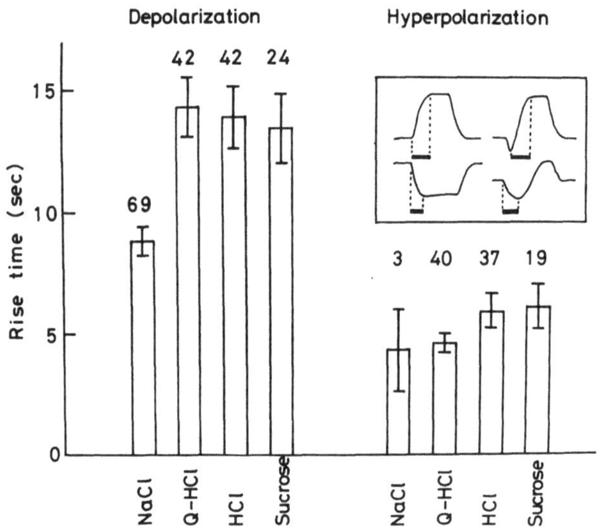

Fig. 5. Rise times of receptor potentials of rat taste cells evoked by application of the four basic taste stimuli (0.5 *M* NaCl, 0.02 *M* Q-HCl, 0.01 *M* HCl, 0.5 *M* sucrose). *Vertical bars* denote SE. *Numerals over columns* denote the number of taste cells sampled in each case. *Inset* shows the time intervals measured as rise times in depolarizing (*upper curves*) and hyperpolarizing (*lower curves*) responses. In the case of type B responses (as shown in Fig. 2B), the rise times are obtained from both the initial hyperpolarizing response and the following depolarizing response. Adapting solution: 41.4 m*M* NaCl. (Sato and Beidler 1982)

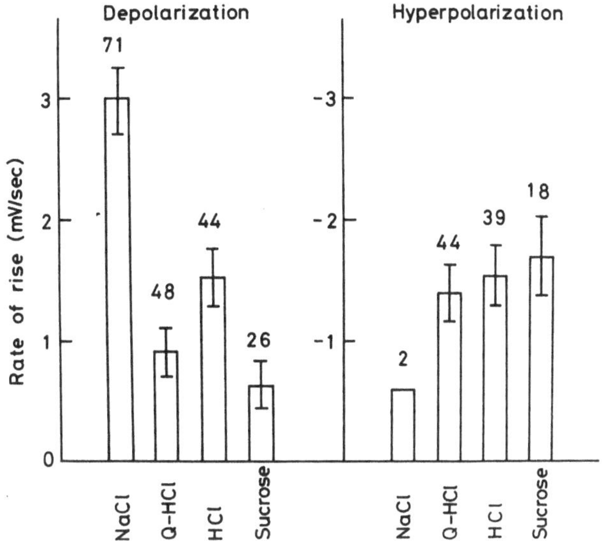

Fig. 6. Rates of rise of receptor potentials of rat taste cells evoked by application of the four basic taste stimuli (0.5 *M* NaCl, 0.02 *M* Q-HCl, 0.01 *M* HCl, 0.5 *M* sucrose). The rate of rise is obtained by dividing the peak amplitude of a receptor potential by its rise time. In the case of type B responses (as shown in Fig. 2B), the rates of rise are obtained from both the hyperpolarizing and the depolarizing components of a receptor potential. Adapting solution: 41.4 m*M* NaCl. *Horizontal bars* denote SE. *Numerals over columns* denote the number of taste cells sampled in each case. (Sato and Beidler 1982)

In the case of depolarizing responses, the rise time of 8.9 s for 0.5 *M* NaCl is much shorter than the rise times of 13.5 – 14.5 s for 0.02 *M* Q-HCl, 0.01 *M* HCl, and 0.5 *M* sucrose. In the case of hyperpolarizing responses, the mean rise times for the four taste stimuli range from 4.3 to 6.1 s, thus showing no remarkable differences. Generally, both the depolarization and the hyperpolarization responses elicited by each of the four basic taste stimuli tend to show a decreasing rise time with increasing response magnitude.

The rates of rise of depolarizations and hyperpolarizations in response to the four basic taste stimuli are illustrated in Fig. 6. The taste stimuli in the concentrations given produced decreasing rates of rise of depolarizations in the order: NaCl > HCl > Q-HCl = sucrose; with respect to the rates of rise of hyperpolarizations, the order is: sucrose = HCl = Q-HCl > NaCl. The relationships between the rates of rise and the amplitudes of receptor potentials in rat taste cells are illustrated by the examples of taste stimulation with 0.5 *M* NaCl and 0.01 *M* HCl in Fig. 7. It is seen that the rate of rise of receptor potentials generally increases with increasing amplitude. The mean rates of rise of the responses to the four basic taste stimuli, whose magnitudes are ± 10 mV, are shown in Table 2. It is apparent that the absolute values of the rates of rise of hyperpolarizing receptor potentials in response to 0.02 *M* Q-HCl, 0.01 *M* HCl, and 0.5 *M* sucrose are larger than those of the depolarizing receptor potentials in response to these stimuli.

The magnitude of the receptor potential in a taste cell usually decreases gradually with time after attaining its peak (Ozeki 1971, Sato 1972a, b). As shown by Sato

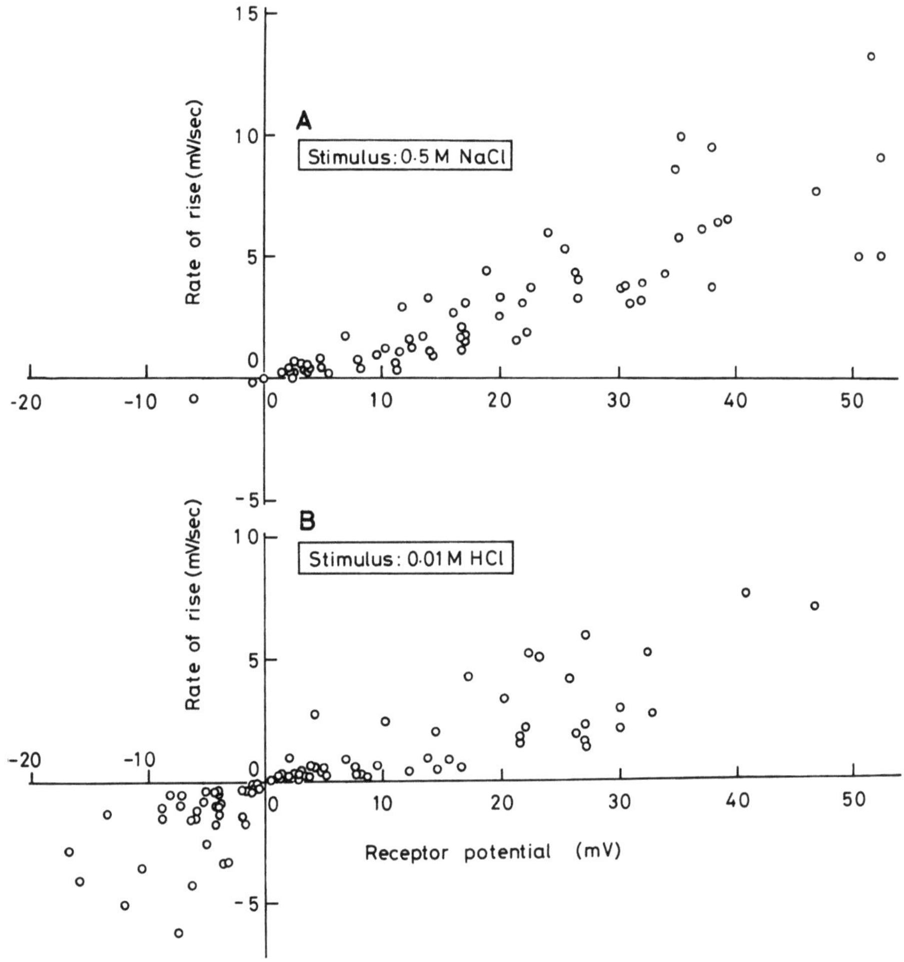

Fig. 7A, B. Relationships between the rates of rise and the magnitudes of receptor potentials in rat taste cells. Taste stimuli: **A** 0.5 *M* NaCl and **B** 0.01 *M* HCl. Adapting solution: 41.4 m*M* NaCl. (Sato and Beidler 1982)

(1972a, b), the mean rate of decrease in the depolarizations of frog taste cells evoked by 0.5 *M* NaCl and 0.016 *M* acetic acid is slower than that in the responses evoked by 0.004 *M* Q-HCl and 0.25 *M* sucrose: The magnitudes of the receptor potentials induced by continuous stimulation with NaCl and acetic acid decreased to 60% – 70% of the peak magnitude within 30 s after the peak time; but the magnitudes of those induced by continuous stimulation with Q-HCl and sucrose decreased to about 40% during the same time.

In the rat taste cell, the rate of decrease in the receptor potential elicited by prolonged stimulation with 0.3 *M* NaCl or 0.5 *M* sucrose is much faster than those produced by continuous stimulation with 0.01 *M* HCl and 0.02 *M* Q-HCl. Further, during NaCl stimulation, the rate of decrease in the receptor potential accelerates with increasing concentration (Ozeki 1971).

Table 2. The rates of rise of receptor potentials of rat taste cells in response to four basic taste stimuli

Taste stimuli	Rate of rise (mV/s)	
	At 10 mV depolarization	At 10 mV hyperpolarization
0.5 *M* NaCl	1.7	−1.5
0.02 *M* Q-HCl	0.8	−3.5
0.01 *M* HCl	1.3	−3.5
0.5 *M* sucrose	1.2	−2.5

The tongue was adapted to 41.4 m*M* NaCl. (Sato and Beidler 1982)

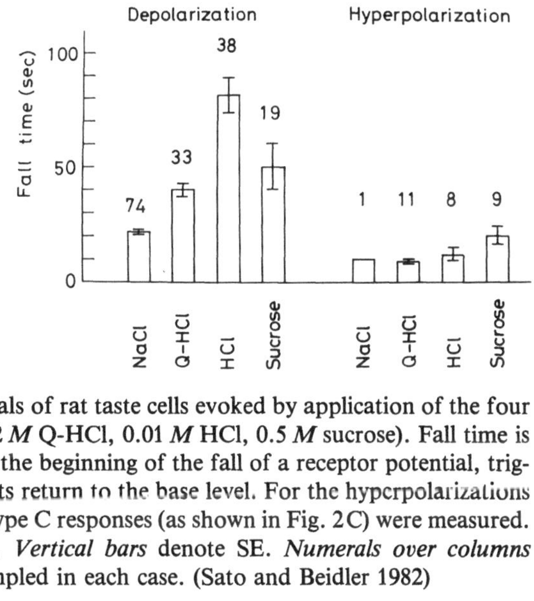

Fig. 8. Fall times of receptor potentials of rat taste cells evoked by application of the four basic taste stimuli (0.5 *M* NaCl, 0.02 *M* Q-HCl, 0.01 *M* HCl, 0.5 *M* sucrose). Fall time is defined as the time interval between the beginning of the fall of a receptor potential, triggered by a rinse of the tongue, and its return to the base level. For the hyperpolarizations in this figure, only the fall times of type C responses (as shown in Fig. 2C) were measured. Adapting solution: 41.4 m*M* NaCl. *Vertical bars* denote SE. *Numerals over columns* denote the number of taste cells sampled in each case. (Sato and Beidler 1982)

The fall times of rat taste cell responses to the four basic taste stimuli are shown in Fig. 8. Fall time is defined as the time interval between the start of a fall of the receptor potential, triggered by a rinse of the tongue, and its return to the state of a resting membrane potential. In these experiments the tongue was adapted to 41.4 m*M* NaCl, stimulated for 20 – 30 s with the taste stimulus, and rinsed with the same adapting solution. For depolarizations, the order of taste stimuli according to the length of fall times they produce is 0.01 *M* HCl > 0.5 *M* sucrose > 0.02 *M* Q-HCl > 0.5 *M* NaCl; for hyperpolarizations, the order is 0.5 *M* sucrose > 0.01 *M* HCl = 0.02 *M* Q-HCl = 0.5 *M* NaCl. For either depolarizing or hyperpolarizing receptor potentials, the fall time is much longer than the rise time. The fall time of HCl-induced depolarizations is very long. This was noticed by Ozeki and Sato (1972). The fall times of depolarizing and hyperpolarizing responses of a rat taste cell elicited by each stimulus increase with in-

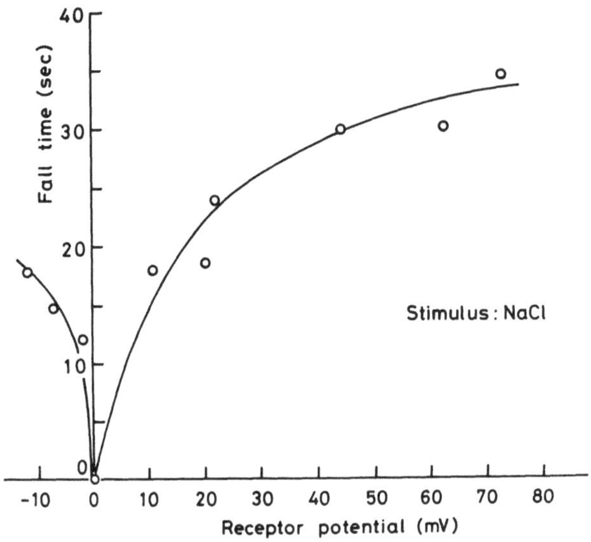

Fig. 9. Changes in the fall times of receptor potentials of a rat taste cell as a function of receptor potential magnitude. Stimulus: 0.003 – 2.0 *M* NaCl. Adapting solution: 41.4 m*M* NaCl. The data are from one cell. (Sato and Beidler 1982)

creasing amplitude. For example, the fall times of responses to NaCl stimulation are illustrated in Fig. 9. At an amplitude of depolarizing receptor potentials of 20 mV, the mean fall times for the four basic taste stimuli were as follows: 25 s for 0.5 *M* NaCl, 60 s for 0.02 *M* Q-HCl. 120 s for 0.01 *M* HCl, and 110 s for 0.5 *M* sucrose.

In a rat taste cell adapted to 41.4 m*M* NaCl, stimulation with water produced a hyperpolarizing response with a rise time of 6.8 ± 1.0 s (mean $\pm$SE, $n = 7$). When that rat taste cell was rinsed with the adapting NaCl solution, the hyperpolarization returned to the base line with a fall time of 12.2 ± 0.4 s (mean $\pm$SE, $n = 4$). In this case, the rise time is similar to those produced by any of the four basic taste stimuli, and the fall time is also similar to those produced by 0.5 *M* NaCl, 0.02 *M* Q-HCl, and 0.01 *M* HCl, but not that produced by 0.5 *M* sucrose stimulation.

5 Membrane Resistance

5.1 Input-Resistance and Current-Voltage Relation in Resting Taste Cell

The input resistance of a taste cell before, during, and after gustatory stimulation can be determined by measuring the voltage drop across the cell membrane in response to constant current pulses applied through an intracellular recording electrode by means of a bridge circuit (Ozeki 1970, 1971, Sato and Beidler 1975, 1982, Akaike et al. 1976). The mean input resistance of taste cells at rest in frog (Sato and Beidler 1975, Akaike et al. 1976), mudpuppy (West and Bernard 1978), and catfish (Teeter and Kare 1974) ranges from 17 to 42 MΩ, whereas that in rat taste cells is 81 MΩ (Ozeki 1971, Ozeki and Sato 1972, Sato 1973). The magnitude of the input resistance of a taste cell is dependent on the kind of adapting solution applied to the tongue surface.

The relationships between injected current and change in the membrane potential in taste cells are mostly linear, indicating that the taste cell membrane behaves in an ohmic manner (Ozeki 1971, Sato and Beidler 1975, Akaike et al. 1976, West and Bernard 1978). A number of investigations have show that regenerative, spike-like potentials are not usually generated in a taste cell during a period of stepwise current injection into the cell (Sato 1980). However, Kashiwayanagi et al. (1983) found that regenerative anode-break potentials are elicited in frog taste cells at the termination of large hyperpolarizing currents. They did not mention whether regenerative, spike-like potentials were produced during a period of depolarizing current injection. In their studies, no spike potentials were observed during chemical stimulation of frog taste cells with 1 m*M* HCl and 1 *M* NaCl. Recently, data quite different from the earlier findings (Sato 1980) have been offered by Roper (1983), who found that taste cells of mudpuppy can produce regenerative spike potentials during the injection of a depolarizing current into the cell. I think that there are some questions about this report; one is the extraordinarily high value of input resistance in a resting taste cell, which on the average was 200 MΩ. A much smaller value of the input resistance would be expected because of the large diameter of the mudpuppy taste cell. In these experiments, taste solutions were not applied to the taste cells. It would be important to learn whether the taste cell can produce regenerative impulses.

5.2 Relation Between Receptor Potential and Resistance Change

The mean membrane resistance changes of rat taste cells under 41.4 m*M* NaCl adaptation in response to the four basic taste stimuli and to 0.1 *M* $CaCl_2$ are given in Table 3. The values were measured at the peaks of the receptor potentials and are expressed as percent of the input resistance magnitudes in the unstimulated state. The mean amplitudes of receptor potentials in response to all taste stimuli were positive. However, the input resistance decreased during the NaCl and $CaCl_2$ stimulation, but increased during the Q-HCl, HCl, and sucrose stimulation. The relationships between the magnitudes of receptor potentials and

Table 3. Change in input resistance of rat taste cells at the peaks of receptor potentials elicited by four basic taste stimuli

Taste stimuli	Resistance (%)[a]	Receptor potential (mV)[b]	No. of cells
0.5 *M* NaCl	62 ± 4	25 ± 2	44
0.1 *M* $CaCl_2$	73 ± 8	24 ± 5	11
0.02 *M* Q-HCl	178 ± 16	9 ± 2	42
0.01 *M* HCl	148 ± 8	11 ± 2	34
0.5 *M* sucrose	138 ± 12	3.4 ± 2.5	26

[a] The values (means ± SEs) are expressed as percent of control input resistance under 41.4 m*M* NaCl adaptation

[b] The values are means ± SEs under 41.4 m*M* NaCl adaptation. (Sato and Beidler 1982)

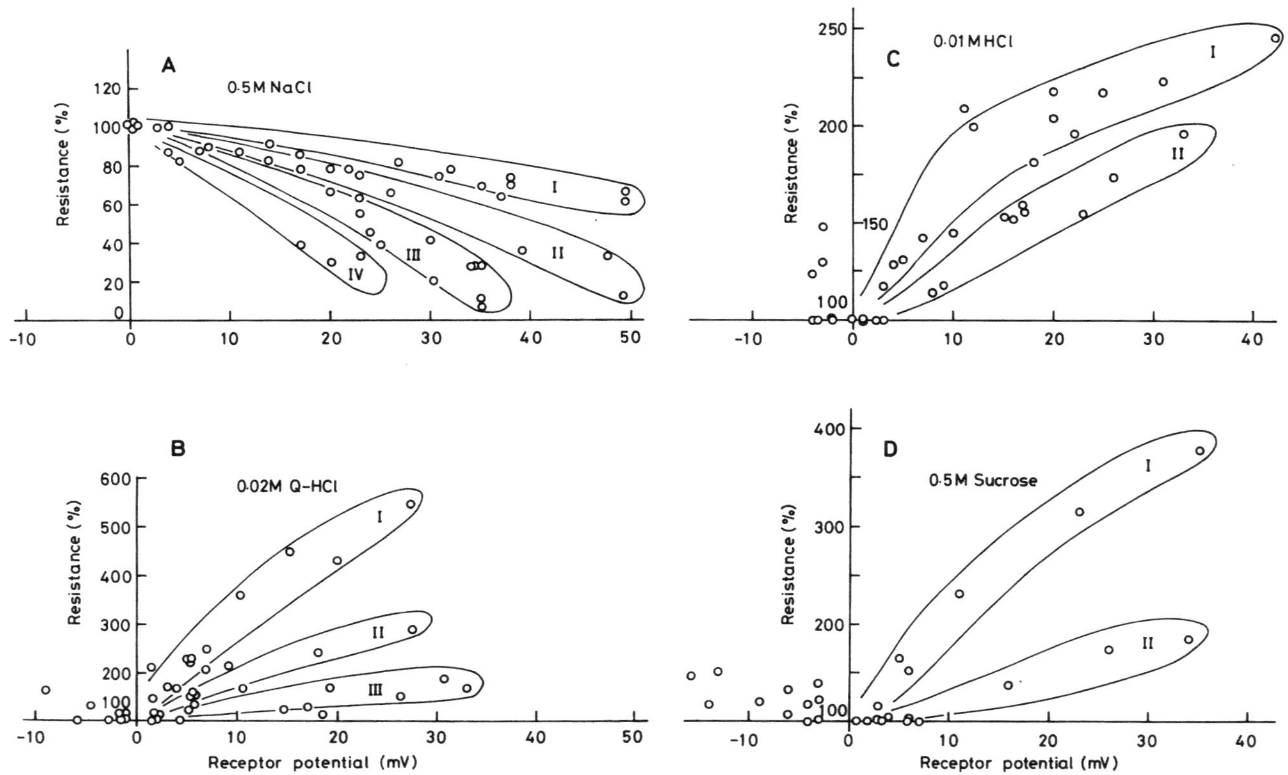

Fig. 10 A – D. Relationships between the magnitude of receptor potentials and the magnitudes of input resistances of rat taste cells. The input resistance of taste cells during taste stimulation is expressed as percentage of the input resistance of the cells at rest under 41.4 m*M* NaCl adaptation. Taste stimuli: **A** 0.5 *M* NaCl, **B** 0.02 *M* Q-HCl, **C** 0.01 *M* HCl, and **D** 0.5 *M* sucrose. ***Elliptic curves numbered I, II, III,*** or *IV* indicate that the relationships with each stimulus can be divided into groups. (Sato and Beidler 1982)

the magnitudes of membrane resistances were examined in many rat taste cells adapted to 41.4 m*M* NaCl (Fig. 10). Stimulation with 0.5 *M* NaCl decreased the input resistances and increased the amplitudes of depolarizing receptor potentials (Fig. 10A). In contrast, stimulation with 0.02 *M* Q-HCl, 0.01 *M* HCl, and 0.5 *M* sucrose increased both the input resistances and the amplitudes of depolarizations (Fig. 10B – D). When the rat taste cells were hyperpolarized by Q-HCl, HCl, and sucrose, the membrane resistance also tended to increase with increasing magnitude of the hyperpolarizations (Fig. 10B – D). As shown by the elliptic curves enclosing the experimental points in the four graphs in Fig. 10, the relations between the input resistance and the receptor potential evoked by a gustatory stimulus can be divided into two to four groups (I, II, III, and IV), depending upon the gradients in the relations. This means that the rat taste cells can be divided into groups, depending on the sensitivity to change in resistance induced by taste stimuli.

The decrease in the input resistance of a taste cell accompanying the depolarization observed after application of concentrated NaCl solutions has also been found in other animals such as frog, mudpuppy, and catfish (Sato and Beidler 1975, Akaike et al. 1976, West and Bernard 1978, Teeter and Kare 1974). The increase in the membrane resistance of a taste cell during a depolarization in response to Q-HCl has been found in frog (Akaike et al. 1976). In contrast to our data for rat taste cells, Ozeki's experiment (1971), also with rat taste cells, shows a decrease in the input resistance during depolarizations elicited by 0.01 *M* HCl and 0.5 *M* sucrose. It is not known at present why the two results are different.

5.3 Time Course of Resistance Change During Taste Stimulation

After the onset of stimulation of a rat taste cell with 0.5 – 2.0 *M* NaCl, the membrane resistance gradually decreases simultaneously with a gradual increase of receptor potential amplitude, and the maximal resistance decrease occurs almost at the peak of the receptor potentials (Fig. 11 C). During the plateau phase of the receptor potential, the maximally decreased resistance is maintained at almost a constant level. After a rinse of the tongue with the adapting solution, the decreased resistance gradually returns to the original control level (Fig. 11 A, B). The recovery rate of changed resistance is generally slower than that of changed receptor potential. This tendency is increasingly noticeable with increasing concentrations of NaCl. Ozeki (1971) found the same recovery process in a rat taste cell. He also found that the rate of adaptation of the receptor potential amplitude during stimulation with 1 – 2 *M* NaCl is more prominent than the rate of increase in resistance originating from the maximally decreased resistance level.

In contrast, during stimulation with 0.02 *M* Q-HCl, 0.01 *M* HCl, and 0.5 *M* sucrose, the membrane resistance gradually increases in parallel with an increase in depolarization and reaches the maximum value at the peak of depolarization. The increased resistance gradually returns to the initial control level after rinsing the stimuli from the tongue. As an example, the responses to sucrose stimulation are shown in Fig. 11. After completion of a 30 – 40-s period of stimulation with sucrose, the resistance usually does not return to the control level, but remains elevated even when the magnitude of the receptor potential returns to the base

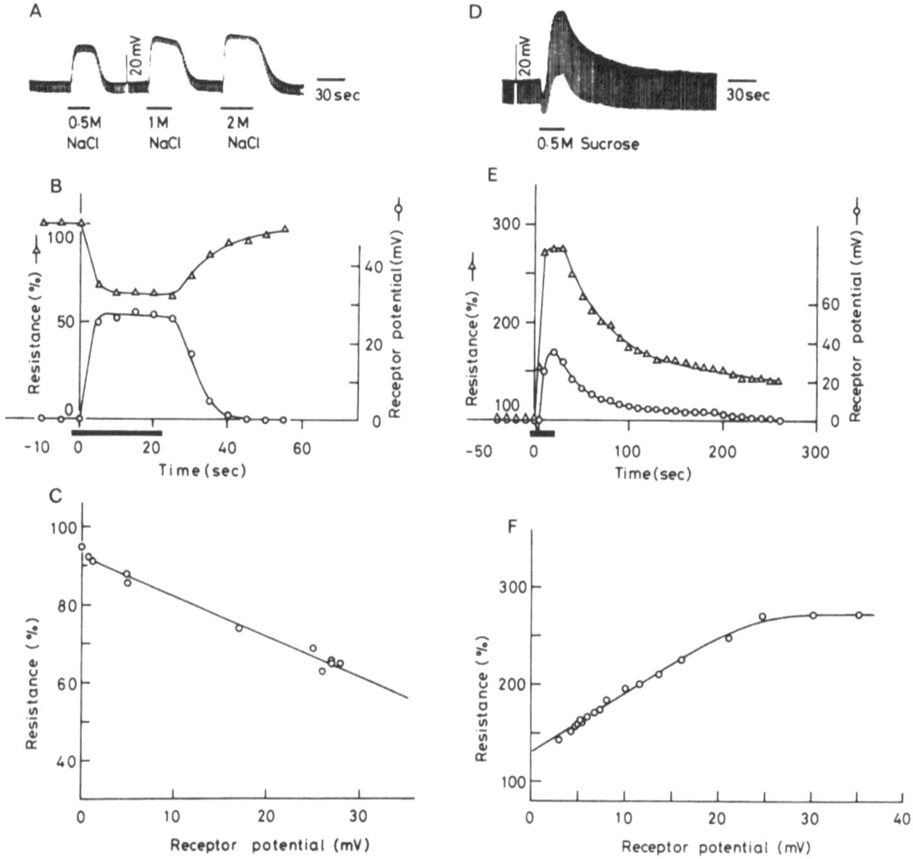

Fig. 11A – F. Relations between the time courses of receptor potential magnitudes and the time courses of input resistance magnitudes in rat taste cells. The input resistance of taste cells during taste stimulation is expressed as percent of the input resistance of the cells at rest under 41.4 m*M* NaCl adaptation. Adapting solution: 41.4 m*M* NaCl. **A** Receptor potentials induced by 0.5, 1.0, and 2.0 *M* NaCl, and, *superimposed*, electrotonic potentials induced by constant hyperpolarizing current pulses applied to the taste cell through a microelectrode. **B** Time courses of change in the magnitudes of receptor potentials produced by 0.5 *M* NaCl (○), and of change in the magnitudes of input resistance (△). The *heavy bar* under the abscissa shows the period of stimulus application. **C** Relation between the time-dependent change in the magnitudes of receptor potentials induced by 0.5 *M* NaCl and the time-dependent change in the magnitudes of input resistance. Legends for **D – F** are the same as those for **A – C**, respectively, except for application of 0.5 *M* sucrose in **D – F**. (Sato and Beidler 1982)

line (Fig. 11E). An analysis of the relation between the magnitudes of resistance and the magnitudes of depolarizing response in a rat taste cell shows that the resistance decreases linearly or exponentially with an increase in the depolarization magnitudes during NaCl stimulation (Fig. 11C), but increases with an increase of the depolarizations during stimulation with 0.02 *M* Q-HCl, 0.01 *M* HCl, and 0.5 *M* sucrose (Fig. 11F).

Table 4. Recovery of changed input resistance of rat taste cells after rinsing the taste stimulus from the tongue

Taste stimuli	Resistance (%)		
	At the peak of response	At 60 s after rinse[a] of tongue	No. of cells
0.5 *M* NaCl	66 ± 4	95 ± 1	40
0.02 *M* Q-HCl	184 ± 18	130 ± 5	39
0.01 *M* HCl	152 ± 8	139 ± 7	33
0.5 *M* sucrose	135 ± 5	108 ± 2	18

All values are means ± SEs

[a] In the case of 0.5 *M* NaCl stimulation, the values are at 30 s after rinse of the tongue. The tongue was adapted to and rinsed with 41.4 m*M* NaCl. (Sato and Beidler 1982)

Table 4 gives the relation between the values of input resistances at the peaks of depolarizations induced by the four taste stimuli and the values of the resistances 30 or 60 s after rinsing the stimuli from the tongue. The recovery of the changed input resistance in response to 0.5 *M* NaCl and 0.5 *M* sucrose is faster than that in response to 0.02 *M* Q-HCl and 0.01 *M* HCl. The order of the rates of recovery of the changed input resistances is: 0.5 *M* NaCl > 0.5 *M* sucrose > 0.02 *M* Q-HCl > 0.01 *M* HCl. The rate of recovery of the changed resistances following rinsing depends on the magnitudes of input resistance change in response to taste stimulation: the larger the input resistance change, the slower the recovery after rinsing (Fig. 12). After NaCl and Q-HCl stimulation, the recovery processes of the input resistances, initiated by rinsing of the tongue, can be divided into fast and slow types (Fig. 12A, B). On the other hand, the recovery processes of input resistances after termination of HCl and sucrose stimulation show only one type (Fig. 12C, D).

6 Mechanism of Generation of Receptor Potential

The membrane which surrounds a taste cell can be divided into at least two parts. One part is the receptor membrane or microvillus membrane, which is situated at the apex of the taste cell and makes direct contact with the chemicals coming into the oral cavity. The rest of the membrane of the taste cell may not make direct contact with taste stimuli.

The following three mechanism hypotheses have been proposed to underlie the generation of receptor potentials in taste cells: (1) After the binding of a taste stimulus to the receptor membrane of a taste cell, a receptor potential may result from a change in the permeability of the taste cell membrane to ions such as Na^+, K^+, and Cl^-, which are commonly found inside and outside the taste cell (Beidler 1954, 1961, 1967, 1971, Beidler and Gross 1971, Sato and Beidler 1975). (2) The receptor potential elicited by monovalent salt stimuli may be due to the direct penetration of monovalent ions, such as Na^+, Li^+, and Cl^-, from the

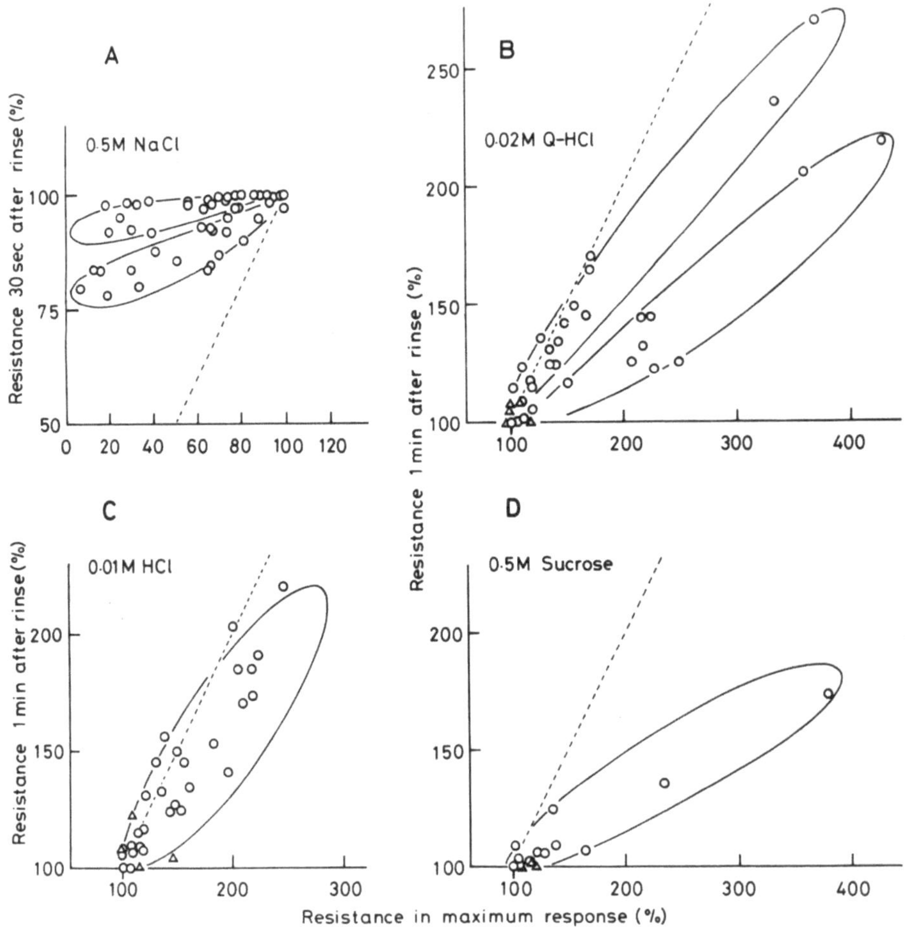

Fig. 12A–D. Recovery of changed input resistance of rat taste cells after rinsing of the tongue. Taste stimuli used: 0.5 *M* NaCl, 0.02 *M* Q-HCl, 0.01 *M* HCl, and 0.5 *M* sucrose (**A**–**D**). Adapting solution: 41.4 m*M* NaCl. The input resistance of the taste cells during taste stimulation is expressed as percent of the input resistance of the cells at rest under 41.4 m*M* NaCl adaptation. *Abscissae* are the magnitudes of input resistance at the response peak. *Ordinate* in **A** is the magnitude of input resistance 30 s after rinsing the stimulus from the taste receptor surface, and *ordinates* in **B**–**D** are those 60 s after rinsing. The resistances were recorded both in the depolarizing (○), and in the hyperpolarizing (△) responses. The *dashed lines* indicate the observed limits of recovery of the changed resistances. (Sato and Beidler 1982)

gustatory stimuli into the taste cell through the receptor membrane (Akaike et al. 1976, Akaike and Sato 1976). (3) The receptor potentials elicited by many kinds of taste stimuli may be due mostly to the change in the phase boundary potential occurring between the taste stimulus solution and the outer surface of the taste receptor membrane. This theory suggests that there are almost no changes in ionic permeability of either the receptor membrane or the cell membrane during the generation of receptor potentials in response to these stimuli (Kurihara et al. 1978).

Although not much evidence to support any of these hypotheses has yet been accumulated, analysis of each theory with intracellular recording techniques has been advanced step by step. By replacing the interstitial fluid surrounding the frog taste cells with various modified Ringer solutions, Sato and Sugimoto (1978) and Sato et al. (1982) found that when the frog taste cell is stimulated with salt stimuli, such as 0.5 *M* NaCl, 0.5 *M* NH_4Cl, and 0.1 *M* $CaCl_2$, the resultant receptor potentials are fairly associated with an increase in the permeability of the taste cell membrane to Na^+ and Ca^{2+} existing in the interstitial fluid. Since the magnitudes of the salt-induced receptor potentials remain at some 40% after the total removal of both Na^+ and Ca^{2+} from the interstitial fluid, it is likely that other mechanisms, such as the phase boundary potential occurring at the taste receptor membrane, are involved in the generation of the taste receptor potential induced by salty taste stimuli. The association of interstitial ions with the salt-stimulus-induced receptor potentials is suggested by such observations as clear decreases in the membrane resistance of a taste cell during response to salt stimuli (Ozeki 1971, Sato and Beidler 1975, 1982, Akaike et al. 1976) and the presence of a reversal potential for the depolarizing receptor potentials induced by salt stimuli (Ozeki 1971, Sato and Beidler 1975, Akaike et al. 1976).

On the basis of extensive studies on the frog taste cell responses, Akaike and Sato (1976) have proposed the following: (1) The change in the membrane potential in frog taste cells elicited by stimulation with monovalent salts originates from the diffusion potential that is due to the permeability of the taste receptor membrane to various monovalent cations and anions of stimulus salts, as well as to Na^+, K^+, and Cl^- ions commonly existing inside the taste cell and just outside the taste receptor membrane. (2) The monovalent anions of stimulus salts also modify, in the order of the lyotropic series, the permeability of the taste receptor membrane to both monovalent cations and anions. If only the diffusion of monovalent ions in taste stimuli into a taste cell through the receptor membrane is involved in the generation of a receptor potential, as has been proposed above, it is likely that the change in ion composition in the interstitial fluid surrounding the taste cells does not seriously affect the generation of receptor potential. As mentioned above, the experimental results by Sato and Sugimoto (1978) and Sato et al. (1982) do not agree with this assumption, so that the monovalent salt-induced receptor potential in the frog taste cell may not be attributed to one mechanism of ion diffusion through the taste receptor membrane. Probably, ionic permeabilities of both the receptor membrane and the basolateral membrane of a taste cell and the phase boundary potential occurring at the taste cell membrane are, to various degrees, related to the receptor potentials that are elicited by salt stimuli applied to the tongue.

To determine the mechanism by which the receptor potential in a taste cell is evoked by varying taste stimuli, it is important to study the electrical and physicochemical properties of the receptor membrane and the basolateral membrane of the taste cell individually. To this end, it is useful to perfuse both the tongue through the lingual artery and the tongue surface. Thus, the ionic compositions of either the outer surface environment of the receptor membrane or the interstitial fluid surrounding the taste cells can be changed separately during recording of intracellular receptor potentials in a taste cell. With this approach, the mechanism underlying the generation of taste cell receptor potentials has been studied recently (Okada and Sato 1982a, b, 1983, Sato and Okada 1984). These researchers have investigated the generative mechanisms of acid-stimulus-induced receptor potential in frog taste cells with both intracellular recording techniques and lingual perfusion techniques. It has been found that the amplitudes of receptor potentials in a frog taste cell in response to acid stimuli are not changed by removing almost any kind of interstitial ion surrounding the taste cell; this indicates that the acid-induced receptor potential is independent from the permeability change of the basolateral membrane of the taste cell. Removal of Ca^{2+}, Na^{+}, or both from the adapting solution covering the taste receptor membrane causes a remarkable reduction in the receptor potential, but some 10% of the control amplitude of the receptor potential still remains. Therefore, it is suggested that the acid-induced receptor potential in frog taste cells is related to both the diffusion potential and the phase boundary potential occurring at the taste receptor membrane surface.

7 Multiple Sensitivity of Taste Cell

7.1 Response Profile

The records shown in Figs. 13 and 14 are examples of receptor potentials produced when rat taste cells adapted to 41.4 m*M* NaCl (Fig. 13) or to water

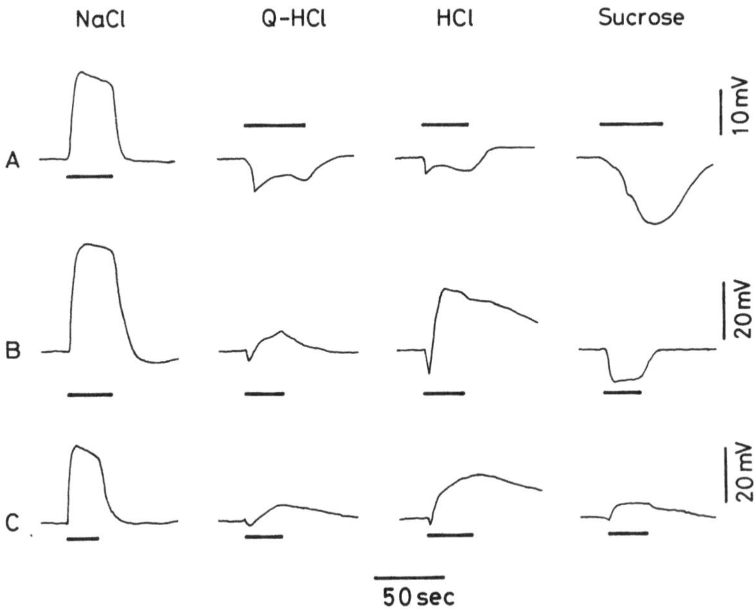

Fig. 13A – C. Intracellularly recorded receptor potentials of three rat taste cells in response to four basic taste stimuli (0.5 *M* NaCl, 0.02 *M* Q-HCl, 0.01 *M* HCl, 0.5 *M* sucrose). *Horizontal bars* under or over the records show the duration of stimulus application. The taste cells were adapted to 41.4 m*M* NaCl. (Sato and Beidler 1983a)

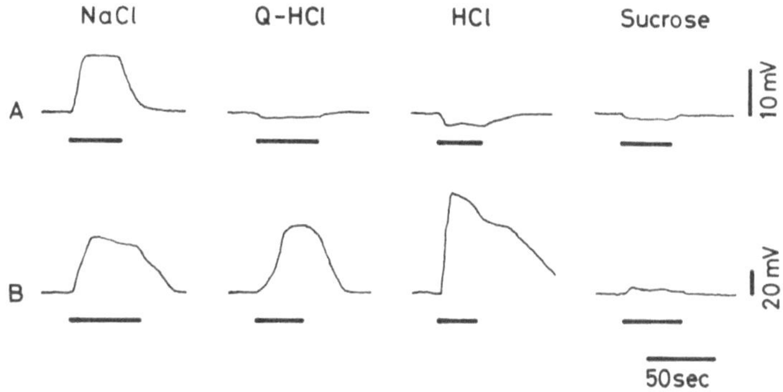

Fig. 14A, B. Intracellularly recorded receptor potentials of two rat taste cells in response to four basic taste stimuli (0.5 *M* NaCl, 0.02 *M* Q-HCl, 0.01 *M* HCl, 0.5 *M* sucrose). *Horizontal bars* under the records show the duration of stimulus application. The taste cells were adapted to water. (Sato and Beidler 1983a)

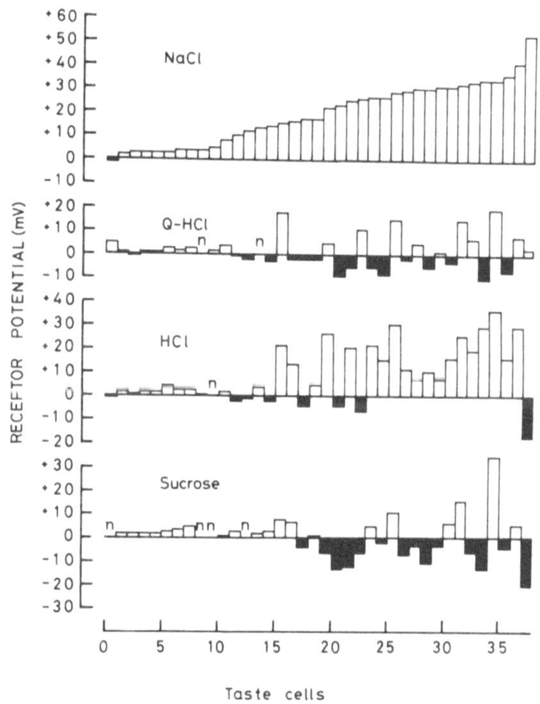

Fig. 15. Profiles of responses of 38 rat taste cells to the four basic taste stimuli (0.5 *M* NaCl, 0.02 *M* Q-HCl, 0.01 *M* HCl, 0.5 *M* sucrose). The taste cells were adapted to 41.4 m*M* NaCl. *Open columns* denote depolarizations alone or depolarizing responses preceded by an initial phasic hyperpolarization; *filled columns* denote hyperpolarizations alone. The columns are arranged in the order of taste cell responsiveness to 0.5 *M* NaCl. The symbol *n* indicates that the resting membrane potential of the taste cell was not changed by the stimulus. The mean resting potential was -29 ± 2 (SE) mV. (Sato and Beidler 1983a)

(Fig. 14) were stimulated with 0.5 *M* NaCl, 0.02 *M* Q-HCl, 0.01 *M* HCl, and 0.5 *M* sucrose. The three types of responses shown in Fig. 2 (Sato and Beidler 1982) are seen here in the taste cells, namely: (1) a depolarization alone; (2) a depolarization preceded by an initial phasic hyperpolarization; and (3) a hyperpolarization alone. Response profiles of 38 rat taste cells, adapted to 41.4 m*M* NaCl, to the four fundamental taste stimuli are shown in Fig. 15. The open columns denote the magnitudes of depolarizing receptor potentials consisting of the first two response types mentioned above, and the filled columns denote the magnitudes of hyperpolarizing receptor potentials. Most of the rat taste cells showed a multiple sensitivity in that single cells responded to various combinations of the four basic taste stimuli with depolarizations or hyperpolarizations. Of the cells represented, 32% were depolarized by all four stimuli; 26% and 32% were depolarized by three and two stimuli, respectively; and 10% were depolarized by only one stimulus. That is, 90% of the taste cells responded with depolarizations to two or more of the four basic taste stimuli.

In rat taste cells, the generative mechanism of the receptor potential has not yet been studied with the tongue perfusion techniques mentioned above. Extensive investigations of this kind with the perfused rat tongue are expected to be carried out in the future.

The profile of the responses of 19 water-adapted rat taste cells to the four basic taste stimuli (Fig. 16) shows that, in spite of adaptation of the tongue to water, a total of seven of the 19 taste cells examined responded to some of these stimuli with hyperpolarizations. This profile again indicates the multiple sensitivity of rat taste cells. The percentage of the hyperpolarizing responses to the NaCl, Q-HCl, and sucrose under 41.4 m*M* NaCl adaptation tends to be larger than that under water adaptation.

The multiple sensitivity of rat taste cells was found by Kimura and Beidler (1961) and Ozeki and Sato (1972). However, hyperpolarizing responses of rat taste cells elicited by the four basic stimuli were not reported in the earlier studies. Similar multiple sensitivities of taste cells have, however, been found in frog (Sato 1969,

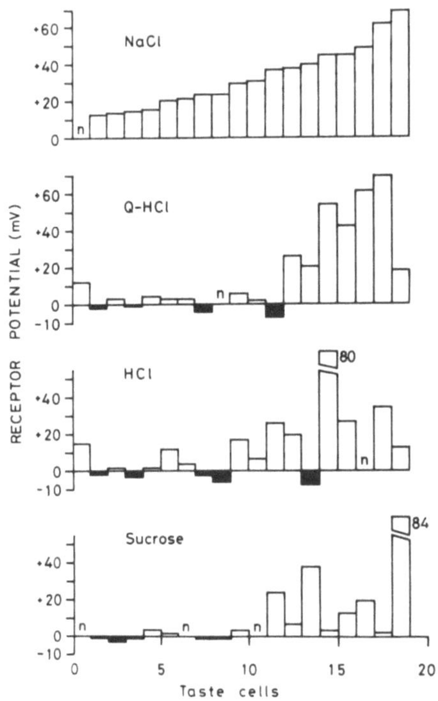

Fig. 16. Profiles of responses of 19 rat taste cells to four basic taste stimuli (0.5 *M* NaCl, 0.02 *M* Q-HCl, 0.01 *M* HCl, 0.5 *M* sucrose). The taste cells were adapted to water. *Open columns* denote depolarizations, and *filled columns*, hyperpolarizations. The columns are arranged in the order of taste cell responsiveness to 0.5 *M* NaCl. The symbol *n* indicates that there was no response. The mean resting potential was -46 ± 6 (SE) mV. (Sato and Beidler 1983a)

1972a, Akaike et al. 1973, 1976), in mudpuppy (West and Bernard 1978), and in fish (Teeter and Kare 1974). These data suggest that most of the individual taste cells in all classes of vertebrates can respond to many different kinds of taste stimuli with depolarizing and hyperpolarizing receptor potentials. On the other hand, there is the possibility that individual taste cells in some or many classes of invertebrates respond to only one kind of taste stimulus, i.e., show single sensitivity, since chemoreceptor cells of flies respond exclusively to substances representing one taste quality (Morita and Yamashita 1959, Morita et al. 1966).

7.2 Random Responsiveness

If the sensitivity of a rat taste cell to any one of the four basic taste stimuli is independent of the sensitivities to the other three stimuli, the probability of eliciting taste cell responses to any pair of the four basic taste stimuli would be given by the product of the probabilities of obtaining a response to each of the two stimuli. The response probabilities for each stimulus can be estimated by the proportion of responding taste cells to the total number of cells examined. Table 5 gives the probabilities for the occurrence of depolarizing responses in rat taste cells to all six pairs of the four basic taste stimuli, and the predicted and observed numbers of rat taste cells depolarized by the pairs. No statistical difference is seen between the predicted and observed numbers (chi-square test, $P > 0.05$). Table 6 shows the probability of occurrence of hyperpolarizing responses to the six pairs of the four taste stimuli, and the predicted and observed numbers of rat taste cells hyperpolarized by the pairs. Again, the two sets of predicted and observed numbers do not significantly differ (chi-square test, $P > 0.05$). Therefore, it is concluded that the depolarization or hyperpolarization responsiveness of rat taste cells to the four basic taste stimuli occurs randomly and independently. The correlation coefficients between the mean magnitudes of receptor potentials of 38 rat taste cells in response to the six pairs of the four basic taste stimuli are

Table 5. Depolarizations in response to all six pairs of the four basic taste stimuli in 38 rat taste cells

Pairs of taste stimuli	Probability	Number of taste cells, depolarized	
		Predicted	Observed
NaCl, sucrose	0.97 × 0.53	19.5	20
NaCl, HCl	0.97 × 0.76	28.0	29
NaCl, Q-HCl	0.97 × 0.53	19.5	19
Sucrose, HCl	0.53 × 0.76	15.3	18
Sucrose, Q-HCl	0.53 × 0.53	10.2	11
HCl, Q-HCl	0.76 × 0.53	15.3	15

The rat tongue was adapted to 41.4 m*M* NaCl. The taste stimuli used are 0.5 *M* NaCl, 0.02 *M* Q-HCl, 0.01 *M* HCl and 0.5 *M* sucrose. Probabilities for the stimuli to produce a depolarization are 0.97 (NaCl), 0.53 (Q-HCl), 0.76 (HCl), and 0.53 (sucrose). (Sato and Beidler 1983a)

Table 6. Hyperpolarizations in response to all six pairs of the four basic taste stimuli in 38 rat taste cells

Pairs of taste stimuli	Probability	Number of taste cells, hyperpolarized	
		Predicted	Observed
NaCl, sucrose	0.03 × 0.37	0.42	0
NaCl, HCl	0.03 × 0.21	0.24	1
NaCl, Q-HCl	0.03 × 0.42	0.48	0
Sucrose, HCl	0.37 × 0.21	2.95	4
Sucrose, Q-HCl	0.37 × 0.42	5.91	8
HCl, Q-HCl	0.21 × 0.42	3.35	5

The rat tongue was adapted to 41.4 m*M* NaCl. The taste stimuli used are 0.5 *M* NaCl, 0.02 *M* Q-HCl, 0.01 *M* HCl and 0.5 *M* sucrose. Probabilities for the stimuli to produce a hyperpolarization are 0.03 (NaCl), 0.42 (Q-HCl), 0.21 (HCl), and 0.37 (sucrose). (Sato and Beidler 1983a)

Table 7. Correlation coefficients between mean magnitudes of depolarizing and hyperpolarizing receptor potentials of 38 rat taste cells in response to six pairs of four basic stimuli

	0.02 *M* Q-HCl	0.01 *M* HCl	0.5 *M* sucrose
0.5 *M* NaCl	0.09	0.44[a]	−0.23
0.02 *M* Q-HCl		0.28	0.58[a]
0.01 *M* HCl			0.40[b]

All taste cells were adapted to 41.4 m*M* NaCl. (Sato and Beidler 1983a)
[a] Correlation is significant
[b] Correlation is probably significant

given in Table 7. Significant correlations are found between 0.5 *M* NaCl and 0.01 *M* HCl and between 0.02 *M* Q-HCl and 0.5 *M* sucrose. The first-mentioned significant correlation has already been reported for rat taste cells (Ozeki and Sato 1972).

8 Contribution of Depolarizing and Hyperpolarizing Receptor Potentials to Gustatory Neural Response

The mean depolarizing responses and the mean hyperpolarizing responses of rat taste cells to the four basic taste stimuli under water and 41.1 m*M* NaCl adaptation are shown in Fig. 17A. Except for 0.01 *M* HCl, the depolarizing responses under water adaptation are significantly larger than those under 41.4 m*M* NaCl. On the other hand, the magnitudes of hyperpolarizing responses to 0.02 *M*

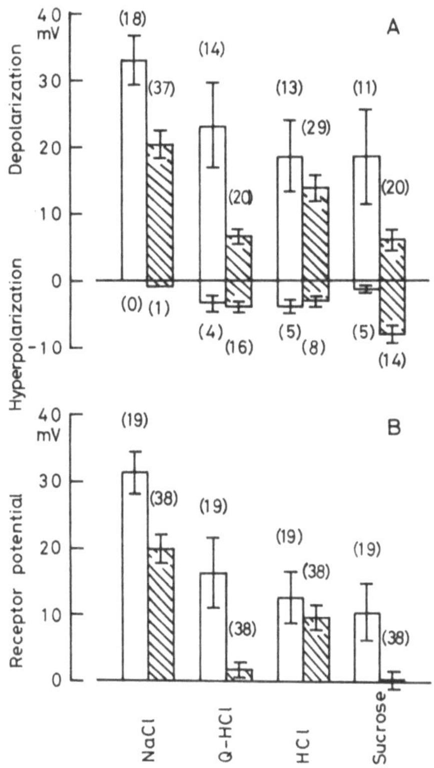

Fig. 17 A, B. Responses of rat taste cells adapted to water (*open columns*) and to 41.4 m*M* NaCl (*hatched columns*) compared to basic taste stimuli (0.5 *M* NaCl, 0.02 *M* Q-HCl, 0.01 *M* HCl, 0.5 *M* sucrose). *Vertical bars* denote SE. *Numerals in parentheses* denote the numbers of taste cells sampled. **A** The mean responses to the four stimuli, divided into depolarizations and hyperpolarizations. **B** The mean receptor potentials obtained by averaging the three kinds of responses (depolarization, hyperpolarization, and unchanged membrane potential) to the four stimuli. (Sato and Beidler 1983a)

Q-HCl and 0.01 *M* HCl stimulation are almost the same under both adapting solutions, while the magnitudes of hyperpolarizing responses to 0.5 *M* NaCl and 0.5 *M* sucrose under 41.4 m*M* NaCl adaptation are larger than those under water adaptation. To obtain Fig. 17B, the amplitudes of depolarizing and hyperpolarizing responses and the cases of no response to each taste stimulus were all combined and averaged. It is seen that the mean taste cell responses to 0.05 *M* NaCl, 0.02 *M* Q-HCl and 0.5 *M* sucrose under water adaptation are larger than those under 41.4 m*M* NaCl adaptation, while the mean responses to 0.01 *M* HCl under adaptation to either solution do not differ significantly.

An example of integrated chorda tympani nerve responses in a rat to lingual stimulation with four basic taste solutions after the tongue is adapted to water or to 41.4 m*M* NaCl is shown in Fig. 18. The mean magnitudes of the phasic and tonic components of gustatory neural responses to the four basic taste stimuli are given in Table 8. The values are expressed relative to the amplitude of the phasic response to 0.5 *M* NaCl. Both the phasic and the tonic responses to each taste stimulus under 41.4 m*M* NaCl adaptation are smaller than those under water adaptation. Particularly, the phasic and tonic components of the neural responses to 0.02 *M* Q-HCl and 0.5 *M* sucrose after adaptation to 41.4 m*M* NaCl are reduced markedly. For example, the magnitudes of the phasic responses to 0.02 *M* Q-HCl and 0.5 *M* sucrose under 41.4 m*M* NaCl adaptation are reduced to 18% and 10%, respectively, of the respective controls under water adaptation.

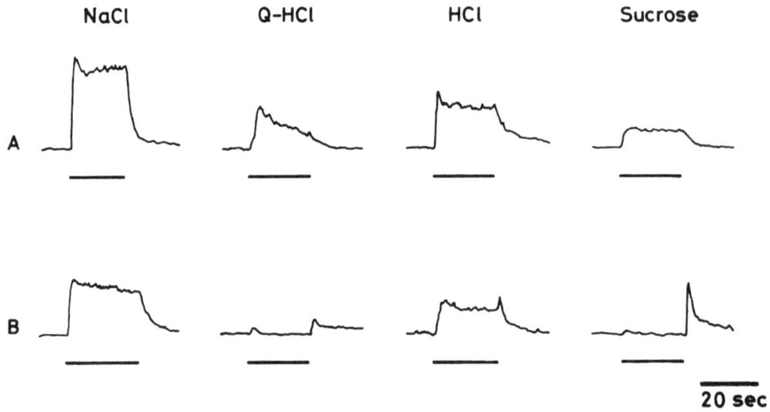

Fig. 18 A, B. Integrated chorda tympani nerve responses of rat to four basic taste stimuli (0.5 *M* NaCl, 0.02 *M* Q-HCl, 0.01 *M* HCl, 0.5 *M* sucrose). *Horizontal bars* denote the duration of taste stimulus application. The tongue was adapted to water in **A**, and to 41.4 m*M* NaCl in **B**. All records were obtained from one and the same nerve. (Sato and Beidler 1983a)

Table 8. Magnitudes of rat chorda tympani nerve responses to four basic taste stimuli

Adapting solution	Response component	Taste stimulus			
		0.5 *M* NaCl	0.02 *M* Q-HCl	0.01 *M* HCl	0.5 *M* sucrose
Water	Phasic	100	49 ± 4	54 ± 2	21 ± 3
41.4 m*M* NaCl	Phasic	72 ± 4	9 ± 3	29 ± 2	2 ± 2
Water	Tonic	78 ± 11	26 ± 6	44 ± 2	17 ± 3
41.4 m*M* NaCl	Tonic	57 ± 6	3 ± 3	21 ± 3	1 ± 1

All values (means ± SEs) are expressed as percent of the magnitude of a phasic component of the neural response to 0.5 *M* NaCl. The data were obtained from three nerves. (Sato and Beidler 1983a)

The mean magnitudes of tonic gustatory responses in the rat chorda tympani nerve elicited by the four basic taste stimuli (Table 8) are compared with the mean, steady-state magnitudes of rat taste cell responses (Fig. 17) in Table 9. This comparison is suggested by the knowledge that there is a linear relation between the magnitude of the steady level of receptor potentials and the frequency of the steady level of neural impulses (Fuortes 1971). When the magnitudes of tonic neural responses are compared with the magnitudes of receptor potentials composed of both depolarizations and hyperpolarizations, it becomes apparent that the magnitudes of the gustatory neural responses and of the gustatory cell responses produced by the four basic taste stimuli are very similar under either water or 41.4 m*M* NaCl adaptation. On the other hand, there is a large difference between the magnitudes of tonic neural responses and the magnitudes of taste

Table 9. Comparison of taste cell responses and chorda tympani nerve responses to four basic taste stimuli in rat

Cell or nerve	Adapting solution	Taste stimulus				Cells (n)	Responses
		0.5 M NaCl	0.02 M Q-HCl	0.01 M HCl	0.5 M sucrose		
Nerve	Water	100	33±8	56±3	22±4	3	Tonic
Cell	Water	100	52±17	40±17	33±4	19	Depolarization plus hyperpolarization
Cell	Water	100	70±19	56±17	56±22	11–18	Depolarization
Nerve	41.4 mM NaCl	100	5±5	37±5	2±2	3	Tonic
Cell	41.4 mM NaCl	100	9±6	48±10	2±8	38	Depolarization plus hyperpolarization
Cell	41.4 mM NaCl	100	32±6	68±10	30±8	20–37	Depolarization

In each row the response magnitude to 0.5 M NaCl is taken as 100%, and the other values are expressed relative to the NaCl response. (Sato and Beidler 1983a)

cell responses composed of only depolarizations. These findings suggest that gustatory neural responses in rat are due to the actions of both depolarizing and hyperpolarizing responses of the taste cells. It is likely that the depolarizing receptor potential of rat taste cells is involved in a generation of gustatory neural impulses, whereas the hyperpolarizing receptor potential is involved in a depression of gustatory neural impulses.

9 Action of Potassium Benzoate

Application to the tongue of low concentrations of some salts, such as potassium chloride, potassium benzoate, potassium formate, reduces the spontaneous discharge frequency of the gustatory nerve (Beidler 1961, Miller 1971, 1975, Ogawa et al. 1972). Application of mixtures of NaCl and low concentrations of K-benzoate reduces gustatory neural responses compared to that of NaCl alone (Miller 1971).

We measured the response characteristics of rat taste cells to various concentrations of K-benzoate to determine its inhibitory mechanism. Figure 19 shows examples of the receptor potentials in a rat taste cell elicited by application of 0.1

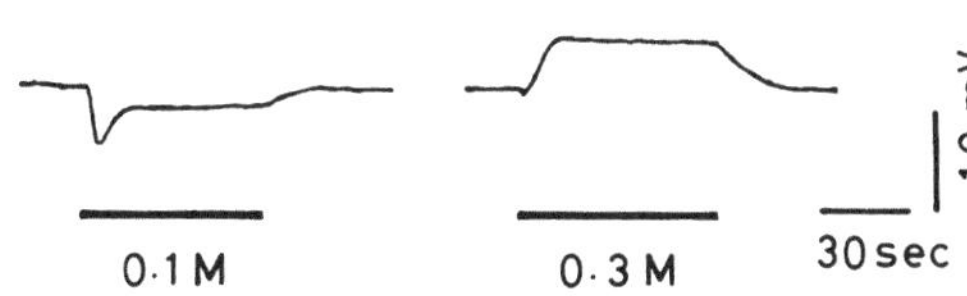

Fig. 19. Intracellularly recorded receptor potentials of a rat taste cell induced by 0.1 and 0.3 M K-benzoate solutions. Adapting solution: water. *Horizontal bars* denote duration of stimulus application. (Sato and Beidler 1983b)

and 0.3 *M* K-benzoate solutions to the tongue adapted to water. Under water adaptation the receptor potential elicited by a low K-benzoate concentration, such as 0.1 *M*, consists of an initial phasic hyperpolarization and a following steady hyperpolarization. Under 41.4 m*M* NaCl adaptation no initial phasic hyperpolarization is seen. On the other hand, under water adaptation the receptor potential elicited by high K-benzoate concentrations, such as 0.3 *M* consists only of a steady depolarization or of steady depolarization preceded by an initial phasic hyperpolarization.

Curves representing the responses of six rat taste cells to different K-benzoate concentrations are shown in Fig. 20. It is seen that the hyperpolarizing receptor potentials are initiated by concentrations below 0.05 – 0.3 *M*, but depolarizing receptor potentials are initiated by concentrations above threshold. The maximum amplitudes of hyperpolarizing responses are obtained at concentrations between 0.03 and 0.1 *M*. As the concentration is reduced, the hyperpolarizations decrease in amplitude. On the other hand, the amplitudes of depolarizations increase with an increase in the concentration.

With either depolarizing or hyperpolarizing receptor potentials, the rise time becomes shorter exponentially with increasing magnitude. The mean rise time of rat taste cells depolarized by 0.1 *M* K-benzoate is 17.6 ± 3.1 s (SE, $n = 14$), while the mean rise time of rat taste cells hyperpolarized by 0.1 *M* K-benzoate is 10.6 ± 1.9 s ($n = 15$). The rate of rise of the receptor potentials in rat taste cells increases as the amplitudes of either depolarizing or hyperpolarizing responses become larger. The slope of the relation between the rate of rise and the amplitude of a receptor potential is steeper with hyperpolarizations than with depolar-

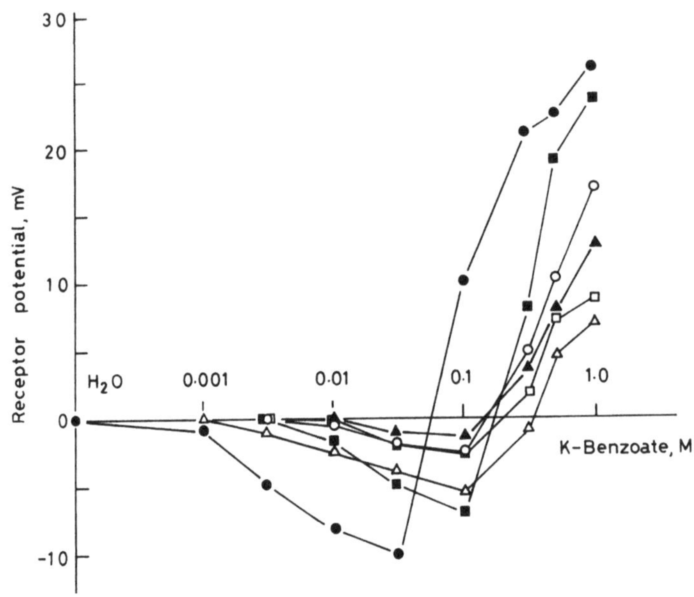

Fig. 20. Responses of six rat taste cells to various concentrations of K-benzoate. Adapting solution: water. The amplitudes of steady hyperpolarizations and steady depolarizations were used as measures of the taste cell responses (Sato and Beidler 1983b)

izations. The fall time of rat taste cell responses after rinsing is 18.8 ± 3.1 s ($n = 12$) in depolarizations evoked by 0.1 *M* K-benzoate and 21.4 ± 5.1 s ($n = 7$) in hyperpolarizations evoked by 0.1 *M* K-benzoate. The fall times of either depolarizing or hyperpolarizing receptor potentials increase with an increase in the magnitude of the receptor potential.

The input resistance of rat taste cells adapted to water decreases during generation of either depolarizing or hyperpolarizing receptor potentials by stimulation with K-benzoate. The input resistance magnitude in response to 0.1 *M* K-benzoate decreases to 87 ± 2% ($n = 21$) of the controls. In this case the mean amplitude of depolarizing and hyperpolarizing responses is −0.3 ± 1.2 mV ($n = 21$). With increasing response amplitude, the input resistance of rat taste cells decreases gradually (Fig. 21).

The effects of the mixtures of various concentrations of NaCl and 0.05 *M* K-benzoate on a rat taste cell are shown in Fig. 22. In this example, 0.05 *M* K-benzoate produces a hyperpolarization. The magnitudes of responses to the mixtures are smaller than of those to NaCl alone. This may be due to a balance of the

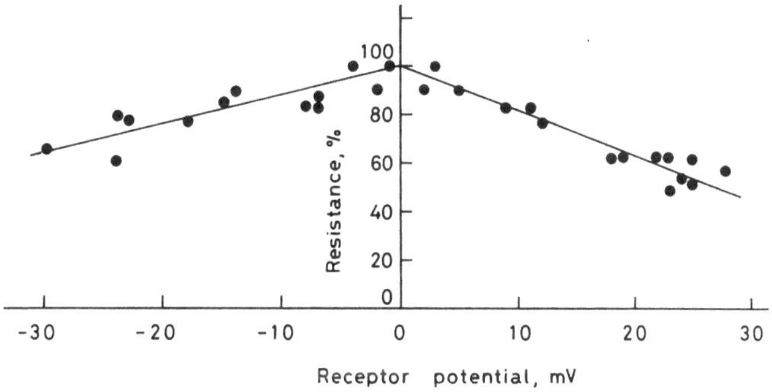

Fig. 21. Relationship between magnitudes of receptor potentials induced by various concentrations of K-benzoate and magnitudes of input resistance in a rat taste cell. The input resistance is relative to the control measured in an unstimulated state. Adapting solution: water. (Sato and Beidler 1983b)

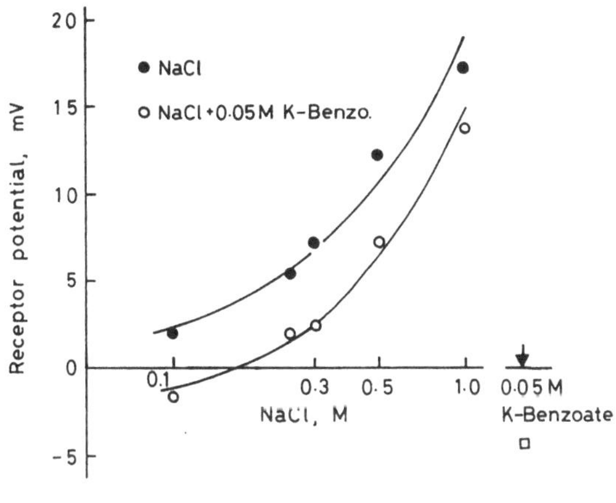

Fig. 22. Amplitudes of receptor potentials induced by various concentrations of NaCl plus 0.05 *M* K-benzoate (○) and various concentrations of NaCl alone (●). □ shows the response to 0.05 *M* K-benzoate. The data are from a rat taste cell adapted to water (Sato and Beidler 1983b)

excitatory action of NaCl and the depressing action of K-benzoate on the rat taste cell.

Miller (1971, 1975) investigated in detail the depressing action of K-benzoate in rats. He found that chemical stimulation of a fungiform papilla with 0.3 *M* NaCl while a low concentration of K-benzoate was applied to the surrounding papillae produced a depressed response of the single nerve fibers to the 0.3 *M* NaCl. Miller (1971) called this phenomenon a lateral taste depression. To interpret this mechanism he proposed a simple model consisting of two adjacent fungiform papillae innervated by two unmyelinated fiber branches of a single myelinated gustatory nerve fiber. In this system it is supposed that a 0.3 *M* NaCl stimulus applied to one papilla produces a depolarization of the axon terminals, while a low concentration of K-benzoate stimulus applied to the other papilla produces a hyperpolarization of the terminals. In addition, it is considered that the generator currents, which flow in two different directions and are initiated at the axon terminals, spread to near the impulse firing site of the gustatory nerve fiber, and that the depression of NaCl-induced impulses by the K-benzoate is due to a summation of the spread generator currents near the impulse firing site.

By combining Miller's idea and other findings including our data, we can explain a mechanism underlying the lateral taste suppression. Figure 23 shows a schematic diagram of two fungiform papillae bearing one taste bud each, and a

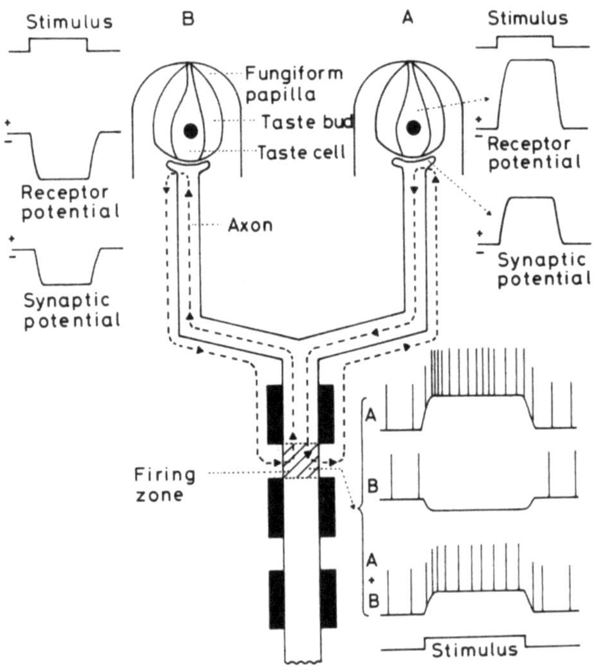

Fig. 23. Schematic drawing of two rat fungiform papillae (*A* and *B*) containing one taste bud each, and a single myelinated afferent nerve fiber with two unmyelinated fiber branches at the base of the papillae. *Dashed lines* show a flow of electrical currents. Inward and outward currents flow through the axon membrane of an impulse firing zone. Further details in the text. (Sato and Beidler 1983b)

single myelinated gustatory fiber with the two unmyelonated branches. In making this diagram, a number of studies mentioned below were helpful. Beidler (1969) reported that more than 90% of the total nerve fibers in a rat fungiform papilla are unmyelinated fibers. Cross-innervation of rat fungiform papilla taste buds has been examined by a light microscope study of chorda tympani nerves (Miller 1974). Takeda and her collaborators (Takeda 1977, Takeda and Kitao 1980, Takeda et al. 1982) have demonstrated histologically that chemical synapses exist between taste cells and gustatory nerve fibers in rats and mice. The existence of a postsynaptic potential in gustatory fiber terminals has been suggested by Nomura and Sakada (1969). The initiation of depolarizing receptor potentials by a NaCl stimulus in rat taste cells has been demonstrated by Kimura and Beidler (1961), Tateda and Beidler (1964), Ozeki (1971), Ozeki and Sato (1972), Sato and Beidler (1979, 1982, 1983a, b).

We assume, as did Miller (1971), that impulses originate at the first node of Ranvier of a gustatory nerve fiber. We assume further that, upon application of 0.3 *M* NaCl only to papilla A in Fig. 23, (1) a depolarizing receptor potential appears in the taste cell, (2) an excitatory postsynaptic potential (EPSP) appears at the fiber terminals, and (3) an outward current flows through the axon membrane of the first node of Ranvier. Intracellular recording of the activity at the impulse firing zone may show that gustatory neural impulses, as superimposed on the decayed EPSP in record A (bottom right, Fig. 23), appear upon application of 0.3 *M* NaCl to papilla A. On the other hand, we assume that, upon stimulation of papilla B with 0.1 *M* K-benzoate, (1) a hyperpolarizing receptor potential appears in the taste cell, (2) an inhibitory postsynaptic potential (IPSP) is evoked at the fiber terminals, and (3) an inward current flows at the membrane of the first node of Ranvier. Usually, the rat gustatory nerve shows a prominent, spontaneous discharge of impulses, so that a hyperpolarization regarded as the IPSP evoked by 0.1 *M* K-benzoate probably results from a depression of the spontaneous depolarization of the postsynaptic membrane at the fiber terminals. As shown in record B (bottom right, Fig. 23), the intracellular response in the impulse firing zone during K-benzoate stimulation might be a decayed hyperpolarization during which an impulse discharge is diminished or abolished. It is supposed that, when papillas A and B are stimulated simultaneously with 0.3 *M* NaCl and 0.1 *M* K-benzoate, respectively, a summation of inward and outward currents occurring at the impulse firing site causes the smaller magnitude of depolarization at that membrane, which results in a depression of impulse frequency (record A + B, bottom right in Fig. 23).

However, it was shown that the frequency of neural impulses elicited in a nerve fiber is modulated at the branch points of the nerve fiber (Spira et al. 1976). Therefore, we cannot rule out the possibility that the frequency of impulses evoked at the fiber terminal by NaCl stimulation of papilla A (Fig. 23) may be modulated at the branch point of the fiber by an IPSP evoked by K-benzoate stimulation of papilla B.

10 Summary

1. The mean resting membrane potential of rat taste cells is −36 mV under adaptation of the tongue to 41.4 m*M* NaCl and −50 mV under water adaptation.
2. The shapes of receptor potentials of rat taste cells in response to the four basic taste stimuli (0.5 *M* NaCl, 0.02 *M* Q-HCl, 0.01 *M* HCl, and 0.5 *M* sucrose) are classified into three types, namely (1) a depolarization alone, (2) a depolarization preceded by a transient hyperpolarization, and (3) a hyperpolarization alone. No regenerative spike potentials are evoked in rat taste cells by chemical stimuli. The amplitude of rat taste cell responses increases with increasing concentration of the taste stimulus. Most of the rat taste cells show a multiple sensitivity in that single cells respond to various combinations of the four basic taste stimuli with depolarizations or hyperpolarizations.
3. The rise and fall times of depolarizing responses to 0.5 *M* NaCl are much shorter than those of depolarizing responses to the other three stimuli. The fall time of depolarization evoked by 0.01 *M* HCl is the longest. The rise and fall times of all hyperpolarizing responses are shorter than those of all depolarizing responses.
4. The input resistance of taste cells decreases during depolarizations elicited by 0.5 *M* NaCl stimulation, but increases during depolarizations and hyperpolarizations elicited by stimulation with 0.02 *M* Q-HCl, 0.01 *M* HCl or 0.5 *M* sucrose. According to the speed at which rinsing of the tongue reduces these resistance changes to the control level, the taste stimuli take this order: 0.5 *M* NaCl > 0.5 *M* sucrose > 0.02 *M* Q-HCl > 0.01 *M* HCl.
5. The ratio of mean magnitudes of depolarizing and hyperpolarizing responses in rat taste cells to the four basic taste stimuli under 41.4 m*M* NaCl adaptation is as follows: NaCl : Q-HCl : HCl : sucrose = 100 : 9 : 48 : 2. On the other hand, the ratio of mean magnitudes of tonic chorda tympani nerve responses in rats for the four basic taste stimuli under 41.4 m*M* NaCl adaptation is: NaCl : Q-HCl : HCl : sucrose = 100 : 5 : 37 : 2. Similarity of the response ratios in taste cell and nerve suggests that the depolarization of rat taste cells is connected with a generation of gustatory neural impulses, and that the hyperpolarization of rat taste cells is connected with a depression of the impulses.
6. The rat taste cells respond to K-benzoate solutions of higher-than-threshold concentrations (0.03 – 0.3 *M*) with a depolarizing receptor potential, but to K-benzoate solutions of concentrations lower than threshold with a hyperpolarizing receptor potential. Application of mixtures of various concentrations of NaCl and 0.05 *M* K-benzoate results in a reduced receptor potential amplitude compared with that evoked by application of NaCl alone. It is concluded that a depression of gustatory neural impulses by low concentrations of K-benzoate is mainly due to the hyperpolarizing receptor potential of taste cells elicited by the K-benzoate solutions. Chemical stimulation of a fungiform papilla with 0.3 *M* NaCl while a low concentration of K-benzoate is applied to the surrounding papillae results in a depression of the single-nerve-fiber response to 0.3 *M* NaCl (Miller 1971). The mechanism of this lateral taste depression is discussed using a simple model consisting of two adjacent fungiform papillae innervated by two unmyelinated fiber branches of a single myelinated gustatory nerve fiber.

Acknowledgements. I am very gratiful to Dr. L. M. Beidler for his helpful suggestions. The assistance of Miss Y. Kitada in preparing the manuscript and the figures has been invaluable.

11 References

Akaike N, Sato M (1976) Role of anions and cations in frog taste cell stimulation. Comp Biochem Physiol 55A:383–391

Akaike N, Noma A, Sato M (1973) Frog taste cell response to chemical stimuli. Proc Jpn Acad 49:464–469

Akaike N, Noma A, Sato M (1976) Electrical responses of frog taste cells to chemical stimuli. J Physiol [Lond] 254:87–107

Beidler LM (1954) A theory of taste stimulation. J Gen Physiol 38:133–139

Beidler LM (1961) Taste receptor stimulation. In: Butler JAV, Huxley HE, Zirkle RE (eds) Progress in biophysics and biophysical chemistry, vol 12. Pergamon, Oxford

Beidler LM (1967) Anion influences on taste receptor response. In: Hayashi (ed) Olfaction and taste, vol II. Pergamon, Oxford

Beidler LM (1969) Innervation of rat fungiform papilla. In: Pfaffmann (ed) Olfaction and taste, vol III. Rockefeller University, New York

Beidler LM (1971) Taste receptor stimulation with salts and acids. In: Beidler LM (ed) Chemical senses, part 2, taste. Springer, Berlin Heidelberg New York (Handbook of sensory physiology, vol 4)

Beidler LM, Gross GW (1971) The nature of taste receptor sites. In: Neff DW (ed) Contributions to sensory physiology, vol 5. Academic, New York

Esakov AI, Byzov AL (1971) Electrical response of the frog's gustatory cells to centrifugal stimulation. Bull Exp Biol Med 72:723–726

Farbman AI (1965) Fine structure of the taste bud. J Ultrastruct Res 12:328–350

Fuortes MGF (1971) Generation of responses in receptor. In: Loewenstein WR (ed) Principles of receptor physiology. Springer, Berlin Heidelberg New York (Handbook of sensory physiology, vol 1)

Graziadei PPC, DeHan RS (1971) The ultrastructure of frog's taste organs. Acta Anat 80:563–603

Hije Y (1969) Gustatory response and preference behavior in alloxan diabetic rats. Kumamoto Med J 22:109–119

Kashiwayanagi M, Miyake M, Kurihara K (1983) Voltage-dependent Ca^{2+} channel and Na^{+} channel in frog taste cells. Am J Physiol 244 (Cell Physiol 13): C82–C88

Kimura K, Beidler LM (1956) Microelectrode study of taste bud of the rat. Am J Physiol 187:610–611

Kimura K, Beidler LM (1961) Microelectrode study of taste receptors of rat and hamster. J Cell Comp Physiol 58:131–140

Kurihara K, Kamo N, Kobatake Y (1978) Transduction mechanism in chemoreception. Adv Biophys 10:27–95

Miller IJ Jr (1971) Peripheral interactions among single papilla inputs to gustatory nerve fibers. J Gen Physiol 57:1–25

Miller IJ Jr (1974) Branched chorda tympani neurons and interactions among taste receptors. J Comp Neurol 158:155–166

Miller IJ Jr (1975) Mechanisms of lateral interactions in rat fungiform taste receptors. In: Denton DA, Cochlan JP (eds) Olfaction and taste, vol V. Academic, New York

Morita H, Yamashita S (1959) Generator potential of inset chemoreceptors. Science 130:922

Morita H, Hidaka T, Shiraishi A (1966) Excitatory and inhibitory effects of salts on the sugar receptor of the fleshfly. Mem Fac Sci Kyushu Univ Ser E (Biol) 4:123–135

Murray RG (1969) Cell types in rabbit taste buds. In: Pfaffmann C (ed) Olfaction and taste, vol III. Rockefeller University, New York
Murray RG (1971) Ultrastructure of taste receptors. In: Beidler LM (ed) Chemical senses, part 2, taste. Springer, Berlin Heidelberg New York (Handbook of sensory physiology, vol IV)
Murray RG (1973) The ultrastructure of taste buds. In: Friedman J (ed) The ultrastructure of sensory organs. Elsevier, Amsterdam
Murray RG, Murray A (1970) The anatomy and ultrastructure of taste endings. In: Walstenholme GE, Knight E (eds) Ciba Foundation symposium on taste and small in vertebrates. Churchill, London
Murray RG, Murray A, Fujimoto S (1969) Fine structure of gustatory cells in rabbit taste buds. J Ultrastruct Res 27:444–461
Murray RG, Murray A, Hellekant G (1972) Fine structure of taste buds of rat fungiform papillae. In: Schneider D (ed) Olfaction and taste, vol IV. Wissenschaftliche Verlagsgesellschaft, Stuttgart
Nomura H, Sakada S (1969) Local potential changes at sensory nerve fiber terminals of the frog tongue. In: Pfaffmann C (ed) Olfaction and taste, vol III. Rockefeller University, New York
Ogawa H, Yamashita S, Noma A, Sato M (1972) Taste responses in the macaque monkey chorda tympani. Physiol Behav 9:325–331
Okada Y, Sato T (1982a) Receptor potential of frog taste cell in response to acid stimuli. J Physiol Soc Jpn 44:436
Okada Y, Sato T (1982b) Mechanisms of receptor potential of frog taste cell in response to acid stimuli. Jpn J Oral Biol 24 [Suppl]:S98
Okada Y, Sato T (1983) Role of Ca ions in receptor potential of frog taste cell in response to acid stimuli. J Physiol Soc Jpn 45:495
Ozeki M (1970) Hetero-electrogenesis of the gustatory membrane in rat. Nature 228: 868–869
Ozeki M (1971) Conductance change associated with receptor potentials of gustatory cells in rat. J Gen Physiol 58:688–699
Ozeki M (1973) Physiological and morphological identification of the gustatory cell in rat. J Physiol Soc Jpn 35:251–252
Ozeki M, Ōura C (1972) Physiological and electron microscopical studies of gustatory responsive cell in the fungiform papilla of the rat. Proc Jpn Acad 48:544–547
Ozeki M, Sato M (1972) Responses of gustatory cells in the tongue of rat to stimuli representing four taste qualities. Comp Biochem Physiol 41A:391–407
Pfaffmann C, Fisher CL, Frank MK (1967) The sensory and behavioral factors in taste preferences. In: Hayashi T (ed) Olfaction and taste, vol II. Pergamon, Oxford
Roper S (1983) Regenerative impulses in taste cells. Science 220:1311–1312
Sato M (1973) Gustatory receptor mechanism in mammals. Adv Biophys 4:103–152
Sato T (1969) The response of frog taste cells (*Rana nigromaculata* and *Rana catesbeiana*). Experientia 25:709–710
Sato T (1972a) Multiple sensitivity of single taste cells of the frog tongue to four basic taste stimuli. J Cell Physiol 80:207–218
Sato T (1972b) The electrical response of the frog taste cells as studied with the intracellular microelectrode. In: Schneider D (ed) Olfaction and taste, vol IV. Wissenschaftliche Verlagsgesellschaft, Stuttgart
Sato T (1976a) Depolarizing and hyperpolarizing receptor potentials of rat taste cells. J Physiol Soc Jpn 38:143–144
Sato T (1976b) Does an initial phasic response exist in the receptor potential of taste cells? Experientia 32:1426–1428
Sato T (1980) Recent advances in the physiology of taste cells. Prog Neurobiol 14:25–67

Sato T, Beidler LM (1975) Membrane resistance change of the frog taste cells in response to water and NaCl. J Gen Physiol 66:735 – 763
Sato T, Beidler LM (1979) Receptor potential of rat taste cell to potassium benzoate. Experientia 35:1203 – 1205
Sato T, Beidler LM (1982) The response characteristics of rat taste cells to four basic taste stimuli. Comp Biochem Physiol 73A:1 – 10
Sato T, Beidler LM (1983a) Dependence of gustatory neural response on depolarizing and hyperpolarizing receptor potentials of taste cells in the rat. Comp Biochem Physiol 75A:131 – 137
Sato T, Beidler LM (1983b) The response characteristics of rat taste cells to potassium benzoate. Comp Biochem Physiol 76A:777 – 781
Sato T, Okada Y (1984) Receptor potential in frog taste cell elicited by acid stimuli. J Dent Res 63:558
Sato T, Sugimoto K (1978) Effects of external ions on resting potential and receptor potential in taste cells. J Physiol Soc Jpn 40:291
Sato T, Sugimoto K, Okada Y (1982) Ionic basis of receptor potential in frog taste cell in response to salt stimuli. Jpn J Physiol 32:459 – 462
Spira ME, Yarom Y, Parnas I (1976) Modification of spike frequency by regions of special axonal geometry and by synaptic inputs. J Neurophysiol 39:882 – 899
Takeda M (1976) An electron microscopic study on the innervation in the taste buds of the mouse circumvallate papillae. Arch Histol Jpn 39:257 – 269
Takeda M (1977) Uptake of 5-hydroxytryptophan by gustatory cells in the mouse taste bud. Arch Histol Jpn 40:243 – 250
Takeda M (1979) Tritiated thymidine autoradiographic study of taste buds in the mouse (in Japanese). Acta Anat Nippon 54:230 – 231
Takeda M (1981) Fine structure of taste buds (in Japanese with English abstract). Adv Neurol Sci 25:320 – 333
Takeda M, Hoshino T (1975) Fine structure of taste buds in the rat. Arch Histol Jpn 37:395 – 413
Takeda M, Kitao K (1980) Effect of monoamines on the taste buds in the mouse. Cell Tissue Res 210:71 – 78
Takeda M, Shishido Y, Kitao K, Suzuki Y (1982) Monoamines of taste buds in the fungiform and foliate papillae of the mouse. Arch Histol Jpn 45:239 – 246
Tateda H, Beidler LM (1964) The receptor potential of the taste cell of the rat. J Gen Physiol 47:479 – 486
Teeter J, Kare MR (1974) Passive electrical properties and responses to chemical stimulation of cutaneous taste bud cells and surrounding surface cells of the catfish. Fed Proc 33:416
Tonosaki K, Funakoshi M (1983) The mouse taste cell responses to some sugar stimuli. Proc Jpn Sym Taste Smell 17:41 – 44
West CHK, Bernard RA (1978) Intracellular characteristics and responses of taste bud and lingual cells of the mudpuppy. J Gen Physiol 72:305 – 326
Yamada T (1980) Chorda tympani responses to gustatory stimuli in developing rats. Jpn J Physiol 30:631 – 643

Functional Properties of the Fish Olfactory System

K. B. Døving

Department of Biology, Division of General Physiology, University of Oslo,
P.O. Box 1051 Blindern, 0316 Oslo 3, Norway

---

1 Introduction

The olfactory system in vertebrates comprises a highly specialized sensory organ for detection and identification of minute quantities of chemicals in the environment. Experimental studies have documented the role played by olfactory information in social relationships, prey or predator recognition, and the search for food. The demands imposed by the aquatic environment have made the olfactory apparatus of fish a sensory system with many specialized features. At present only a few properties of the fish olfactory organ have been investigated.

Several comprehensive reviews have been published on the general properties of the olfactory system in vertebrates (Ottoson 1963, Ottoson and Shepherd 1967, Moulton 1976, Holley and MacLeod 1977). Numerous reviews are available on the fish olfactory system (Hara 1971, 1975, Teichman 1962, Barnett 1977, Liley 1982) together with monographs (Kleerèkoper 1969, Grant and Mackie 1974, Hara 1982). Most of this material has focused on the anatomy of the olfactory organ or the behavior of the fishes. Problems related to the physiology of the olfactory system are less represented.

The aim of this review is to present some fundamental facts and current ideas on the function of the olfactory system in fish. The presentation of morphological and physiological information follows the chain of events in the olfactory pathways. As demonstrated in Fig. 1, the olfactory system in fish is distinct and separated from other nervous structures, and is readily accessible. All students of chemoreception, whether biochemists, physiologists, or ethologists can find suitable material, preparation, or subjects for study among the multitude of anatomical adaptations of the olfactory system in fish. I hope that the reader will extract from this report the information which is of particular value for his own field of interest.

2 The Peripheral Olfactory Organ

2.1 Gross Anatomy

The olfactory organ of fishes varies in gross as well as fine morphology. In spite of the diversity in appearance, there are certain general traits that are common to most species.

In cyclostome fishes the olfactory epithelium is situated in the snout, with an anterior opening at the most rostral part of the head. The nasal tube has a single opening on the dorsal surface of the head, both in *Petromyzon marinus* (Kleerekoper and van Erkel 1960) and in *Lampetra fluviatilis* (Thornhill 1967). The tube is continous with the nasal cavity containing the sensory epithelium. There is a valve at the entrance to the cavity. Behind the nasal cavity is found a nasopharyngeal sac. In the hagfish the nasal valve is located in a similar position as in the lampreys (Theisen 1973, 1976). The pouch is lacking, however, and the nasal cavity opens directly to the nasopharyngeal duct (Fig. 2).

In elasmobranchii the paired olfactory cavities are found at the ventral surface of the head. There is only one opening to the cavity, but an intricate fold of skin divides the cavity in two chambers (Døving et al. 1977).

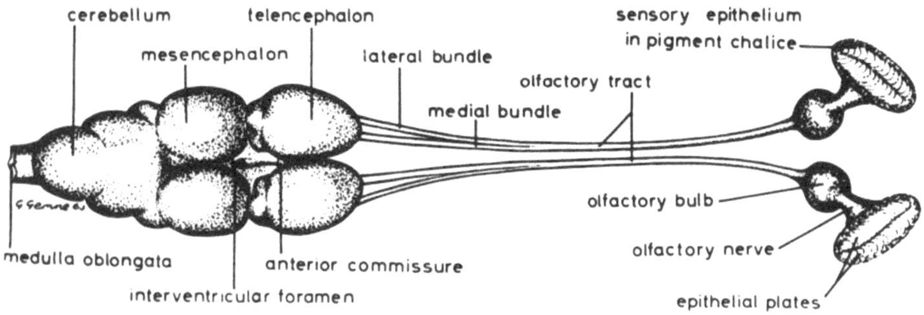

Fig. 1. Dorsal view of the brain and olfactory system of burbot, *Lota lota*, showing the sensory epithelium, the stalked olfactory bulbs, and the divided tracts. (From Døving and Gemne 1965)

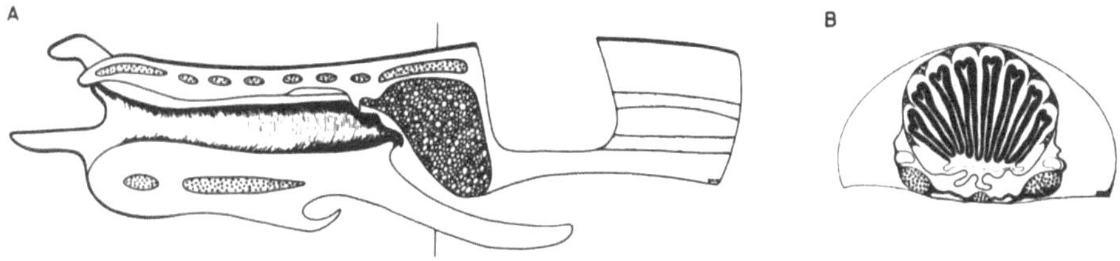

Fig. 2. A Medial section of the anterior dorsal part of the hagfish, *Myxine glutinosa*, showing the naris leading into the olfactory cavity. The position of the transverse section (**B**) through the olfactory organ is indicated by the *vertical line* in **A**. (From Døving and Holmberg 1974)

In teleosts the olfactory epithelium is found as a paired organ on the dorsal side of the head above the ethmoidal region. Usually there are two openings to the olfactory cavity, an anterior and a posterior naris.

A thorough description of the comparative anatomy and different forms of olfactory organ can be found in the zoological literature on species from the European continent (Burne 1909, Liermann 1933, Faure 1944, Devitsyna and Malyukin 1977). A multitude of Indian species have also been described (Singh 1978, Ojha and Kapoor 1972, 1973, 1974, Kapoor and Ojha 1974, Holey and Belsare 1973, Goel 1978a, b, 1980a, b, Rahmani and Khan 1980, 1981, Shahab et al. 1981). A few fishes have been reported to have an internal naris, for example, the ratfish *Chimaera monstrosa* (Holl 1973b), Antarctic fish *Gymnodraco acuticeps* (Jakubowski 1975), the *Astroscopus* (Dahlgren 1908). These findings have led to lengthy discussions on the origin of the choana in higher vertebrates. In some fish there is only one opening to the olfactory chamber (monotrematous olfactory organs), for example, in the eel pout *Zoarces viviparus* (Pipping 1926), *Pholis gunellus, Liparis montaqui* (Liermann 1932), *Aphanopus carbo* (Holl and Meinel 1968), and fifteen-spined stickleback, *Spinachia spinachia* (Fig. 3; Theisen 1982). In the pike *Esox esox* (Burne 1909) and garfish *Belone belone* (Theisen et al. 1980), the olfactory epithelium is placed in a shallow groove at the skin surface.

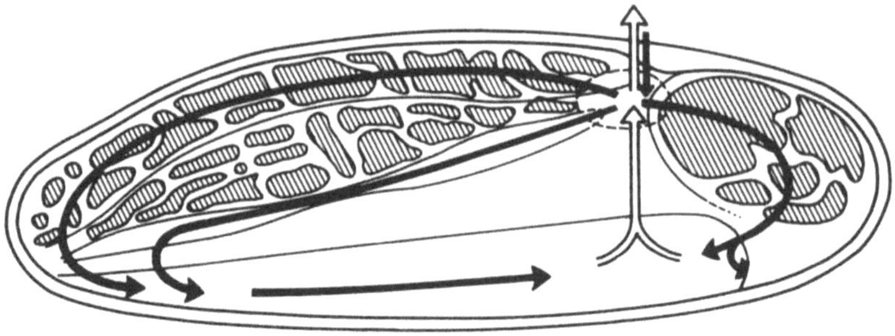

Fig. 3. Drawing of the olfactory organ in the monotrematous fish, the fifteen-spined stickleback, *Spinachia spinachia.* The sensory epithelium (*hatched*) is distributed in small ilets anterior and posterior to the opening (*stippled line*). The inward waterflow (*black arrows*) covers the sensory epithelium, whereas the outward water current leaves directly from the underlying cavity (*open arrows*). (Redrawn from Theissen 1982)

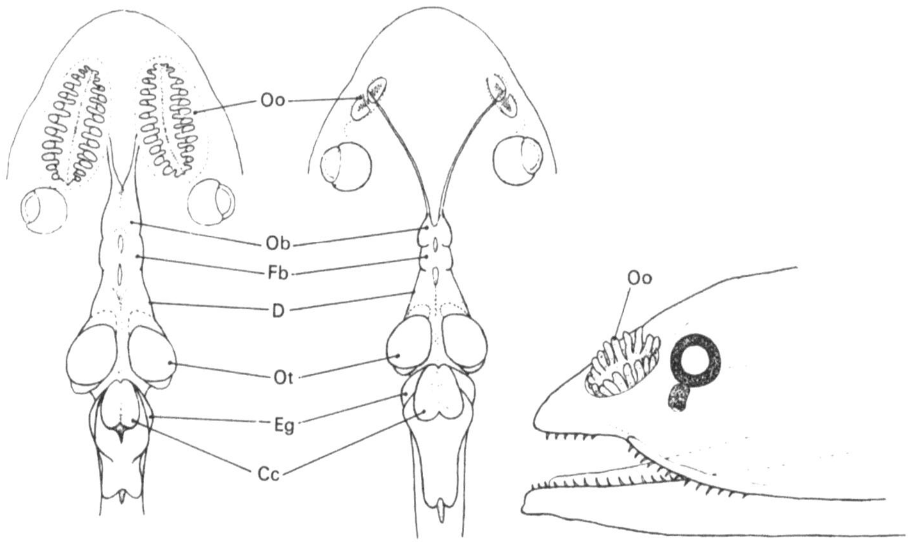

Fig. 4. Olfactory organs and brain of the bathypelagic fish *Cyclothone microdon*; *left*, male; *right*, female. *Cc*, corpus cerebellum; *D*, diencephalon; *Eg*, eminentia granularis; *Fb*, forebrain; *Ob*, olfactory bulb; *Oo*, olfactory organ; *Ot*, optic tectum. (From Marshall 1979)

Most of the bathypelagic fishes, i.e., species that live at depths of between about 1000 and 4000 m, a marked sexual dimorphism has evolved with regard to size of the olfactory system (Marshall 1979). The males have large olfactory organs with more than 20 olfactory lamellae. In the females the olfactory organs are regressed and only relics of the lamellae are detectable. There are corresponding differences in the size of the olfactory nerves, bulbs, and forebrain (as shown in Fig. 4).

2.2 Water Sampling

The mechanism for water sampling by the olfactory organ in fish has evolved in various directions, and the evolutionary history of fishes shows both convergent and divergent lines of development. A large variety of functional designs have been evolved to propel water through the olfactory organ. The olfactory organs of fish achieve the propelling action by two different, but not necessarily mutually exclusive, mechanisms. One mechanism is based on cilia (isosmates), the other on pumping (cyclosmates).

Isosmates. In some fishes the water sampling by the olfactory organ is based upon the action of cilia. These fishes have been given the name *isosmates* (from Greek *iso*, "equal" and *osmare*, "who smells") (Døving et al. 1977). In a typical isosmate fish the major part of the olfactory cavity is occupied by the olfactory rosette. When describing the water current, I shall keep to the architectural nomenclature. Figure 5 A shows a transverse section of the olfactory organ of the eel *Anguilla anguilla* (Døving and Thommesen 1977). From the entry, which is always the anterior naris, a central vestibule opens above the raphe. From this compartment water can reach all the corridors arranged at right angles to the raphe. The water passes from the vestibule through the corridors of the olfactory rosette to the peripheral compartment called the gallery. In the gallery the water is collected and leaves to olfactory cavity via the posterior naris.

The water current is created by the beating of the kinocilia found on the lamellae of the olfactory rosette. The efficiency of the cilia beat is dependent upon the length of the cilia, the spacing of the corridor, and the direction of the effective stroke. Synchronized beating of kinocilia may even occur in such a fashion as to cause peristaltic pumping (H. Westerberg, personal communication 1983). The cilia are so arranged that the water current will travel from the central aspect of the olfactory rosette down towards the basement of the rosette, leave it at the periphery of the lamellae and flow up to the gallery, as shown in Fig. 5 A. From the gallery the water is expelled to the outside via the posterior naris. In the eel, a dorsal skin fold is formed along the whole length of the olfactory rosette, creating two compartments, the vestibule and the gallery. The skin fold is essential for the unidirectional flow of water through the olfactory organ. When the roof is cut away the water flow becomes irregular and abnormal (see Teichmann 1959).

There are also cilia lining the epithelium of other parts of the olfactory cavity. In the carp family a velum extends down from the anterior naris covering the posterior parts of the olfactory rosette (Døving et al. 1977). The peripheral part

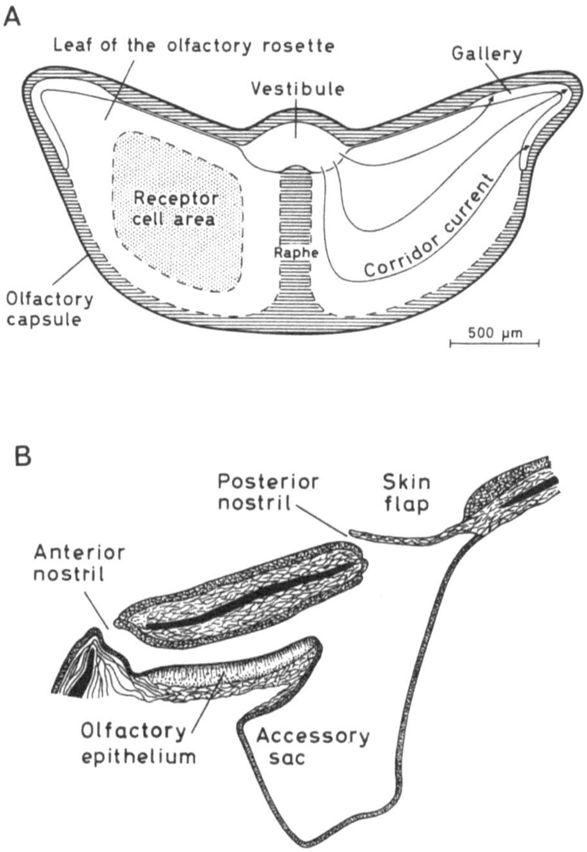

Fig. 5A, B. Examples of olfactory organs in fish. **A** Transverse section through the olfactory organ of an isosmate, the eel, *Anguilla anguilla*, showing the vestibule, corridors, and the gallery which are the compartments through which the water passes, in sequential order. (From Døving and Thommesen 1977). **B** Sagittal section of the olfactory organ of a cyclosmate, the cyprinodont with an accessory sac which pumps water, and the skin flap serving as an unidirectional valve. (From Zeiske 1974)

of the lamellae in the olfactory rosette expands into the posterior opening. One result of this functional design in isosmates is that water traverses a corridor only once. There is no recurrent action, which would make it possible for an odorant to be effective a second time to receptors of another corridor. This point has been clearly stressed, e.g., by Bashor et al. (1974) who investigated the ciliary action and normal movement of odorant wave fronts of the garfish *Lepisosteus osseus*. Other fishes where the same principle of water current is used, are the European catfish *Silurus glanis L.* (Jakubowski and Kunysz 1979), Indian catfish *Clarias batrachus* (Ojha and Kapoor 1972, Goel 1978a, b), freshwater carp *Labeo rohita* (Ojha and Kapoor 1972), and murenae *Rhinomurhena ambonensis* (Holl et al. 1970). Probably, of the fish which can be called isosmates, the one with the most highly developed and voluminous olfactory organ is the African bichir, polypteridae (Figs. 6 and 7; Pfeiffer 1968, Schulte and Holl 1971).

The anterior nares of isosmates are often shaped like tubes. They can be small protusions, as found in carps (Døving et al. 1980), or large trumpet-formed appendages, as seen in the murenae *Rhinomurhena* (Fig. 8; Holl et al. 1970). The tubes of the anterior naris possibly function as a means of ensuring water sampling above the boundary layer of the skin.

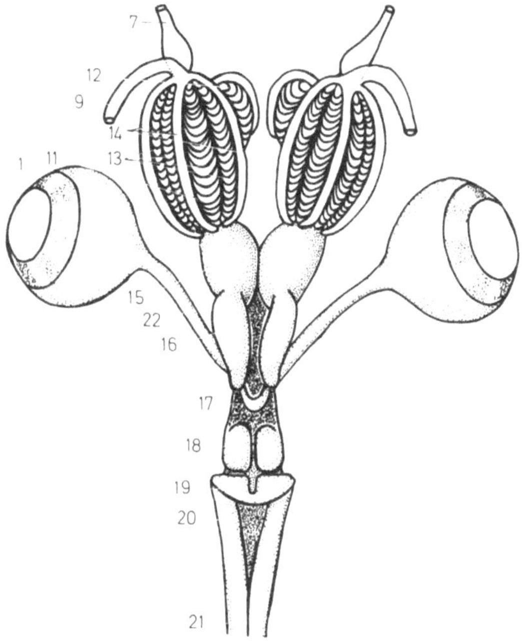

Fig. 6. The olfactory organ of an African bichir, *Polypterus ornatipinnis*, dorsal view. *1*, eye; *7*, anterior naris; *9*, posterior naris; *11*, olfactory chalice; *12*, accessory olfactory organ; *13*, olfactory lamellae; *14*, gallery channels; *15*, olfactory bulbs; *16*, telencephalon. (From Pfeiffer 1968)

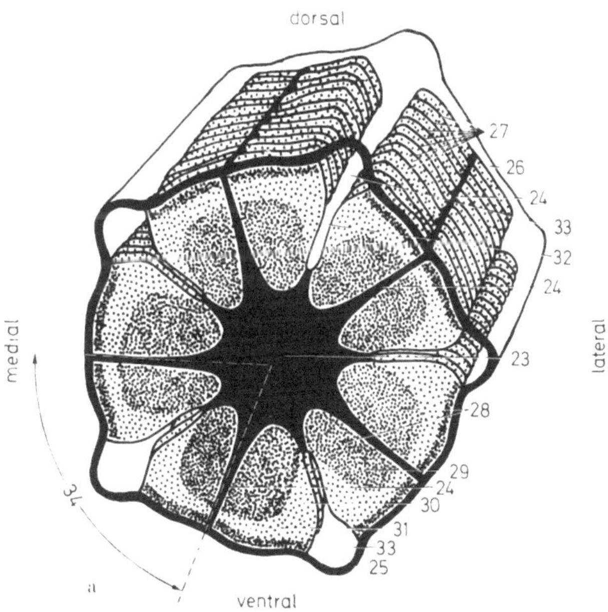

Fig. 7. Transverse section of the olfactory organ of a *Polyperus* showing the fivefold symmetry of the olfactory lamellae. (From Pfeiffer 1968)

Cyclosmates. In many fishes the olfactory cavity connects to one or two accessory sacs (Burne 1909). The olfactory rosette lies typically close to the anterior naris, and a valve is frequently found at the posterior naris. An example of the design of the olfactory cavity in a cyclosmate is given in Fig. 5B. In the wrasses *Labridae*, the rosette is situated on a sail (the velum), and the rosette itself acts as

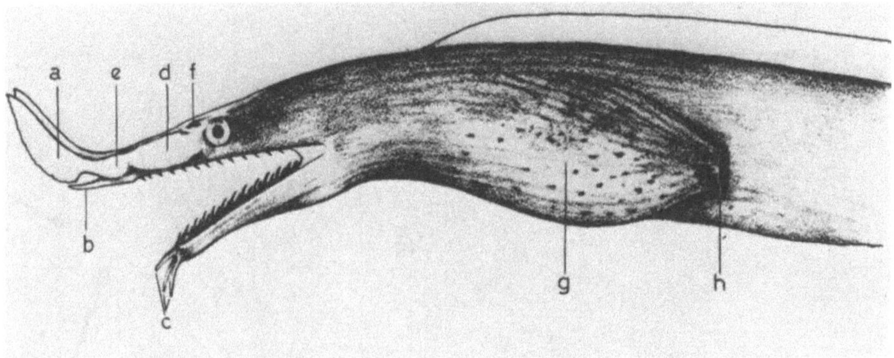

Fig. 8. The head of the murenae, *Rhinomuraena ambonensis*. *a*, trumpet-formed skin; *b*, rostrum; *c*, mandibular tentacles; *d*, olfactory chalice with sensory epithelium; *e*, anterior naris; *f*, posterior naris; *g*, gill pouch; *h*, gill opening. (From Holl et al. 1970)

a valve. The propulsion of water is caused by varying the volume of the accessory sacs, which again is a result of the movements of the mouth and/or respiratory apparatus (Døving et al. 1977). A concise account of the morphology and method of water sampling in the olfactory cavity of the black rockfish *Sebastodes melanops* is given by Johnson and Brown (1962).

Studies in *Bedotia geayi* (Melinkat and Zeiske 1979) and in *Lampris quttatus* (Oelschläger 1976) have described in detail the functional morphology which ensures efficient control of the volumes of the accessory sacs. Melinkat and Zeiske (1979) stressed in their observations that the lymphatic cavities in the head region of the fish play an intermediate role in connection with variations in the volume of the accessory olfactory sacs and, thus, the ventilation of the olfactory organ.

The fishes which have this form of water sampling mechanism have been termed cyclosmates (from Greek *kyklos*, "circle", and *osmate*, "who smells"), focusing on the recurrent pumping action that leads to intermittent water sampling (Døving et al. 1977). This mechanism is clearly different from that found in *isosmates*; thus, in the cyclosmates there are no structures resembling the vestibule and gallery. The water flow is made unidirectional by the hydrodynamic properties of the olfactory cavity or by the arrangement of valves in relation to the water current. Water enters the anterior and leaves at the posterior naris; thus, in the cyclosmates, too, there are mechanisms which prevent recycling of the water already sampled.

In fishes with monotrematous olfactory organs, water, enters and leaves by the one opening. Theisen (1982) has recently described the hydrodynamic properties of the olfactory organ of the fifteen-spined stickleback *Spinachia spinachia*. She demonstrated that the entering water spreads out over the olfactory epithelial sheet, but when it leaves, the water is "squeezed" out as a jet without coming into contact with the sensory epithelium (see Fig. 3). This observation once again demonstrates that there are mechanisms which decrease the probability that the water which has already flushed the olfactory epithelium is being recycled.

There are several intermediate forms between cyclosmates and isosmates. Two of these are the Indian snake head *Channa punctatus* and Indian carp *Labeo rohita*.

In both species, Kapoor and Ojha (1972) demonstrated that a continuous passage of water is created by ciliary activity. In addition, propulsion of the water is caused by a variation in volume of the accessory sacs. The possible function of the accessory sac in the olfactory organ of the catfish *Siluris glanis* was studied by allowing a liberal amount of mud particles to enter the anterior naris (Døving et al. 1977). A few particles passed the organ and appeared at the posterior naris fastened to mucus. After a time the fish opened its mouth widely, as if yawning; when it closed its mouth a ball of mud particles embedded in mucus was expelled from the posterior naris. When the olfactory organ was examined, no particles were found. This observation indicates that the accessory olfactory sac of the catfish aids in removing mud particles from the olfactory organ. This conclusion is in agreement with the observations made by Kapoor and Ojha (1972) on the Indian carp. It is of interest to note in this connection that both species are bottom feeders in turbid water.

It has been the common belief that the water passes through the olfactory cavity and over the olfactory sensory epithelium as the fish swims. This may be correct in some fishes, like the pike *Esox esox* and the garfish *Belone belone*, studied by Theisen et al. (1980). It is of interest to known that the fast-swimming fishes, like the skipjack tuna *Katsuwonus pelanes* (Gooding 1963) and mackerel *Scomber scombrus* (Døving unpublished obervations) have well-developed olfactory organs with accessory sacs. These fishes are also equipped with valves at the posterior naris. Gooding (1963) concluded that a continuous stream of water flows slowly through the olfactory capsule of the swimming skipjack tuna, entering through the anterior naris and leaving through the poterior naris. Furthermore, an additional increment of water is possible via the anterior naris when the mouth is open, and is ejected via the posterior naris when the mouth is closed. The swimming speed of the fish does not seem to influence the water sampling. It is evident, however, that by its swimming the fish ensures that each sample is taken afresh.

Some of the variations of the mechanisms of water sampling by the olfactory organ have been discovered only recently, for example, in the African lungfish *Protopterus annectens* (Derivot et al. 1979a), *Mastecembelidae* (Holl 1974), and ratfish *Chimaera monstrosa* (Holl 1973a, b).

The speed of water passage through the olfactory organ is of interest for understanding the exposure time of the olfactory apparatus. In the isosmate *Lepisosteus osseus*, the stream of water, estimated using dye experiments, had a mean velocity in the corridors of 2.2 mm/s. A lamella surface in this species would be traversed in 400 – 600 ms (Bashor et al. 1974). In the cyprinodont cyclosmate, the water velocities were observed using laser technique. The peak velocities were 12 – 17 mm/s at the entrance and 30 – 35 mm/s at the exit (Kux et al. 1978). The olfactory organ of the latter species was about 1.5 mm in length. The time needed to traverse the olfactory cavity will be dependent upon the cross-sectional area of the water-ways. If the cavity cross section equals that of the posterior naris, water will pass the olfactory cavity in less than 0.1 s. Experiments on the functional implications of the speed of water sampling do not appear to have been published.

In the cyclosmate there seems to be a higher speed of water sampling than in the isosmate. Such a feature combined with the ability of voluntary control of the sampling speed and frequency, i.e., a sniffing action, may be important for the fish comparing the water quality on the left and right sides. This allows a spatial mechanism for comparison of water quality or stereoolfaction as proposed by Stoddart (1979).

2.3 Olfactory Epithelium

The epithelium sheet on which the sensory receptors are found varies greatly in shape, from the flat surface as seen, for example, in the cyprinodonts (Zeiske 1974), to the intricate lamellar folding of the crossopterygian *Latimeria chalumnae* (Millot and Anthony 1957) and actinopterygian, *Calamoichtys* sp. (Pfeiffer 1968, Schulte and Holl 1971). In the latter species the olfactory organs are very large and occupy most of the volume of the snout. The paired organ has a five-fold symmetry around the vestibule and water leaves via tubes at the periphery of the organ, which join before leaving through the posterior naris. Pfeiffer (1968) has made a comparative study of the polypteridae. In a fish measuring 41 cm the total olfactory area of the lamellae was estimated to be about 3200 mm^2 (Fig. 7; Pfeiffer 1968).

The size of the olfactory organ is interesting from the functional point of view because the larger surface can contain a larger number of receptors. Teichmann (1954) has made quantitative measurements of the area of the olfactory epithelium and related it to the retina surface. Such measurements are ambiguous, since the receptor-cell surface can vary in proportion to the epithelium sheet as shown by Zeiske (1974), who measured this relation in cyprinodonts. He also used scanning electron microscopy, which has given new insight into the organization of the sensory epithelium. Using this technique Yamamoto and Ueda (1977, 1978a – f, 1979a – c) have described a great many freshwater and saltwater fishes local to Japan. The fishes most intensively studied are the cyprinids (Branson 1963, Breipohl 1974, Breipohl et al. 1973, 1974, Ichikawa and Ueda 1977, Appelbaum 1981) and salmonids (Pfeiffer 1963a, Jahn 1972, Bertmar 1972a – d, 1973, 1978, Yamamoto and Ueda 1977, Thommesen 1983). The olfactory organs have also been described in hagfish (Theisen 1973, 1976), garfish (Theisen et al. 1980), rainbow fish (Zeiske et al. 1979, Breucker 1979, Breucker et al. 1979), cyprinodonts (Zeiske 1974, Zeiske et al. 1976a, b), silurids (Jakubowski and Kunysz 1979, Caprio and Raderman-Little 1978), gadids (Devitsyna 1972, 1977, Gemne and Døving 1969), a Pacific fish species (Pfeiffer 1964), dipnoid (Derivot et al. 1979a), and lantern fish (Lawry 1973).

Detailed reviews of the forms of the olfactory rosette in different fishes are given by Burne (1909), Liermann (1932), and Faure (1944). The different forms of the epithelium sheet have induced several attempts to classify the fishes according to the shape of this sensory organ (Burne 1909, Yamamoto and Ueda 1979c, Holey and Belsare 1973). Burne (1909) distinguished four types of olfactory organs in fish. Yamamoto and Ueda (1979c) classified the organs into eight groups according to the presence and development of lamellae. According to their scheme, the cyprinodonts fall into group A, with only one flat sheet of sensory epithelium, while the eels, cods, and carps are found in group H, in which the lamellae are numerous and arranged transversely to a central raphe.

2.4 Receptor Cells

Trujillo-Cenóz (1961) described the olfactory epithelium in fish, and using electron microscopical techniques identified the ciliated receptor cells. Bannister (1965) found in *Phoxinus* sp. and *Gasterosteus* sp. three forms of receptor cells: ciliated, microvillous, and rod-tipped. The rod cells vary considerably in number from species to species and among different individuals of the same species. Yamamoto and Ueda (1979c) refer to unpublished work on the salmonid *Oncorhynchus masou*, in which they claim that the number of rod-tipped receptor cells increased during the smoltification period. Intermediate forms of ciliated and rod-tipped receptor cells are found in the flying fish and in three species of *Perciformes* (Yamamoto and Ueda 1979a). The authors suggest that the rod-tipped cells can be placed in the same category as the ciliated receptor cells, the latter being transformed into the rod-tipped cells for some unknown reason. Ichikawa and Ueda (1977) found that in the goldfish, two types of cells disappeared from the sensory epithelium 6 – 7 days after sectioning the olfactory nerve. By this method they seem to have settled a long dispute on what kind of epithelial cells were primary olfactory neurones. Their experimental findings indicate a retrograde degeneration of two types of receptor cells. The cells that showed retrograde degeneration were:

1. The ciliated cells, in which the dendritic process bears cilia. The dendritic portion of the cell protrudes slightly above the surface of the neighboring supporting cells, forming a terminal swelling (olfactory knob). From this swelling a varying number (5 – 10) of radially oriented cilia are found. The length of the cilia varies around 7 μm, and the diameter is about 0.2 μm. The dendritic position is characterized by longitudinally arranged microtubuli and mitochondria. Each receptor cell is isolated from the surrounding receptor cells by the supporting (sustentacular) cells. The cilia of the sensory cells seem to lack dynein arms, a property associated with immotility (Afzelius 1976).
2. The microvillous receptor cell, in which the apical dendrite is equiped with a large number of microvilli (30 – 80). The length of the microvilli is about 2 μm and the diameter is around 0.08 μm. A centriole has often been seen in the apical dendrite, indicating that this receptor cell developed from a ciliated precursor cell (Ichikawa and Ueda 1977). There seems to be general agreement that both types of cells are present in the olfactory epithelium in teleosts. This has been found to be the case in the eel *Anguilla anguilla* (Schulte 1972), garfish *Belone belone* (Theisen et al. 1980), rainbow fish (Zeiske 1974), catfish (Jakubowski 1981), gadids (Lowe 1974, Lowe and MacLeod 1975), and *Opticephalus* sp. (Schulte and Riehl 1977). Cancalon (1983) described a receptor cell in the catfish *Ictalurus punctatus* where the microvillae formed a crown at the surface.

The two types of olfactory receptor cells are illustrated in Fig. 9. These receptor cells have also been found in the hagfish *Myxine glutinosa* (Theisen 1973, 1976) and African lungfish *Protopterus annectens* (Derivot et al. 1979b), but not in the Australian lungfish *Neoceratodus forsteri* (Theisen 1972), which is described as having only microvillous receptor cells. Schulte and Holl (1971) claimed that there is only one type of receptor cell in *Calamoichtys calvaricus*. This fish is

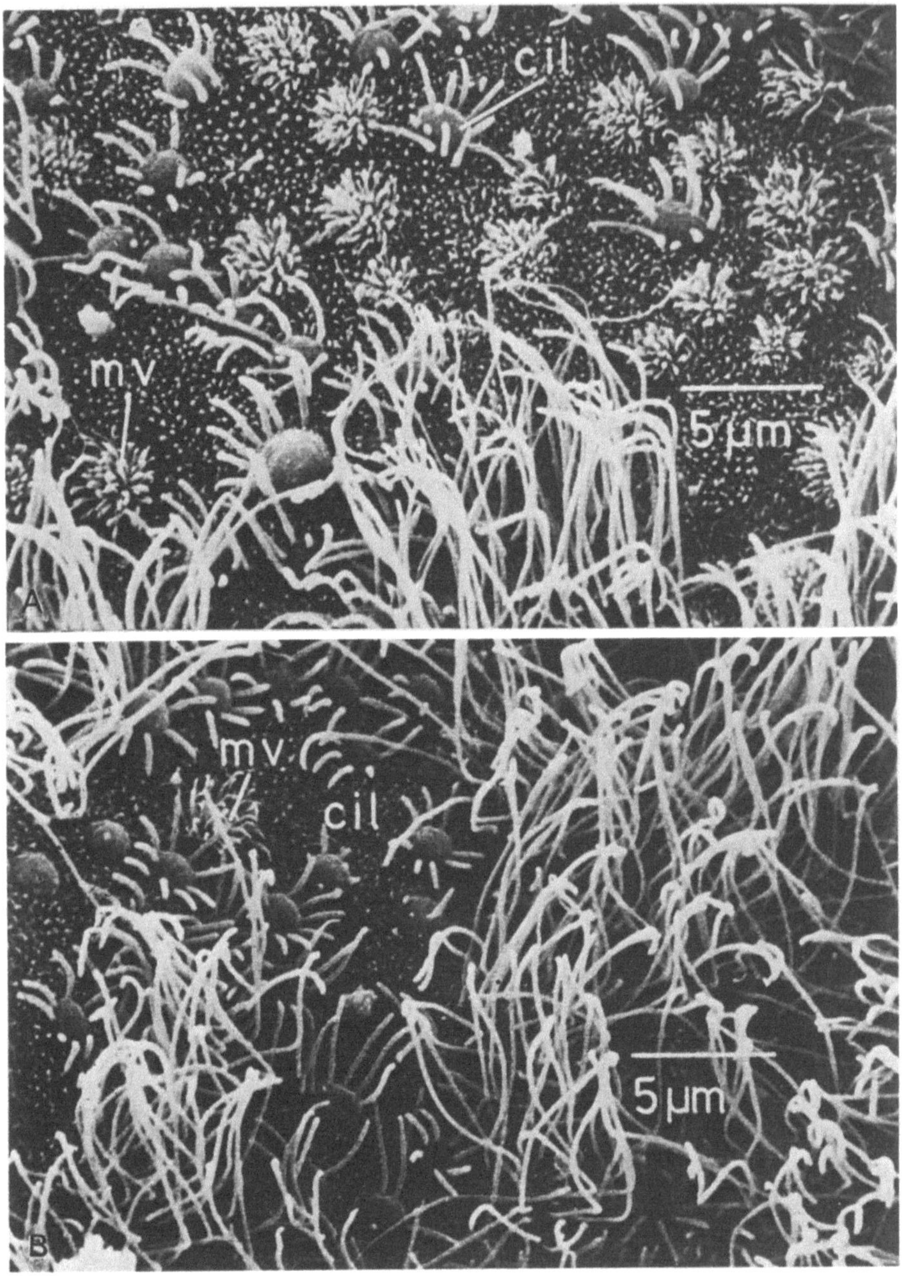

Fig. 9 A, B. The receptor surface of the olfactory lamella of a rainbow trout showing the microvillous (*mv*) and the ciliated receptor cells (*cil*). Photo **A** is taken from the central part of the lamella close to the raphe and shows a high density of microvillous receptor cells. Photo **B** is taken from the peripheral part of the lamella and shows a high density of ciliated receptor cells. The kino cilia of the supporting cells are sometimes so dense that they obscure the view of receptor cells. (From Thommesen 1983)

equiped with cilia at the olfactory knob, but microvillae could also be seen. In the ratfish *Chimaera monstrosa*, Holl (1973a, b) describes receptor cells with microvillae and another naked receptor cell with rod-shaped, free appendages. In their review of 130 species of 18 orders of teleosts, Yamamoto and Ueda (1979c) state that the two types of receptors are present in all species examined. Reese and Brightman (1970) reported that only microvillous receptor cells were present in the olfactory epithelium of the Elasmobranchii, nurse sharks *Ginglyomostoma cirratum*, and guitar fish *Rhinobatus lentiginosus*. It should be noted that the number of individuals which have been examined is small in relation to the total number of elasmobranch species. The functional implication of the two types of receptor cells must be clarified in order to understand the evolution of receptor cells (Vinnikov 1982).

Holl (1980, 1981) has developed a promising method for selective staining of olfactory receptor cells. He exposed the olfactory organ for 30 min with procian brilliant yellow. Thereby he stained the majority of receptor cells and the nerve fibers. The method offers new possibilities in studies of the functional anatomy of the olfactory system in fish.

Receptor-Cell Density. For the electrophysiologist it might be advantageous to search for receptor-cell activity in species in which the receptor-cell density is high. The variations are great from species to species. It seems that in olfactory organs where the kinocilia are few (cyclosmates), the receptor density is higher than in the fishes equiped with a high density of kinocilia (isosmates). In salmonid fishes the mean receptor-cell density is about 50000 receptors mm^{-2} (Teichmann 1954, Holl 1965, Thommesen 1983). In patches of 100 μm^2 the receptor-cell density may reach 120000 receptors mm^{-2} (Thommesen 1983). In cyprinodonts, which are cyclosmates, the densities varied from 200000 to 500000 receptors mm^{-2}. Zeiske et al. (1976a, b, 1979) and Theisen et al. (1980) studied the development of receptor-cell density.

Receptor-Cell Distribution. The receptor cells can be arranged in different ways on the epithelium sheet. Yamamoto and Ueda (1979c) made a classification in four groups (Fig. 10). On the lamella surface the receptor cells can be found I, in continuous areas except for the margin (eels and cods), II, in areas separated by indifferent epithelium (salmonids); III, in irregular patches (herrings), IV or (d) in discreet islets separated by indifferent epithelium (cyprinodonts). Thommesen (1983) showed that the density of the two types of receptor cells in the olfactory epithelium of migrating salmonids varied in a consistent manner (Fig. 9). The microvillous receptor cells had highest density at the central region close to the raphe, the ciliated receptor cells had highest density in the peripheral part of the olfactory lamellae. The mean ratio of receptor-cell density at the peripheral vs the central regions was 0.56 for the microvillous and 2.25 for the ciliated receptor cells. Thommesen found that the total number of ciliated cells in the olfactory organs of salmonid species examined was about twice that of the microvillous receptor cells. Pyatkina (1976) found similar rates in two species of other migrating fish, the sturgeon *Acipenser güldenstadti* and sevruga *A. stellatus*. It is remarkable that in the stationary species *A. ruthenus* she found twice as many microvillous cells as ciliated receptor cells.

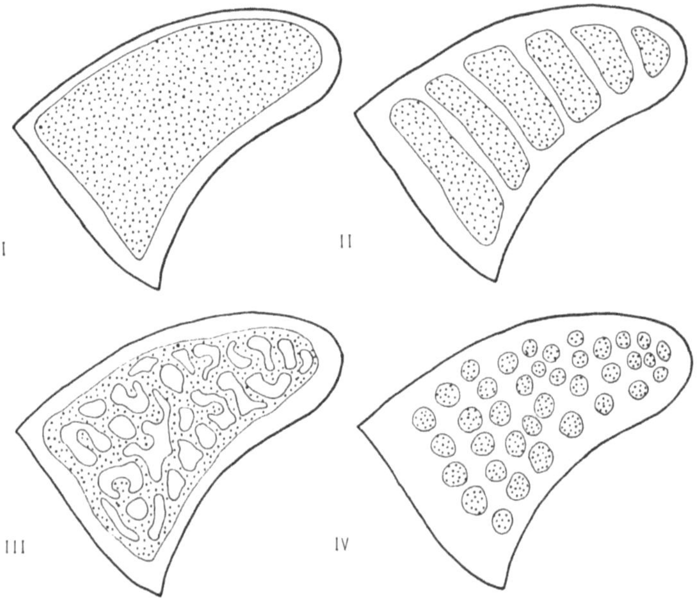

Fig. 10. Schematic representation of the principal types of arrangement of the sensory epithelium on the olfactory lamellae of fish; see text. (From Yamamoto and Ueda 1979c)

Biochemical Studies. Histochemical investigations on the olfactory epithelium have been mainly concerned with the composition of the mucus. According to the reports, the mucus is free of enzymes and contains neutral and acid-sulfated mucopolysaccharides (Doroschenko and Pinchuk 1974, Bianchi 1978, Herberhold 1968).

In a preparation of membranous elements from the olfactory rosette of the rainbow trout *Salmo gairdneri*, Rhein and Cagan (1980) assessed binding properties of this membrane preparation to functionally important amino acids. The authors state that their preparation was rich in membranous elements of isolated cilia and that binding properties were correlated to the olfactory potency of the amino acids. They also propose that the odorant recognition sites are integral parts of the cilia. The authors do not distinguish between the cilia of the olfactory receptor cells and those of the sustentacular cells. Their preparation may also contain microvillae of receptor cells. The correlation between biochemical and physiological properties may be fortuitous.

2.5 Physiological Activity

Early behavioral studies showed that the fish have a remarkably keen sense of smell (Parker 1910, Parker and Sheldon 1912). Physiological studies have confirmed most of the results from behavioral experiments.

As in other vertebrates, a slow negative, monophasic potential from the exposed side of the olfactory epithelium can be recorded when this is stimulated with

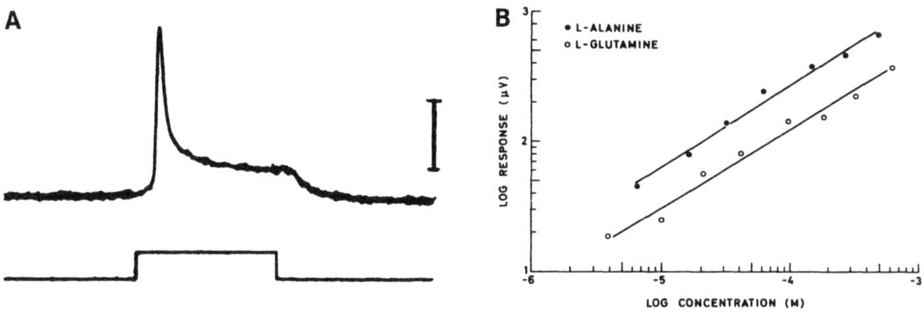

Fig. 11. A Response of the olfactory organ of *Myxine glutinosa* to stimulation with 1.4 10^{-4} *M* L-alanine. Stimulus duration indicated by *lower trace* is 7.5 s. *Vertical bar* 0.2 mV. **B** Relation between peak amplitude of the olfactory response and stimulus strength for two amino acids. Each point is the mean value obtained by four stimuli. Note the logarithmic scales. (From Døving and Holmberg 1974)

potent odorants. This potential, called the electroolfactogram (EOG) was first analyzed in detail by Ottoson (1956) in studies of the olfactory organ in the frog (for review, see Ottoson 1963, 1970, Ottoson and Shepherd 1967). An example of the wave form of the EOG recorded in the hagfish *Myxine glutinosa* is shown in Fig. 11 A (Døving and Holmberg 1974). The electrode in contact with the water flowing past the olfactory epithelium recorded a slow negative monophasic potential when the organ was stimulated with amino acids, confirming the results found in terrestrial animals. EOG responses have also been recorded in the Atlantic stingray *Dasyatis sabina* and sea catfish *Arius felis* (Silver et al. 1976, Silver 1979, Caprio 1980), salmonids (Thommesen 1982, 1983), and eel *Anguilla anguilla* (Silver 1982). All the species examined respond to a large concentration range of amino acids (between four and six log units). The EOG amplitude increased exponentially as a function of the amino acid concentration (Fig. 11 B). The relationship between response and intensity of the stimulus fitted a power function with an exponent varying between 0.18 and 0.29. The thresholds varied from 10^{-7} to 10^{-9} *M* concentration. When the olfactory nerve of rainbow trout is cut, there is a loss of receptor cells and a reduction in EOG amplitude (Evans et al. 1982).

Silver (1982) compared responses from the olfactory epithelium either as EOG or multiunit activity to those obtained by recording from a twig of the olfactory nerve. The recordings from the nerve gave lower threshold values than those obtained by the other two techniques. In all three types of recordings, power functions described the dose/response relationship. The exponent was lower, however, in the case of nerve twig (0.1) than in the case of EOG (0.15) and multiunit activity (0.19).

Response Specificity. By recording the EOG in salmonids, Thommesen (1982) showed that the responses were elicited by two different types of chemicals, amino acids and bile salts (see Sect. 4, Olfactory Stimuli). His results on cross-adaptation experiments demonstrated that the responses were functionally independent, and suggested that the responses originated in two different populations of receptor cells.

Response Distribution. Thommesen (1982, 1983) also demonstrated that the EOG responses, when recorded simultaneously from two sites, the central and peripheral part of the olfactory rosette in salmonids, were largest for the bile salts at the periphery and largest for the amino acids at the central electrode. On the basis of the physiological experiments and the anatomical investigation of the distribution of receptor cell types in salmonids, Thommesen (1983) put forward the hypothesis that the ciliated receptor cells respond to bile salts, while the receptor cells with microvillae respond to amino acids.

Activity from Single Receptors. From the olfactory epithelium it is also possible to record action potential activity from single receptor neurons (A. Holley 1980, personal communication). Most published work has been obtained by studying the multiunit activity. Thus, Suzuki and Tucker (1971) found that the amino acids were potent olfactory stimuli for the freshwater catfish *Ictalurus catus*, and using similar techniques, Sutterlin and Sutterlin (1971) found that this was also true for the Atlantic salmon *Salmo salar*.

Suzuki (1978) has shown that when a calcium-free solution is applied to the olfactory epithelium of the lamprey *Entosphenus japonicus*, receptor activity is completely blocked. He suggests that the Ca^{++} ions play an important role in the generation of the olfactory receptor potential, probably by carrying the inward current through the apical parts of the receptor neurons. The requirement of calcium for the function of olfactory organs is probably the reason for the results obtained by Bodznick (1978c), who showed that a calcium concentration above $10^{-6}\,M$ was needed to induce responses from the olfactory organ of sockeye salmon *Onchorhynchus nerka*.

Suzuki (1977) has carried out intracellular recordings on the receptor cells of the lamprey. He showed that the resting potential varied from -37 to -60 mV. Variations of the external environment did not greatly influence the resting potential, though calcium was essential for the response to odorants. The insensitivity of the receptors towards the outer ionic environment may be the reason for the ability of the olfactory organ of the salmonid fishes to function perfectly upon transfer from a freshwater to a saltwater environment and vice versa (Lorentzen 1983).

Intracellular studies conducted by Suzuki (1977, 1982) on the lamprey receptor cells revealed that prolonged depolarization gives repetitive spike activity. The absolute spike amplitude decreases with increasing depolarization. Thus, the intracellular recordings confirm the decrease in spike height, seen at high frequency burst in extracellular recordings. The lamprey olfactory receptor membrane had time constants of 3.4 and 0.34 ms. The results indicate that the receptor-cell membrane is heterogeneous and composed of at least two compartments with different electrical properties.

Suzuki (1977) also determined the passive membrane characteristics of the olfactory receptors. He found the input impedance of the receptor membrane to have a mean value of 108 Mohm. This value is lower than that obtained from intracellular recordings of the olfactory receptor cells of the salamander *Salamandra salamandra* (Trotier and MacLeod 1983). These cells have membrane potentials in the order of -56 mV. Olfactory stimulation evoked a slow depolarization up to 28 mV. The input resistance of the cell fell from about 180 at rest to 90 Mohm at the peak of depolarization.

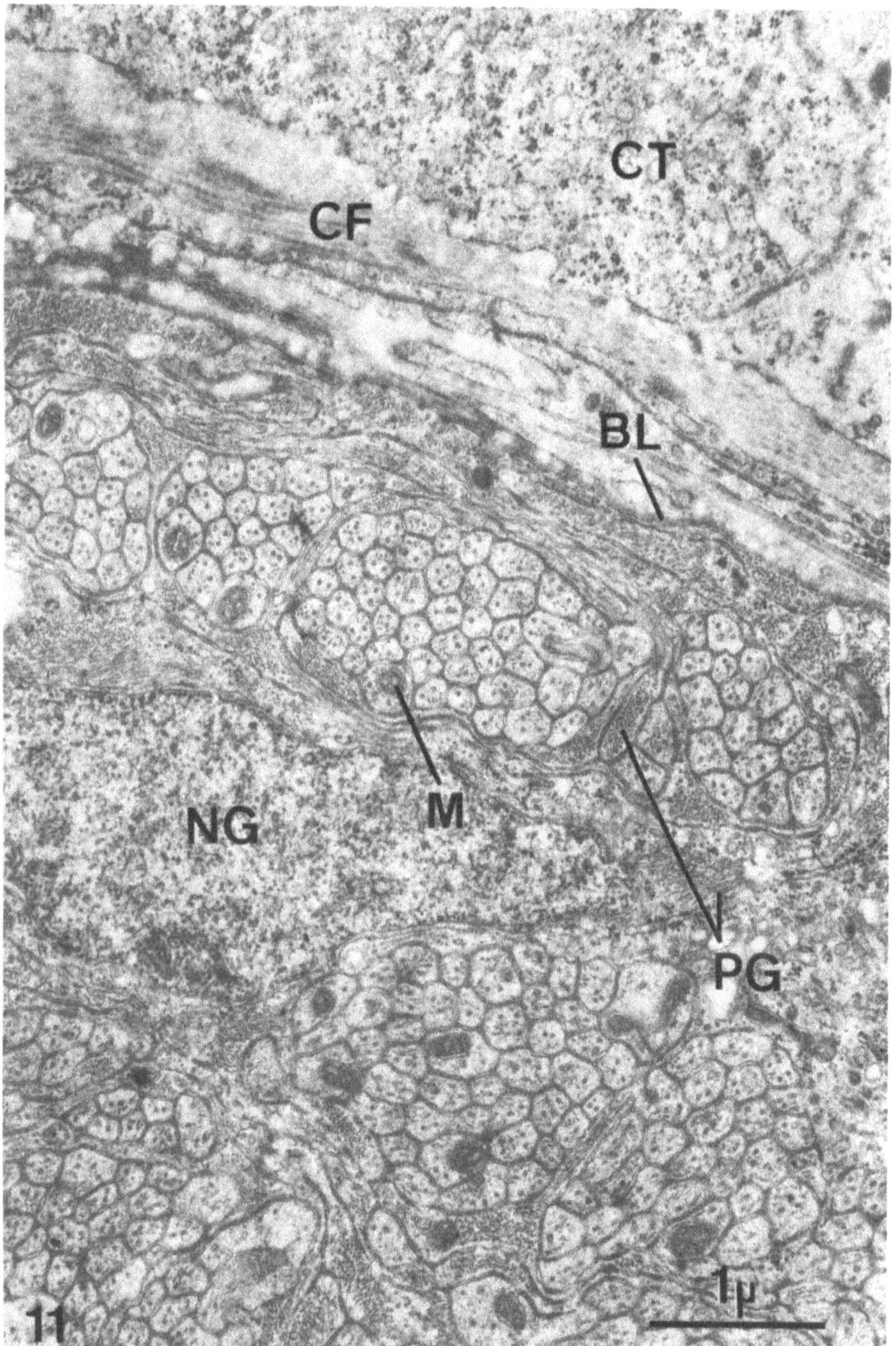

Fig. 12. Part of the olfactory fila formed just under the basal lamina of the olfactory epithelium. *BL*, basal lamina; *CF*, collagen fibers; *CT*, connective tissue cell; *M*, mitochondrion; *NG*, nucleus of glia cell; *PG*, glial cell process. (From Gemne and Døving 1969)

2.6 Ultrastructure of the Olfactory Nerve

The sensory neurons of the olfactory epithelium are primary receptors, which means that a receptor cell forms an axonal projection, terminating in the olfactory bulb. The axons form bundlets (filae olfactoria) which cluster to make up the olfactory nerve. The fibers are unmyelinated, with numbers up to 1000 within one mesaxon of a glial cell (Fig. 12). This feature would seem to favor ephapses and "cross-talk" between receptors, but no evidence for such properties has been found. Since the axons are so thin, there must be an increase in their diameters in order to make room for mitochondria. Each axon contains a varying number of microtubules, usually about 4, and 12-nm neurofilaments which are often found in clusters of up to 30 (Kreutzberg and Gross 1977). The axon diameter is thinner than found in the nerve fibers of any other group of vertebrates. In the burbot *Lota lota*, the mean ± one standard deviation was found to be 0.14 ± 0.04 μm (Døving and Gemne 1965; Fig. 13), in garfish *Lepisosteus osseus*, 0.24 ± 0.05 μm, and in pike *Esox esox*, 0.2 ± 0.04 μm (Kreutzberg and Gross 1977, von Muralt et al. 1976). In these three studies, the total number of axons within one nerve was estimated to be 10^7, 1.3×10^7, and 5.1×10^6, respectively. The total surface of axolemma per cm of nerve has been calculated to be 320 cm^2 (Kreutzberg and Gross 1977).

Physiological Properties. Due to its unique properties, the olfactory nerve became a tool in early neurophysiological research. Because of its length and its

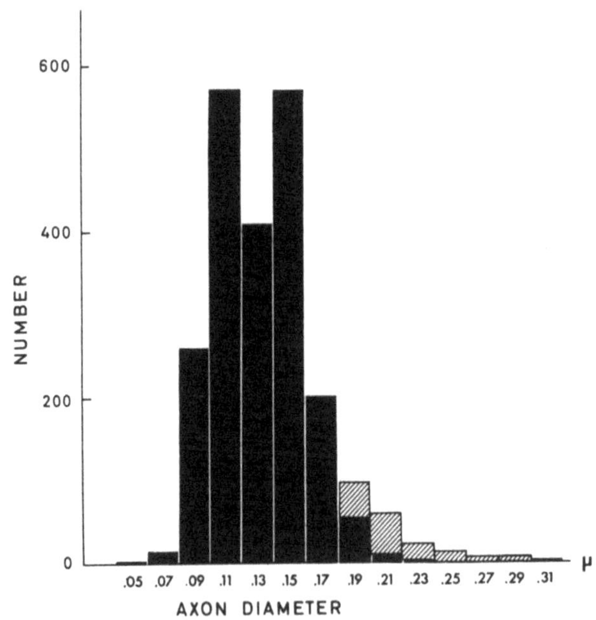

Fig. 13. Histogram showing the distribution of axon diameters of the olfactory nerve in burbot, *Lota lota* ($N = 2240$). The axons containing mitochondria are indicated by *hatched columns*. (From Gemne and Døving 1969)

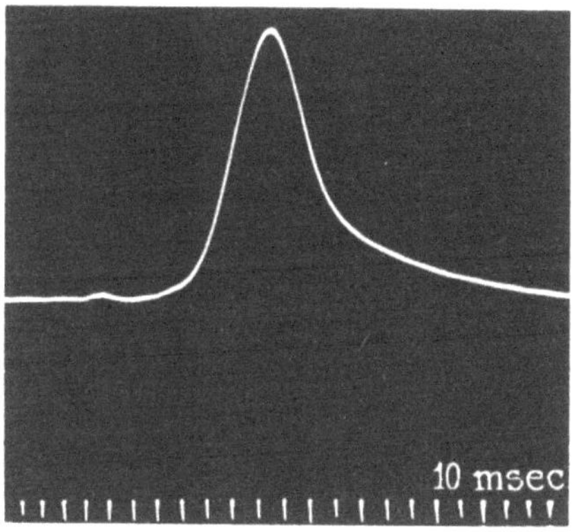

Fig. 14. Action potential of the olfactory nerve of the pike. Conduction distance, 6.5 mm, 21 °C. (From Gasser 1956)

slowness of conduction, which was compatible with the slow action of the capillary electrometer used at that time, the olfactory nerve of the pike *Esox esox* was found useful in the first measurements of the compound nerve action potential by Kühne and Steiner (1879). Garten (1900) made an extensive study of the properties of the olfactory nerve in pike, and Gasser (1956) investigated the anatomical and functional properties of the nerve of the pike, as well as of the pig (Fig. 14). Studies of the fish olfactory nerve can give information about the physiological, biochemical, and biophysical properties of the axolemma, which is abundant in this preparation. Studies by Kühne and Steiner (1879) und Garten (1900) showed that conduction of nerve action potential was slow, in the order of 0.2 m s^{-1}, as can be expected from the dimensions of the axon diameters (Gasser 1956). The spike is followed by a negative and then a positive afterpotential. The excitability cycle extends through relative refractoriness to supernormality and then to subnormality, and conforms with the properties of other nerve fibers, except the C fibers of the dorsal root.

The action potential of the olfactory nerve has been measured in the garfish *Lepisosteus osseus* (Matsumoto and Tanner 1975), lesser fork-beard *Raniceps raninus* (Døving 1967), and pike *Esox esox* (Howarth et al. 1975). In the garfish, the Q_{10} for conduction velocity varied from 2.5 – 1.7 in the range from 5° – 25 °C (Matsumoto and Tanner 1975). In a fish with a 50-mm long nerve, a message will take 1 s from the olfactory epithelium to the olfactory bulb, where neurones conducting at higher velocities can take over the information transmission. This would be the case for a fish like a returning salmon at a temperature around 5 °C, when the conduction velocity approximates 50 mm/s.

During the passage of an action potential, a small quantity of heat is released in the phase of depolarization. Howard et al. (1975) found that the positive heat at 0 °C was 44.2 µcal/g impulse. The intracellular potassium concentration in the pike olfactory nerve was estimated to be 108.8 mmol/kg fiber water (Howarth et al. 1975). Hendersson and Strichartz (1974) and Ritchie and Straub (1975) have studied the influences of potassium over the garfish olfactory nerve membranes. Studies have also been conducted (Kracke and Chako 1979, Kracke et al. 1981) of the (Na^{+} + K^{+}) ATP-ase activity in a preparation of the plasma membrane of the garfish olfactory nerve. Ritchie and Straub (1975) calculated that the nerve membrane axolemma has 350 pumping sites per square micron.

Axoplasmic Transport. The homogeneity of the fiber diameter in the olfactory nerve makes it ideally suited for the study of the axoplasmatic flow of proteins. The role of such transport is to supply the various cell components with necessary material for maintaining proper structure and function. Application of radioactively labeled amino acids to the olfactory epithelium results in rapid uptake and incorporation into proteins and the export of labeled material along the olfactory nerve. The peak of rapid transport at 19 °C is about 127 mm/day in the pike *Esox lucius* (Gross and Kreutzberg 1978) and 149 mm/day in the garfish *Lepisosteus osseus* (Gross and Beidler 1973, 1975). The amino acids are also transported as free substances at the same velocity as the peak of rapid transportation, together with the rapidly transported proteins (Weiss 1980). The amount of axonal transport was found to increase dramatically in the regenerating nerve fibers; the speed of transport remained the same as in the normal nerve (Cancalon and Beidler 1975, Cancalon and Elam 1980a, b).

3 The Olfactory Bulb

The olfactory bulb is the first relay station in the olfactory system of vertebrates. Here the axons of the sensory cells terminate, and the transmission of olfactory information is conveyed by a new set of secondary neurones called relay cells. In some fishes the bulb can be found close to the olfactory epithelium, connected to the brain by a long central tract. This is the case, for example, in the codfish, as drawn in Fig. 1. Such an anatomical feature is also found in the catfish and carp. In many other fishes the olfactory nerve is long, for example, in the pike, garfish, salmon, and perch. It is obvious that such features should be considered when selecting the most useful animals for functional studies. Uchihashi (1953) has published a comparative essay on these anatomical properties.

3.1 Anatomy

The early works on the fish brain are landmarks in the descriptive anatomy of the brain and of the olfactory system of fishes. The most extensive works are those by Sheldon (1912) on the carp *Cyprinus carpio*, Holmgren (1920) on the smelt *Osmerlanus erlanus*, and Jansen (1930) on the hagfish *Myxine glutinosa*. Later studies have been conducted on the paddlefish *Polydon spathula* (Story 1964), sturgeon *Scaphirhyhynchus platorhynchus* (Adair 1964), African lungfish *Protopterus* sp. (Schnitzlein and Crosby 1967), and ratfish *Chimaera monstrosa* (Faucette 1969a, b). Reviews of the comparative anatomy have been published by Nieuwenhuys (1967) and Andres (1970).

Starting from the surface, the layers in the olfactory bulb can be distinguished as: (a) the olfactory nerve layer, (b) the layer of olfactory glomeruli, (c) the layer of relay cells, and (d) the internal cell layer. The latter has also been called the granule cell layer or anterior olfactory nucleus. However, granule cells have not been definitively described; thus, the name suggested by Oka et al. (1982), the internal cell layer, seems appropriate. The term anterior olfactory nucleus is misleading, since it causes association with the nucleus of the same name in mammals, with its distinct anatomical features (Kosaka and Hama 1979).

Most of the cellular elements which constitute the olfactory bulb in higher vertebrates are found in the bulb of fishes. Their organization is less distinct, but still appreciable even to the untrained eye. The fibers coming from the olfactory epithelium are distributed in the superficial layer of the anterior part of the bulb. Here the fibers are sorted, as indicated in Fig. 15. The axons terminate in restricted regions of the bulb, so-called glomeruli, structures devoid of cell soma where axons make synaptic contact with dendrites of secondary neurones of the bulb. In contrast to the relay cells of the mammalian olfactory bulb, i.e., the mitral and tufted cells, the dendrites of these cells in the fish branch to reach several glomeruli (Fig. 15). The secondary neurones constitute a third layer inside the layer of glomeruli.

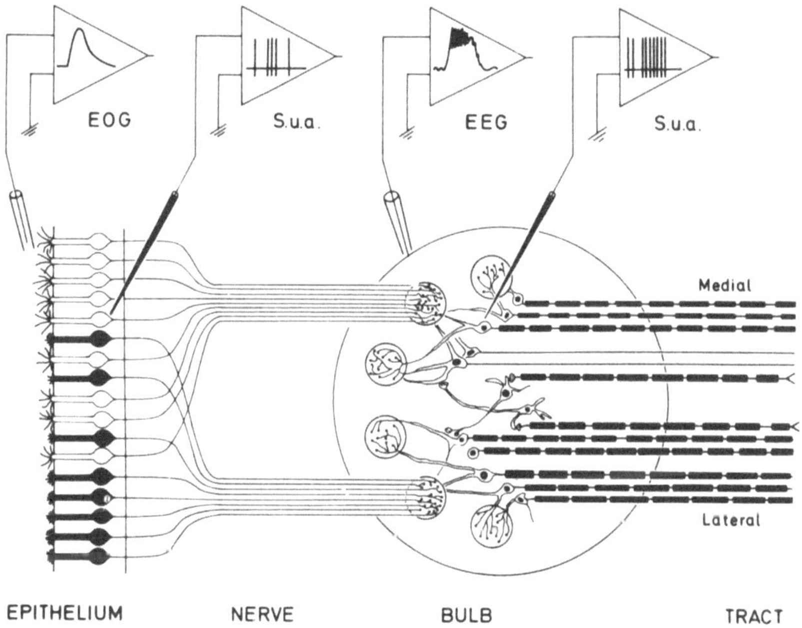

Fig. 15. Schematic and tentative diagram of the olfactory system in fish, illustrating the nervous elements and the methods used for recording electrical activity. In the epithelium, neurones drawn in *black* indicate microvillous receptor cells; neurones drawn with *simple line* indicate ciliated receptor cells. The uneven distribution of neurones in the epithelium is exaggerated. The neurones are tentatively sorted to terminate in the medial and lateral parts of the bulb. From the epithelium can be recorded the receptor potential, electro-olfactogram (*EOG*) or the discharges from single receptors, single unit activity (*S.u.a.*). The secondary neurones with cell soma in the bulb have dendrites extending into the glomeruli. Only two glomeruli are drawn filled with receptor axon terminals; the rest of glomeruli are drawn filled with dendrites of secondary neurones. Centrifugal fibers end on interneurones at the internal cell layer of the bulb. A division of the tract into four bundlets as seen in the cod, *Gadus morhua*, is indicated, as well as the presence of unmyelinated fibers in the lateral part of the medial bundle. Examples are shown of the slow wave activity (*EEG*) that can be recorded from the bulb surface and the single unit activity (*S.u.a.*), from the bulbar neurones

Scattered among the mitral cells are ruffed cells. They were first discovered and described by Kosaka and Hama (1979, 1981) in the goldfish *Carassius auratus*, and have also been described by Kosaka (1980) in the catfish *Prasilurus ascotus* and sea eel *Conger myriaster* (Kosaka and Hama 1980). The ruffed cells have a dendritic nest about 50 – 100 μm from the cell body. Many elaborate protrusions arise from the initial portion of the axon. The protrusions make synaptic contacts with cells believed to be granule cells. The ruffed cells do not appear to receive input from the receptor axons. The function of these cells is unknown. In the glomerular region are found mixed-synapse cells, and around the nest region of the ruffed cell are found perinest cells (Kosaka and Hama 1983).

Synaptic Organization. Few studies have been conducted on the ultrastructure of the fish olfactory bulb. Apart from the studies on the ruffed cells, Ichikawa (1976) has described the fine anatomy of the goldfish *Carassius auratus*.

In the glomeruli, asymmetrical synapses are found, characterized by a dense thickening of the postsynaptic membrane and spherical synaptic vesicles. These synapses are thought to be excitatory. Dendrodendritic synapses were also observed in the glomeruli. Pictures taken by Ichikawa (1976) show that a sensory receptor axon terminates onto several dendrites of secondary neurones. If these dendrites belong to different cells, this implies that the receptor cells branch to make contact with several relay cells, as shown in the mammalian olfactory bulb (Pinching and Powell 1971). In the relay cell layer, dendrodendritic synapses between relay cells and cells thought to be granule cells are observed. According to the authors, the polarity of the synapses is such that the mitral cells are excitatory to the granule cells and the granule cells inhibitory to the relay cells.

Both asymmetrical and symmetrical synapses are found in the internal cell layer. The neuronal elements, however, were not identified. The centrifugal fibers originating in the telencephalon terminate in this layer (Ichikawa and Ueda 1979). In the degenerating terminals both types of synapses are described. The similarity of the synaptic organization to that of the mammalian olfactory bulb is stressed in these studies.

In a recent study on the synaptic organization of the teleost olfactory bulb, Kosaka and Hama (1983) stressed the discrepancies between the functional organization of the teleost bulb and that of mammals. The mixed-synapse cell is postsynaptic both to the relay and the granule cells, and receives asymmetrical synapses from dendrites of the relay cells and symmetrical synapses from granule cells. The perinest cell, not found in the mammalian bulb, has protrusions in the nest region and receives symmetrical synapses from the granule cell only. The output from these two types of cells is still unknown. Based on their recent findings, Kosaka and Hama made a diagrammatic representation of the synaptic connections in the olfactory bulb of teleosts; this diagram is reproduced in Fig. 16.

Convergence Ratio. The number of axons making synapses in the olfactory bulb is in the order of 10^7, while the number of fibers in the olfactory tract is in the order of 10^4 (Døving and Gemne 1965). Thus, the nominal convergence ratio about 1000 : 1. As shown by van Drongelen et al. (1978), this convergence ratio increases the probability of response in the relay cells. Since there is reason to

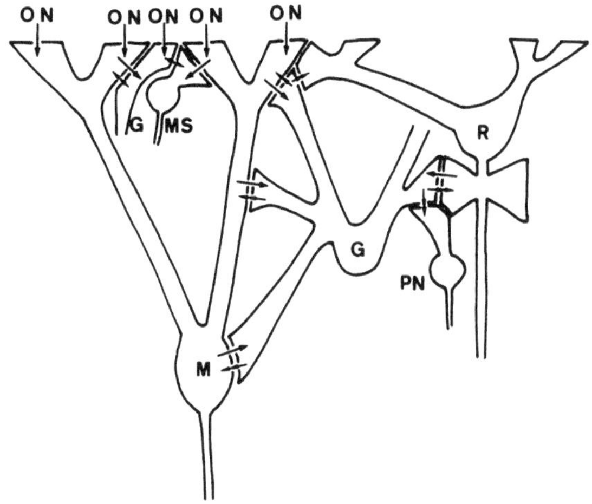

Fig. 16. Schematic drawing of the neurones identified in the fish olfactory bulb and the arrangements of the synaptic connections. *G*, granule cell; *M*, mitral cell; *MS*, mixed-synapse cell; *ON*, olfactory nerve; *PN*, perinest cell; *R*, ruffed cell; →, asymmetrical synapse; ↠, symmetrical synapse; ||, gap junction. (From Kosaka and Hama 1983)

believe that the axons branch and make synapses with several secondary neurones, the functional ratio will be greater than indicated by the counted number of receptor and relay cells, which further increases the response probability.

An increase in axonal branching and terminal sprouting results in an increased sensitivity or "gain" by the olfactory system. There will be greater sensitivity towards those odorants to which the sprouting receptors respond. Such mechanisms may explain a specific odor preference exhibited by animals that have previously been exposed to the odor. Achieved preferences, called learning or imprinting, are known, e.g., from feeding experiments in cod (Solemdal and Tilseth 1978) and from homing studies in salmonids (see Sect. 5.5).

3.2 Bulbar Surface Activity

A convenient method for recording electrical activity of the olfactory system is to place electrodes at the surface of the readily accessible olfactory bulb (Gerard and Young 1937). The electronencephalogram (EEG) recorded is illustrated in Figs. 15 and 17. The surface EEG activity in the fish olfactory bulb (Thommesen 1978) resembles that seen in the mammalian and amphibian bulbs (Ottoson 1959a, b). Stimulation of the sensory epithelium with odorants gives rise to two bulbar components, a slow shift in DC potential upon which are superimposed regular oscillations, frequently referred to as "induced waves" (Adrian 1950). Both the amplitude of the slow potential and the induced waves increase with increasing concentrations (Thommesen 1978, Døving et al. 1980). The amplitude of the slow component may reach a few millivolts. The induced waves are usually of smaller amplitude.

The origin of the slow component of the response obtained with odor stimulation is most probably due to the depolarization of the terminals of the receptor axons and the dendrites of the relay neurones, since antidromic stimulation of the relay neurones effectively blocks the induced waves, but leaves the slow DC shift intact

(Ottoson 1959c). The induced waves are of relatively constant amplitude during continuous stimulation, or they can appear as spindle-shaped bursts of waves (Ottoson 1959a, Thommesen 1978). The dominant frequency of the induced waves is temperature dependent (Dupé and Godet 1969, Graystone et al. 1970, Døving and Belghaug 1977, Thommesen 1978). The origin of the induced waves is presumably the synaptic elements in the layer of relay cells, most probably the synapses between the relay cells and interneurones of the bulb. The functional implication of the temperature dependence of the frequency of induced waves will be considered below (Sect. 3.3).

Any olfactory stimulus influences the olfactory bulb in such a way that a synchronized oscillating activity is produced in a specific population of neurones. By recording and mapping the bulbar surface activity, one can determine the localization of the neuronal pool which is activated by the stimulus. Thommesen (1978) has shown that a certain odorant induces activity in a specific region of the olfactory bulb. Thus, as seen from Fig. 17, amino acids induce activity in the lateral part of the bulb, while bile salts induce activity in the medial part (Døving et al. 1980, Thommesen 1982, 1983).

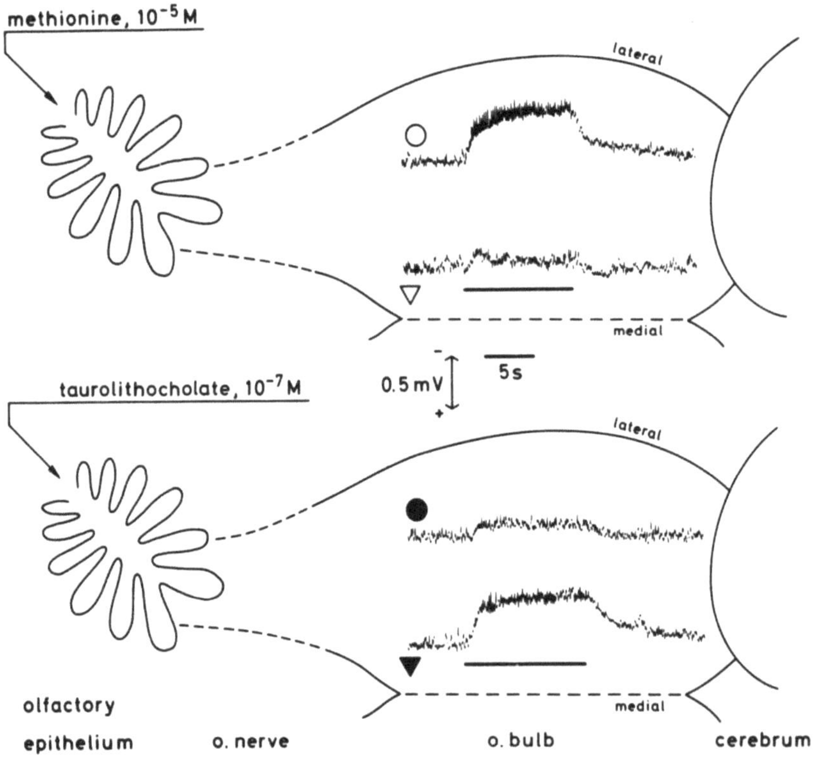

Fig. 17. Diagram of the olfactory epithelium and bulb of the right side of a salmonid, showing the characteristic distribution of responses recorded from the medial (▽, ▼) and lateral (○, ●) part of the bulb when stimulated by an amino acid (*open symbols*) and a bile salt (*filled symbols*). The two substances applied to the receptor epithelium induce responses in two different neuronal pools of the olfactory bulb. (From Døving et al. 1980)

When the electrical activity of the bulbar surface is recorded with capacity-coupled amplifiers, the DC shift is not apparent. Technically, the latter method is in many ways convenient to apply, but such methods may result in the loss of some significant information. Frequently EEG recordings have been made by placing *both* recording electrodes on the bulb surface (Hara 1982). Such recordings show the difference in potential at the two recording sites, e.g., the medial and lateral part of the bulb, but not in which part of the bulb any induced activity originates. The EEG activity is remote from the primary events in the olfactory receptors; thus, only limited inference is possible concerning the stimulating properties of the odorant and more information can be obtained using other techniques. Kudo and Ueda (1980) developed a method for recording the EEG activity of the olfactory bulb in the free-swimming carp *Cyprinus carpio* which may become important when relating the fish behavior to the odors of the environment.

Electrical stimulation of the olfactory nerve induces a series of complex wave patterns in the olfactory bulb of mammals (Shepherd 1963). The field potential recorded from the olfactory bulb of the rainbow trout *Salmo gairdneri* (MacLeod and Lowe 1976) and cod *Gadus morhua* (MacLeod 1976b) shows great similarity to those described for other vertebrates. The results favor the existence of dendrodendritic inhibition between interneurones (granule cells) and relay cells.

Olfactory Sensitivity. The olfactory system might gain an increased sensitivity by stimulus exposure during a certain period of an animal's life cycle. This proposition raises several interesting aspects of altering the olfactory function.

It is plausible that the olfactory system becomes more sensitive by increasing the number of molecular receptor sites bound to the receptorcell membrane. This increase in sensitivity can occur by an increase in the number of receptor cells, or can be take place by increasing the number of receptor sites on a given number of receptor cells.

The olfactory system is characterized by a large number of receptor cells terminating on a much smaller number of secondary neurones (Sect. 3.1). The sensitivity of the olfactory system is dependent upon the convergence of primary-onto-secondary neurones (van Drongelen et al. 1978). The contact between the axon terminals and the dendrites of the secondary neurones takes place in the olfactory glomeruli. It seems reasonable to assume that sensitivity can increase by increasing the number of axon terminals on the secondary neurones, thereby increasing the convergence ratio. This process has been termed sprouting. The latter alternative for sensitivity augmentation towards odors can only be demonstrated by recording from the olfactory bulb as no changes in the peripheral olfactory system are involved.

3.3 Activity of Single Neurones

The first measurements of the electrical activity of the fish olfactory system induced by odorants were made by Adrian and Ludwig (1938), who recorded the discharges of the olfactory tract in catfish. From the brain of a decapitated animal it was possible to record electrical activity for 30 min. Besides mechanical stimuli, which the authors found highly stimulatory, the most potent materials used as test stimuli were human saliva, frogs' eggs, and extracts of liver and blood. Extracts of frog muscle and of cheese were not particularly effective. The authors also showed that the discharges were enhanced upon stimulation of the olfactory epithelium with solutions of different kinds of food known to be eaten by the fish.

In later studies, the activity of the bulbar neurones has been recorded either by isolating a single active fiber from the nerve bundle of the peripheral stump of the olfactory tract, or by using microelectrodes for recordings from the bulb itself; the latter method is illustrated in Fig. 15.

Single-unit studies on the burbot *Lota lota* (Døving 1966, Døving and Hyvärinen 1969), char *Salmon alpinus* (Døving et al. 1974), rainbow trout *Salmo gairdneri* (MacLeod 1976a), sockeye salmon *Oncorhynchus nerka* (Bodznick 1978a), goldfish *Carassius auratus* (Meredith and Moulton 1978), and cod *Gadus morhua* (Kjøstolfsen 1983), all confirm the pioneer studies of the 1930s. A group of bulbar neurones are found to be activated by exposing the olfactory epithelium to a specific stimulus. In another group of neurones the activity is suppressed and in some inhibited to silence throughout the stimulation period by the same stimulus. A majority of the cells are left indifferent to the stimulus in question. The efficiency in evoking a specific response in a bulbar neurone by means of an olfactory stimulus increases with increasing concentration (Meredith and Moulton 1978).

Gusel'nikova et al. (1971) have made intracellular recordings from neurones of the carp olfactory bulb. The membrane potential was found to vary between 25 and 60 mV, with the spike amplitude never exceeding the membrane potential. The bulbar neurones were of two kinds. In one type of bulbar neurone, antidromic and orthodromic electrical stimulation induced one, seldom two, spikes. The other type of neurones responded to such stimulation with rhythmical potential variations of a frequency of 26–33 Hz. In both neurones the activity was followed by an early and late inhibitory postsynaptic potential (IPSP). The late IPSP has been shown by Nowycky et al. (1981) to be blocked by bicuculline in studies of the turtle olfactory bulb.

The results obtained by recording the surface potentials are qualitatively different from those obtained by single-unit studies. Only the latter are closely related to the input messages that the fish brain receives. This point has been missed several times in the past. In some discussions of physiological studies different techniques have been considered together as EEG recordings (see e.g., Hahn 1976, Quinn et al. 1983), despite differences in the level of precision.

Discrimination. In more recent years, when discussing the discriminatory properties of a sensory system, attention has focused upon the activity of single neurones rather than on gross EEG activity. In studies of sensory systems the cellular recordings have revealed new and important information about the nervous substrate for behavior. The cellular approach has been formulated by Barlow (1972) in five "dogmas". The first reads: "To understand nervous function one needs to look at interactions at cellular level, rather than either a more macroscopic or microscopic level, because behavior depends upon the organized pattern of these intercellular interactions." According to this theory, the perception of a signal corresponds to the activity of a small selection of very numerous high-level neurones. Another of the dogmas is that the high impulse frequency in the evoked neurones corresponds to a high degree of certainty that the trigger feature is present.

In most sensory systems, stimulus parameters (frequency of sound, wavelength of light, depth of mechanical indentation) have been used to study the responses of the sensory neurones. The situation is reversed in the olfactory system, where the neural recordings have been used to obtain information about the odorants, i.e., the neural responses of olfactory pathways have been explored in a search for physical/chemical parameters determining odor quality.

Variations in the odorous environment of the fish can release behavior patterns and will evoke activity in different samples of olfactory neurones. These variations can be monitored by recording the activity of single neurones. Single-unit activity can be obtained from all parts of the olfactory system. Recordings from bulbar neurones have shown that the olfactory system can distinguish between different fish populations as in the char *Salmo alpinus* (Døving et al. 1974) and between a number of amino acids as in the burbot *Lota lota* (Døving 1966), rainbow trout *Salmo gairdneri* (MacLeod 1976a), and goldfish *Carassius auratus* (Meredith 1981).

As mentioned above, in the case of salmonids, it has been shown that amino acids evoke activity in the lateral part of the dorsal aspect of the bulb, while bile salts give rise to activity in the medial part (Fig. 17). In the olfactory bulb of salmonids the two groups of chemicals, at present represented by bile salts and amino acids, evoke responses in distinct neural pools. Even gross EEG recording can be used to demonstrate this difference in odor quality. As will be discussed below, the medial and lateral parts of the olfactory bulb project to different parts of the central nervous system (CNS). The implications are that fish have a definite need to distinguish the two groups of chemicals. Further studies will show whether there are other groups of chemicals which activate other neuronal pools. When recordings are made with both recording electrodes placed on the bulbar surface at the medial and lateral parts (Hara 1982), spatial information is lost, and it is impossible to determine from which recording site the activity originates.

The frequencies of induced waves have been related to discriminatory mechanisms (Hara et al. 1973, Kaji et al. 1975, Satou and Ueda 1975, Kudo et al. 1972). The frequency is a local and temperature-dependent phenomenon (Dupé and Godet 1969, Graystone et al. 1970, Døving and Belghaug 1977), and it is hardly reasonable to associate the discriminatory properties of the neurones of the olfactory pathways with their dependence on ambient temperature.

Discrimination of odors is of major importance for many functions in animals. It is essential that sensory discrimination can take place in different environments where one variable is, for example, temperature. It seems unlikely that the oscillatory waves of the bulb have a more profound implication for the discriminatory ability of poikilotherm animals. If "neural coding" of olfactory "language" is reflected in the spectral patterns of the olfactory response (Kaji et al. 1975), meaning that the frequency of the oscillatory waves is important for the animal's ability to determine the quality of an odor, it follows that the quality of the sensation would change with the temperature of the animal's environment.

Based on studies on the bream *Abramis brama* and goldfish *Carassius auratus* (Nanba et al. 1966), it has been suggested that the time sequences of impulse activity in the relay cells may transmit information about the chemical quality of the odorants. Using similar recording techniques on the burbot *Lota lota*, Døving and Hyvärinen (1969) could not reveal any specific mode of activity connected to a particular odorant. Differences in activity patterns were usually greater between the cells than under different experimental conditions for one cell. The authors concluded that it was unlikely that differences in external conditions could be coded as variations in the time structure of the impulse sequence. This conclusion is strengthened by the observation of spatial separation of the responses of the olfactory bulb to bile salts and amino acids, and the induction of specific behavior when stimulating different parts of the olfactory tract in cod (Døving and Selset 1980).

3.4 Adaptation

A fish following an olfactory clue, either searching for a mate, home site, or food, will be exposed to the pertinent chemicals for some considerable time. An odor trail may extend for kilometers in a specific layer in the ocean. A fish following such a layer would be at a loss if the sensory adaptation prevented it from obtaining information about the quality and concentration of the stimulus.

Studies of the function of the olfactory apparatus in amphibians and mammals (Ottoson 1956, 1959a, b) have shown that the primary sense organ adapts slowly to a continuous stimulus or pulsed stimulation repeated over a considerable time. Also the secondary neurones in the olfactory bulb of the frog respond to repeated stimulation with a burst discharge of nerve impulses (Døving 1964), which can remain at the same level for periods of several minutes. Satou (1971) studied the adaptation process by recording the induced waves in the bulb of the carp, and found that the lower frequency components remained at a constant level throughout stimulation.

In the cod *Gadus morhua* the relay cells (secondary neurones, recorded in the olfactory tract) responded to exposure of the olfactory epithelium to dilute concentrations of amino acids and bile salts. Continuous stimulation with the same substance for a period of up to 30 min revealed that some secondary neurones maintained the same discharge frequency at the end as in the initial part of the stimulation period (Kjøstolfsen 1983). The last mentioned study is the first to conclusively demonstrate that there are neurones in the olfactory system in the fish which are able to give sufficient information over time for guiding the animal in a homogeneous chemical environment. It follows that there is no need for the fish to leave a chemical environment to "disadapt" in order to be able to perceive the odorant.

4 The Olfactory Tract

The "stalked" olfactory bulb (Fig. 1) is an anatomical feature of the olfactory system found in fishes of the families *Gadidae, Siluroidae, Cyprinidae*, and

Myrmoridae. This property offers a unique entrance, not only to the olfactory system but also to a central tract, for anatomical, physiological, and behavioral studies.

The olfactory tract fibers are situated around the lumen of the lateral ventricle (rhinocele), which extends out to the olfactory bulb. Its inner surface is lined with a layer of ependymal cells. The tract fibers divide into a medial and a lateral bundle (Figs. 1 and 15), each of which in codfish can be divided into smaller bundlets. This anatomical feature, which is most easily seen just as the tract

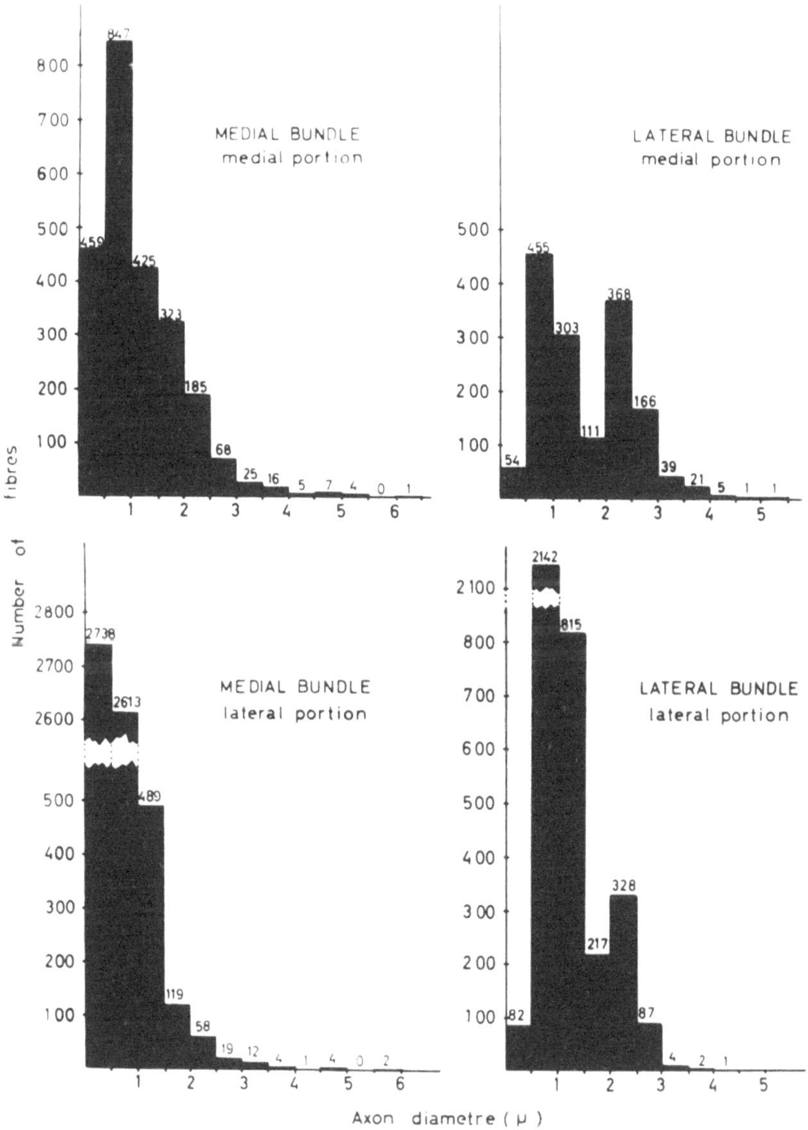

Fig. 18. Histograms of the axon diameters of the fibers of the four bundlets in the olfactory tract of burbot, *Lota lota*. (From Døving and Gemne 1965)

enters the telencephalon, has a functional implication which is vital for understanding the connection of the olfactory system to the centers commanding the behavior patterns in the CNS.

4.1 Axon Properties

In the burbot *Lota lota*, the myelinated tract fibers are less than 6.5 μm in diameter. There are two groups of myelinated fibers, one group with mean axon diameters of 2.2 – 2.4 μm and one group with diameters around 1.0 μm; see Fig. 18 (Døving and Gemne 1965). As seen from Fig. 19, the compound action potentials of the bundlets reflect the composition of the tract fibers. Two fast components have peak conduction velocities of 2.5 – 3.0 m s^{-1} and 1.2 – 1.8 m s^{-1}. In the carp *Carassius carassius*, the largest fibers were about 1 μm and the mean axon diameter in the lateral tract is about 0.6 μm (Westerman and Wilson 1968). These findings are reflected in electrophysiological measurements of several other species of carp (Døving 1967, Westerman et al. unpublished results).

The tract contains a large number of unmyelinated fibers with a diameter ranging from 0.1 – 0.5 μm. This is revealed by electron microscopic studies and in electrophysical records. The slow component in the compound action potential has a peak velocity of 0.14 m s^{-1} at 10 °C, which is consistent with the properties of unmyelinated fibers found in other fish nerves (Døving and Gemne 1965). In the burbot the unmyelinated fibers dominate the lateral aspect of the medial olfactory tract. In the carp, unmyelinated fibers are more common throughout all bundlets of the tract.

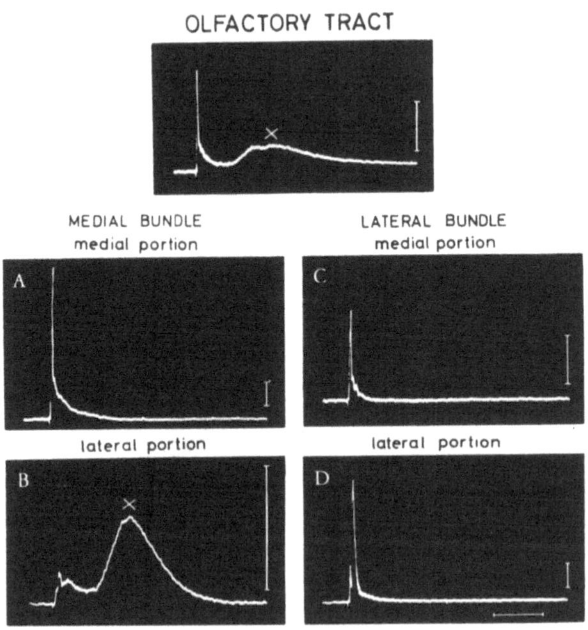

Fig. 19 A – D. The compound action potential of the olfactory tract (*upper trace*) in burbot, *Lota lota*, reflecting the two fast components and a third slow component marked *x*, which is found in the lateral portion of the medial bundle only (**B**). The traces **A** to **D** were obtained by recording from single bundlets corresponding to the axon diameter histograms in Fig. 18. Conduction distance in all cases, 20 mm. *Vertical bars*, 0.4 mV. *Time bar*, 100 ms. (From Døving and Gemne 1965)

Total Number of Fibers. The number of myelinated fibers is in the order of 10^4 in the burbot *Lota lota* (Døving and Gemne 1965). The number of unmyelinated fibers was not counted in the burbot, but Westerman and Wilson (1968) estimated that the lateral tract in the carp *Cyprinus carpio* contained about 5 650 and the medial bundle about 59 900 unmyelinated fibers.

4.2 Organization of Tract Fibers

The pathways of the olfactory tract fibers were first studied by Sheldon (1912) in the goldfish *Carassius auratus*. The tract contains myelinated and unmyelinated fibers of central and peripheral origin. In the present review, I shall discuss the origin and termination of the neurones of the olfactory bulb and then the origin and termination of the fibers that run from the CNS out to the olfactory bulb.

Bulbar Origin of Olfactory Neurones. The neat subdivision of the olfactory tract into bundlets (Sheldon 1912) and the clear subdivision of the responses of the bulb potential in response to natural odorants (Døving et al. 1980) makes it reasonable to ask whether this organization reflects the organization of the cell soma of the bulbar neurones.

Electrophysiological studies on the carp *Cyprinus carpio* (Satou et al. 1979) and tench *Tinca tinca* (Dubois-Dauphin et al. 1980) have shown that the majority of neurones situated in the lateral aspect of the olfactory bulb have axonal projections in the lateral olfactory tract, while neurones situated in the medial part of the olfactory bulb have axons that project to the medial tract. This property of projection is indicated in Fig. 15. In both studies mentioned above, a few units were found which were antidromically activated by stimulation of both bundlets of the tract, showing the bifurcation of secondary neurones which sends branches to both the medial and the lateral olfactory tract.

The activity of bulbar neurones picked up by microelectrode recordings could be suppressed by tract stimulation of either tract (Dubois-Dauphin et al. 1980). However, most of the units which showed suppression by tract stimulation were cells that were excited by the other tract portion. Since there is a clear topographical projection of neurones in the bulb to a tract bundle, stimulation of the lateral olfactory tract inhibited units at the medial part of the bulb and vice versa. This subdivision of the bulb was confirmed in a recent study using intracellular techniques (Satou et al. 1983).

Termination of Olfactory Neurones. Following sectioning of the olfactory tract in the shark, degenerating fibers could be traced in three distinct areas. The lateral olfactory area, the olfactory peduncle, and a more caudal region called the lateral portion of the area superficialis basalis (Schnitzlein 1966, Heimer 1969, Ebbeson and Heimer 1970, Ebbeson 1972, Bodznick and Northcutt 1979). A more extensive distribution of tract fibers was described in the shark *Scyliorhinus canicula* and ray *Raja clavata* (Smeets 1983). Three major terminal fields have also been found within the forebrain of teleosts: (a) lateral, (b) medial, and (c) central-posterior. A tentative diagram of the terminal fields is shown in Fig. 20. For the reader interested in the nomenclature of the different

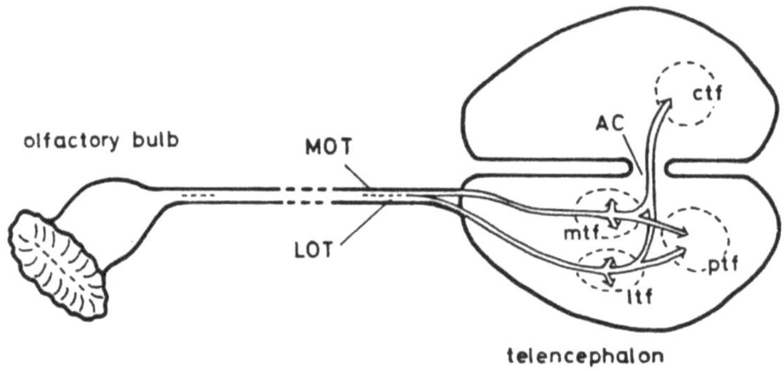

Fig. 20. Schematic diagram of the terminal fields of the secondary olfactory pathways in goldfish. *AC*, anterior commissure; *ctf*, contralateral terminal field; *ltf*, lateral terminal field; *mtf*, medial terminal field; *ptf*, posterior terminal field. (From Oka, Ichikawa and Ueda 1982)

areas, Finger (1975) has made a useful tabulation of the various terms and compared them with the mammalian brain. The terminations of the tract fibers in different areas have been confirmed in the goldfish *Carassius auratus* (Ichikawa 1975a, b, Ichikawa and Ueda 1979), bullhead catfish *Ictalurus nebolosus* (Finger 1975), channel catfish *Ictalurus punctatus* (Bass 1981), moray eel *Gymnothorax funebris* (Scalia and Ebbeson 1976), and *Eugerres* sp. and *Holocentrus* sp. (Vanegas and Ebbeson 1976). In the bichir *Polypterus palmas* and *Polypterus ornatipinnis*, the following target areas were identified: the lateral subpallium nucleus and the dorsal and ventral endopeduncular nuclei (Bradford and Northcutt 1974).

Rhinoendocrine Connections. Finger (1975) has shown that in the bullhead catfish *Ictalurus nebolosus* neurones of the olfactory bulb terminate in the hypothalamic nucleus, nucleus preopticus. Kandel (1964) has demonstrated that electrical stimulation of the olfactory tract induces orthodromic discharges in neuroendocrine cells of the preoptic nucleus, which were identified by antidromic stimulation of the hypophyseal stalk. The delay at which the orthodromic response was evoked indicates that this connection is formed by the unmyelinated fibers of the tract. Long-term electrical stimulation of the olfactory tract in the catfish *Clarias batrachus* for 5 min resulted in total degranulation of the endocrine neurones of the pars magnocellularis (Rao and Dabhade 1973). These findings indicate a direct effect by the olfactory system on the endocrine function in fish. Future studies might yield exciting new information in the borderland between the olfactory sense and hormones.

Centrifugal Fibers. The olfactory tract contains fibers that transmit impulses from the telencephalon to the olfactory bulb. Such fibers have been traced in the goldfish, *Carassius auratus* (Oka 1980). By staining the tract fibers at one side the stained cell bodies have been found ipsilaterally to the injected side in area ventralis, and in the ventral part of the area dorsalis in the telencephalon.

In the same anatomical study, Oka (1980) described the centripetal fibers from the bulb and found that these secondary bulbar neurones terminated in regions of the telencephalon overlapping those containing the cell bodies of centrifugal fibers. Oka (1980) claimed that the medial bundle contained most of the centrifugal fibers to the bulb. This finding is not consistent with electrophysiological studies on the burbot (Døving and Gemne 1966), in which centrifugal fibers to the bulb were found to be present in all bundlets of the olfactory tract.

Nervus Terminalis. Sheldon proposed that the olfactory tract contained fibers of the nervus terminalis, i.e., fibers from the primary receptors of the vomeronasal organ (organ of Jacobson). However, the olfactory nerve is composed of another type of unmyelinated fibers. The axon diameters of the olfactory tract fibers are larger and the corresponding conduction velocity higher than in the fibers of the olfactory nerve. Another argument dismissing the possibilities of nervus terminalis is the disappearance upon asphyxia of the neurophysiological response in the olfactory tract, when stimulating the olfactory nerve with electrical pulses. As would be expected with a postsynaptic potential, the response from the olfactory tract diminished rapidly as the blood flow stopped. If the olfactory tract fibers were stimulated directly, however, a normal compound action potential existed (Døving, unpublished observation).

Interbulbar Connections. Sheldon (1912) questioned the existence of interbulbar fibers in the carp. The studies by Oka (1980) and Døving and Gemne (1966) have confirmed Sheldon's doubt. The latter authors demonstrated, however, a close connection from one bulb to another, mediated via synaptically coupled neurones. The interbulbar connections are interesting from a functional point of view and indicate one way in which comparisons of the information received by the right and left olfactory organs may be interpreted by the fish.

4.3 Olfactory Tract Recordings

The stalked olfactory bulb is a convenient target for recording devices. Single and multiunit recordings made from the tract can readily be related to the stimuli flushed through the olfactory organ. When the olfactory tract is cut and recordings made from the bulbar stump, the activity originates from the bulbar neurones, of which most are probably secondary neurones.

The first to demonstrate the convenience of the preparation were Adrian and Ludwig (1938), who used the catfish. It took over two decades before such experiments were continued (Boudreau 1962, Døving 1966, Nanba et al. 1966, Satou and Ueda 1978). Westerman and von Baumgarten (1964) showed that the neurones of the olfactory tract presumably grow back to their original target areas after being cut, because the fish perform the same behavior patterns as before sectioning. The results of the different studies on the olfactory tract will be discussed in more detail in their appropriate connection in this review.

5 Olfactory-Induced Behavior

Studies on olfactory function in fish have shown that chemical substances can induce specific and socially relevant behavior. The olfactory system mediates this

behavior. Since the olfactory tract is separated into bundlets and the relay neurones end in discreet terminal fields in the telencephalon, it is plausible that the different bundlets serve different functions. The role played by the telencephalon in the promotion of behavior has been thoroughly reviewed by Hollis and Overmier (1978). In the present review, I shall concentrate on the effects incurred by manipulating the olfactory input. Figure 21 gives a summary of the behavior induced by electrical stimulation of the olfactory bundlets in the cod *Gadus morhua.*
In the past doubts have been expressed as to the validity of discriminating between smell and taste in fish (for review, see Atema 1977). In some fish the olfactory system seems to dominate the chemosensory pathways, in others, e.g., the catfish, the gustatory system plays an important role in chemical detection, especially for food. At present we do not know exactly which chemicals evoke the different behavior patterns, neither do we know which sensory system is involved. Frequently the idea has been presented and also substantiated that it is the mixture of substances rather than single components that are important for induction of a particular behavior. I would like to advocate an opposite view, namely, that the substances are more elusive than we imagine at present and the mixtures are efficient because the component substances act synergistically on the receptor sites, just as any small shoe will fit into a large galosh.

Fig. 21. The cod in a resting position at the bottom (*middle*), and the various behavior patterns observed by electrical stimulation of the four olfactory tract bundlets. Stimulation of the lateral part of the lateral olfactory tract induces a benthic food search with head down, the fish moving backwards (*lower left*). Stimulation of the medial part of the lateral tract induces swimming and snapping with the jaws (*upper left*). Stimulation of the lateral part of the medial tract induces swimming at the surface (*upper right*). When the medial part of the medial tract is stimulated, the fish presses to the bottom and changes colour (*lower right*). (From Døving 1982)

5.1 Feeding Response

Electrical stimulation of the olfactory tract in the goldfish *Carassius auratus* was carried out by Grimm (1960). He showed that fish in which electrodes were implanted responded by a complete feeding performance, arousal, or orienting behavior. The hungry fish, when encountering a familiar food odor, would start swimming around. This behavior was followed by a phase in which objects on the aquarium floor were selectively attacked. A nonobjective "seeking" phase continued with the fish head-to-sand, indiscriminately sampling large areas of the sand for food. Stacey and Kyle (1983) showed that in the goldfish the feeding response was less affected by medial than lateral olfactory tract section. The reason for the remaining feeding behavior after complete olfactory tract sectioning may be due to the well-developed gustatory system of the goldfish.

The situation in the cod favors subdivision of the tract into bundlets. This feature was put to advantage in a study on the cod *Gadus morhua* (Døving and Selset 1980), where behavior was observed when stimulating discreet bundlets of the olfactory tracts. When stimulating the lateral part of the lateral olfactory bundle, the cod took a tilted position with head down, swimming backwards (Fig. 21). The behavior could be induced by shock intensities of 0.3 to 6 μA. The behavior was seen with frequencies as low as 4 Hz, but higher shock frequencies increased the speed at which the cod adopted the head-down position. Increasing the frequency above 40 Hz did not induce further changes in the release of behavior pattern.

The head-down and backward movement seen in cod can be associated with a behavior pattern adapted to finding benthic prey. The input from the sense organs of the barbel and the first rays of the pelvic fins, which are dragged along the bottom during the feeding behavior, give the fish additional information for locating the prey. The behavior seen when stimulating this particular part of the olfactory tract is equivalent to that described as a feeding response in the free-swimming cod (Brawn 1969, Pawson 1977).

Feeding Substances. The use of the olfactory sense by fishes to find food was demonstrated in a series of experiments described by Bateson (1890), Sheldon (1911, 1912), and Parker (1910, 1912). The first attempts to find the natural substances from the prey's odor, releasing a specific response in the predator, were apparently those by Kleerekoper and Morgensen (1963). They showed that the body odor of trout stimulates an unconditioned parasitic lamprey *Petromyzon marinus* to increase its locomotor activity. Substitution of the body odor with a mixture composed of nine out of the eleven amino acids present in the body odor indicated that the responses could be elicited by a single component only. The substance has later been said to be isoleucine methyl ester (Kleerekoper, referred to by Bardach and Villars 1974).

The substances evoking feeding response in a series of West Indian fish species were isolated from a mollusc *Arca zebra* and a snail *Strombus gigas*. The substances, called arcamine and strombine, were identified as hypotauryl-2-carboxyglycine and C-methyliminodiacetic acid, respectively (Haynes et al. 1967, Sangster et al. 1975).

Sutterlin (1975) investigated the responses of a series of marine species to extracts of bait and found that of the amino acids present, glycine was the most effective.

In the same study he reports on the attraction of the hagfish *Myxine glutinosa*. The amino acid composition of the bait was reconstituted in the right proportions and tested. In the attraction studies *no* hagfish were caught. Pawson (1977) suggests that glycine may be responsible for the behavioral responses, i.e., attracting the whiting *Merlangus merlangus* and cod *Gadus morhua*. Mackie et al. (1980), studying the Dover sole *Solea solea* found that an imitation of the preferred diet, mussel *Mytilus edulis*, containing the complete chemical mixture of L-amino acids, did not possess the feeding stimulant activity. However, feeding behavior was evoked by glycine betain together with amino acids (Mackie et al. 1980). Chemical fractions of the natural prey odor for the yellow-fin tuna *Thunnus albacares* and synthetic mixtures of amino acids were less effective than the whole natural rinse (Atema et al. 1980). The authors suggest the existence of a chemical search image, which may be a useful working hypothesis in forthcoming studies from that research group.

The feeding stimulants for the rainbow trout *Salmo gairdneri* were claimed to be L-amino acids, since the extract of squid was stimulatory as well as a complete synthetic mixture of known composition; but omission of the amino acids abolished the activity (Adron and Mackie 1977). Freeze-dried clam extract induces feeding behavior in the red hake *Urophycis chuss*, at concentrations of 10^{-8} g/l and higher (Olla, Pearson and Miller 1980). The feeding behavior is a complex phenomenon involving both olfactory and gustatory pathways. The feeding behavior and the substances involved have been reviewed by Carr (1982), Hidaka (1982), and Mackie (1982).

Amino Acids. The behavioral studies mentioned above indicate that amino acids are potent olfactory stimulants for fish. Amino acids are also potent gustatory stimulants, and in some cases the sensitivity to these substances is higher in the gustatory than in the olfactory system.

Behavioral studies using classical conditioning techniques have confirmed the high sensitivity of the fish to amino acids in the catfish (Holland and Teeter 1981), char *Salmo alpinus* (Jònsson 1977), and cod *Gadus morhua* (Johnstone 1980). Johnstone (1980) demonstrated a dependence of the threshold level of glycine on the background concentration of the same substance. He stressed the importance of monitoring the background levels of chemicals when performing threshold determinations.

Structure Activity Relationship. The induced waves, which can be recorded from the olfactory bulb, have been used by Hara and coworkers for studying the molecular configuration of the receptor sites in the sensory cells. As discussed in the section on the olfactory bulb, EEG responses are dependent upon the recording site and the quality of the chemical. The studies have been performed on the rainbow trout *Salmo gairdneri* (Hara 1973, 1976a, b, c, 1982), brook trout *Salvelinus fontinalis*, and white fish *Coregonus clupeaformis* (Hara et al. 1973b). A hypothetical receptor site was proposed from the above data (Hara 1975, 1976a, Brown and Hara 1982). A review of these studies is published by Hara (1982).

Valuable information has been obtained on the olfactory receptors recording the receptor potential (EOG) using the method of cross-adaptation studies. Caprio

(1977, 1978, 1980, 1982) compared the properties of the gustatory and olfactory systems in the catfish *Ictalurus punctatus* by recording EEG nerve activity and bulbar responses. Cross-adaptation experiments suggest that at least three different receptor sites are involved in amino acid detection. Three groups of stimuli show a relative lack of cross-adapting effectiveness: the basic amino acid L-arginine, the 3-carbon sulphur-containing amino acid L-cysteine, and the 5-carbon amino acid L-methionine or L-norleucine. In these studies by Caprio, arginine seems to have a receptor function that is independent of the other amino acid receptor mechanisms, both in the gustatory and the olfactory systems.

By recording multiunit activity from the olfactory tract, the thresholds to some odorants were determined in the carp *Cyprinus carpio* (Goh and Tamura 1978a). The same authors have studied the effect of amino acids on the olfactory system in the Red Sea bream *Chrysophyrys major*, conger eel *Conger myriaster*, and mullet *Mugil cephalus* (Goh and Tamura 1980, Goh et al. 1979). The olfactory spectra were about the same for the saltwater as for the freshwater fish.

Zippel and Voigt (1982) did not find a high sensitivity to amino acids in the olfactory system of the goldfish *Carassius auratus* and state that the amino acids are probably of limited biological relevance for the olfactory behavior of this species. Threshold determination of the amino acids was performed on the char *Salmo alpinus* by recording the EEG activity in a descending series of concentrations until the responses were no longer seen (Belghaug and Døving 1977). The results of these studies are congruent with the measurements of the olfactory efficiency measured at a given concentration. For a more detailed discussion of the various studies on amino acids as gustatory and olfactory stimuli, the reader is referred to the review by Caprio (1982).

5.2 Orienting Reaction

Electrical stimulation of the medial bundlet of the lateral olfactory tract in the cod *Gadus morhua* (Døving and Selset 1980) induced an orienting reaction which was characterized by swimming movements and searching up, down, and around the aquarium. The orienting reaction was followed by snapping movements of the jaws which were so fast that they were completed within 100 ms (Fig. 21).

Bell and Tobin (1982) have published a very useful review of the orientation of animals exposed to chemicals in the environment. They point out the very correct notion that studies in chemoorientation must begin to include the combined effects of the chemical and nonchemical sensory information available to the organism. Another point which may be useful to note is that a single brief stimulation, either chemical or electrical, can release an appropriate behavior, such as an orienting reaction that outlasts the actual duration of the stimulus.

In behavior experiments, taurocholate – one of the most common bile salts found in fish – has been shown to evoke an orienting reaction and a snapping reaction in the cod *Gadus morhua* (T. Hellstrøm, personal communication). The reaction is only partially dependent upon an intact olfactory system, since cutting of the olfactory tracts increased the threshold for eliciting the reaction by a factor of 100 from 7 n*M* to 700 n*M*. Most probably the gustatory system can evoke the same type of behavioral reaction as the olfactory organs in this fish.

5.3 Reproductive Behavior

Stimulating the lateral part of the medial bundle with electric shocks induces swimming with the snout above the water surface in the cod. At high shock intensities the fish quivered, and rapid uncoordinated muscle contractions could be seen down the side of the body (Døving and Selset 1980). In a comprehensive description of spawning behavior, Brawn (1961b) noted that the upward swimming is an invitation to the female from the male, to spawn (Fig. 22). The spawning involves a horizontal swim with stiff body and quivering movements. Quivering movements were observed in the cod on electrical stimulation, but ejection of egg or milk was never induced. If mature cod are taken to aquaria in the spawning season the fish swims with the snout above water (S. Tilseth, personal communication). In the goldfish *Carassius auratus*, Stacey and Kyle (1983) found that bilateral sectioning of the medial olfactory tracts reduced male sexual behavior. The authors state that the medial part of the medial olfactory tract may be more important than the lateral part for courtship expression. Demski et al. (1975) found that the sperm-release mechanisms in the male green sunfish *Lepomis cyanellus* included a system extending from the preoptic region to the spinal cord, passing through the medulla, the hypothalamus, midbrain, tectum, and basal midbrain. Lesions in the medial nuclei of the ventral telencephalon blocked the initiation of spawning behavior in the male goldfish *Carassius auratus* (Kyle and Peter 1982, Kyle et al. 1982). This region binds sex steroids (Kim et al. 1978), a finding which may be relevant to the steroid-dependent release of spawning behavior in the goldfish. Reproductive behavior could be evoked by stimulation of the area dorsalis telencephali pars centralis and the preoptic area in the male sunfish *Lepomis cyanellus* (Demski 1978) and bluegill *Lepomis macrochirus* (Demski and Knigge 1971).

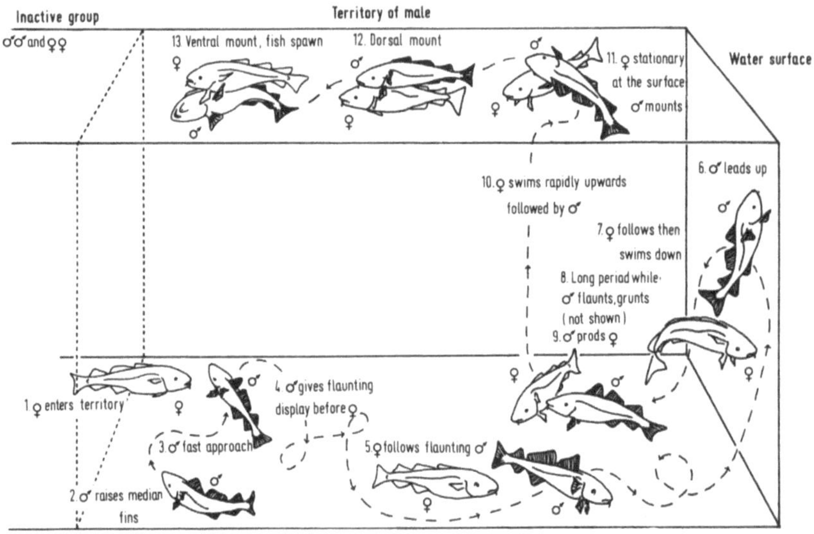

Fig. 22. Simplified representation of the courtship and spawning behavior of the cod, *Gadus morhua*. The *dotted lines* indicate the path followed by the male. Note the position of the fin rays of the male. (From Brawn 1961b)

Pheromones in Reproduction. It has been shown that sexually ripe female fish attract males of the same species. This is the case in the rainbow trout *Salmo gairdneri* (Newcombe and Hartman 1973, Emanuel and Dodson 1979, Honda 1980), goldfish *Carassius auratus* (Partridge et al. 1976, Yamazaki and Watanabe 1979), and chiclid, *Haplochromis burtoni* (Crapon de Caprona 1980). Males secrete substances that attract females in species of blennids, *Hypsoblennius jenkinsi, H. robustus* and *H. gentilis* (Losey 1969), *Brachidanio rerio* (Bloom and Perlmutter 1978), and *Gobius jozo* (Colombo et al. 1982). The last mentioned authors have demonstrated that the prominent mesorchial gland synthesizes mainly 5-beta-reduced androgen conjugates, particularly ethiocholanolone glucoronide. The latter substance has also been found to attract females of the goldfish *Carassius auratus* and guppies *Poecilia reticulata.* Colombo et al. (1982) concluded that the substance is specific with respect to sex and chirality, but not to one particular species. Intact, but not anosmic male zebra fish, *Brachydanio rerio*, are attracted to the sexually mature females by means of steroid glucuronides secreted from the ovaries. The olfactory stimuli induce the males to start a series of courtship behaviors (van den Hurk and Lambert 1983).

5.4 Alarm Reaction

Von Frisch (1941a, b) was the first to describe the alarm reactions of the minnows *Phoxinus phoxinus* to extracts from the skin of their own species. Typically, the fish, when exposed to skin extract, form a tight school, refuse food, and seek hiding places at the bottom. The heart rate decreases upon stimulation (Pfeiffer and Lamour 1976). The coloration on the sides changes (E. Fabricius, personal communication). The behavior is mediated via the olfactory system, as shown by the loss of reaction in fish recovering from extirpation of the olfactory bulbs. It should be remembered that the minnows belong to the carp family and have a stalked olfactory bulb. In the cod (Døving and Selset 1980), stimulation of the medial part of the medial olfactory tract induced lifting of the pelvic fins, which made the fish lie down in close contact with the bottom. The respiratory movements decreased in frequency, and the fish changed color; they usually acquired white spots on the dorsal skin and dark areas on the side of the body (Fig. 21). Sometimes a snapping of the jaws was elicited, followed by a movement of retreat. The reactions are similar to those seen in cod frightened by mechanical stimulation of the aquarium.

The alarm reaction induced by skin extracts is found in the order of Ostariophysi fishes. The reaction is species specific as shown by Schutz (1956); Fig. 23 is a diagram illustrating his main results. As seen, the skin extracts from individuals of the same species induced the most potent reactions. The skin of the fishes in which this reaction can be demonstrated contains characteristic goblet cells (Pfeiffer 1967). The literature on the alarm reaction has been reviewed by Pfeiffer (1962, 1963a, b, 1982) and Smith (1982).

Alarm Substances. Pfeiffer and Riegelbauer (1978) showed that the dorsal light response of the black tetra *Gymnocorymbus ternetzi* varied when exposed to the skin extract of the minnows. The isolated "pure" alarm substances, together with

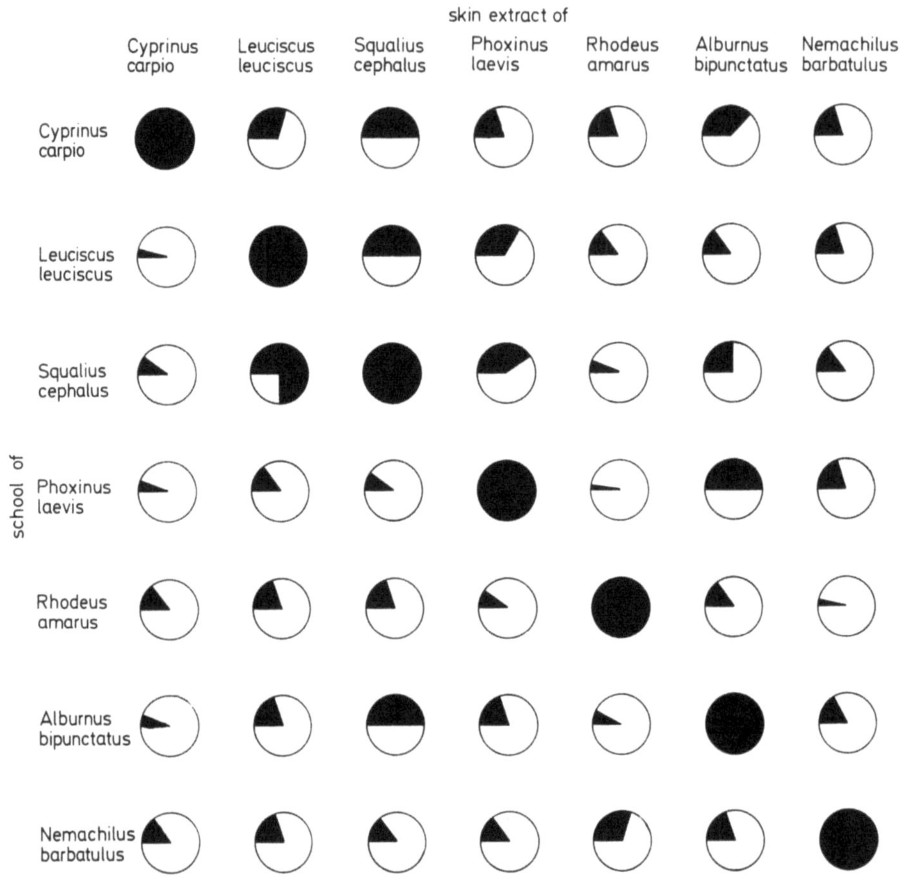

Fig. 23. Intensity of interspecific fright reaction shown diagrammatically by the *blackened area* of the circles. The intraspecific response is always most intense as shown by the *completely blackened circles.* Other reactions are rated in terms of extract dilution required to give comparable response; thus, skin extract of *Squalius cephalus* tested on schools of the same species produces full intensity reaction (*circle 100% black*), while skin extract of *Alburnus bipunctatus* must be four times as concentrated to produce the same intensity of reaction in *Squalius cephalus* and the response is rated as 1/4 (*blackened areas of the circle, 25%*). (From Schultz 1956 and Pfeiffer 1962)

isoxanthopterin and 6-acetonyl-isoxanthopterin, were highly effective in evoking the dorsal light response in the black tetra; the article did not state the concentrations used. Work on the alarm substances (reviewed by Pfeiffer 1982) has revised the previous statements, and the suggestion is now that the alarm reaction in cyprinids is due to hypoxanthine-3(N)-oxide. It is difficult to see how this simple substance can account for the variability and specificity of the alarm reaction in cyprinids (Fig. 23). Soviet physiologists have analyzed the skin extract of the minnow *Phoxinus laevis* and found that the substances inducing alarm have a mol. wt. of about 1100 (Kasumjan and Lebedeva 1979). They state that the alarm substance is most probably a hydrocarbon with amino groups.

Tucker and Suzuki (1974) studied the alarm substances by recording EEG and multiunit receptor responses from the olfactory epithelium of the catfish (species not stated). Only modest enhancement of biological activity was obtained by chemical purification procedure of the skin extract that induced an alarm reaction.

5.5 Homing

Since the pioneer studies of Gerking (1953), the ability of the fish to return to its home site after being relocated by floods or curious scientists has repeatedly been confirmed in a number of fish species. The return in salmonids is associated with an intact olfactory sense, as impairment of the olfactory sense drastically reduces the probability of correct return (Wisby and Hasler 1954, Groves et al. 1968, Toft 1975).

Most of the behavioral and physiological experiments have been performed on salmonid species. The behavioral work has been reviewed repeatedly (Harden Jones 1968, Hasler 1966, Stasko 1971, Groot 1982, Hasler and Scholz 1983). The experimental work comprising electrophysiological techniques has not been adequately evaluated as the differences between and limitations of the various methods have been ignored (Hara 1970).

Several experiments on the effect of transplanting fish have been conducted in an attempt to understand the mechanisms for homing in salmonid migration (see Donaldson and Allen 1957, Hasler and Scholz 1983 for review). The conclusions drawn from these experiments are that the memory of the home-stream "odor" is not inherited, and secondly that a rapid learning process "imprinting" takes place at the time of seaward migration, i.e., in the smoltification period.

Imprinting. The sucess of salmonid homing behavior led to the proposal of the imprinting hypothesis (Hasler and Wisby 1951, Hasler 1966). According to this hypothesis, the salmon smolt learns the specific river odors when descending the river on its outward migration. If follows from the imprinting hypothesis that the salmon smolt would home to the river of release, which in experimental situations could differ from the parental river.

Imprinting comprises a group of learning situations in which a stimulus comes to evoke a response in the animal. The stimulus starts a set of processes involving acquisition, storage, and retrieval of information. In the imprinting process it is not essential that the stimulus and response bear a close temporal relationship (Horn 1981).

It might well turn out that the salmonid behavior will fall within this description of imprinting. Up to the present time, several field experiments have been made that appear to support the imprinting hypothesis (for reviews, see Hasler and Scholz 1983). It is noteworthy that the evidence for odor imprinting is circumstantial. When the imprinting hypothesis for salmon homing was presented, only a few experiments indicating its odor dependence had been made (Craigie 1928, Wisby and Hasler 1954). Several later studies have demonstrated that the olfactory system is important for homing. However, no conclusive experiments have been conducted to show that the olfactory system is operating in the imprinting process of salmonid migration.

If the imprinting hypothesis for salmonid migration is to gain wider acceptance and become a useful tool in the stocking and management of salmon (or other fish species), several well-controlled experiments have to be performed. For example, one must know what substances are involved in the natural homing behavior. The fish pheromones might be promising candidates, but so far no salmon pheromones have been isolated and identified.

It has been demonstrated that certain chemicals can induce attraction to the salmon, as e.g., morpholine (Cooper et al. 1976, Scholz et al. 1978, Johnsen and Hasler 1980) and certain body odors attract the char (Høglund and Åstrand 1973, Selset and Døving 1980). But these attraction studies have not been knit to the function of the olfactory system. To demonstrate that the olfactory system is involved in an imprinting process of salmonid homing, a series of ablation experiments have to be done.

The transplanting experiments induced an extensive search for chemicals which could be used for inducing the "imprinting memory" of the fish. The choice fell on morpholine (C_4H_9NO), tetrahydro-1,4-oxazine, a heterocyclic compound, a strong base (pK, 9.61) with a molecular weight of 87, and the pH of a $10^{-1}\,M$ solution close to 10.5.

Behavioral Studies. Hasler's group conducted release experiments with the Pacific salmon and trout in the Great Lakes of North America. The work has been reviewed by Hasler and Scholz (1983). Fingerlings and smolt of the coho salmon *Oncorhynchus kisutch* and rainbow trout *Salmo gairdneri* (Scholz et al. 1973, 1978, Cooper et al. 1976, Cooper and Scholz 1976) were exposed to morpholine added to basin water, to give a concentration of $5.7 \times 10^{-10}\,M$ (1.5×10^{-5} mg/l) and released in Lake Michigan. During spawning migration in the lake, increased numbers of exposed fish were captured at the estuary of a creek scented with morpholine.

The stream movements of the adult coho salmon *Oncorhynchus kisutch* which had been exposed for 4 weeks to morpholine or phenylethyl alcohol in their smoltification period were studied by Johnsen and Hasler (1980). Morpholine was added to the right or the left half of the stream. Upstream movements of adult fish were observed in the case of 13 out of 18 experimental fish exposed to that substance as young, while only four of 19 fish exposed to phenyl ethyl alcohol showed positive rheotaxis under the same conditions. Since the dye mapping showed uniform distribution over the length of the study area, it is unlikely that the fish were following a gradient. Most probably the chemical acted as a sign stimulus inducing positive rheotaxis in the fish. It has been suggested that morpholine is a nonspecific attractant for salmonids (Mazeaud 1982), but the experiments mentioned above do not support this suggestion.

In ultrasonic tracking experiments (Scholz et al. 1973) migration was typically within 30 m of the shore. After the exposed fish encountered the line of morpholine, i.e., the region of the estuary of the river scented with morpholine, the fish stopped migrating and milled in the estuary for up to 4 h before continuing migration. The control fish not exposed to morpholine did not behave in this manner. The authors conclude that the fish exposed to morpholine recognized the chemical and used it for homing in an experimental situation. Another proposition is that the fish exposed to morpholine were conditioned for a considerable time to the substance and associated it with food. If the fish encountered such a familiar chemical environment during the migration phase, it would explore the environment scented with this chemical. No experimental evidence has been published showing that it is the olfactory system of the fish that senses morpholine in this situation.

The results of the above experiments show that, in the absence of natural spawning populations, Pacific salmon are able to home to tributaries scented with morpholine if they have been exposed to this substance as smolt. Nordeng (1971, 1977) has proposed that in natural populations pheromones are the significant clues.

Electrophysiological Experiments. Experiments have been performed with the notion that the EEG recordings could be used as a bioassay for home-stream recognition in migrating salmon to see if the previous history of the fish influences the magnitude of physiological response. Fingerlings of the rainbow trout *Salmo gairdneri* and coho salmon *Oncorhynchus kisutch* were divided into two groups. One served as a control, the other was exposed to morpholine.

As in the behavior work, the concentration of morpholine in the basin water was 1.5×10^{-5} mg/l (5.7×10^{-10} *M*). The concentrations used in the electrophysiological tests were 1%, 0.1% and 0.01% (that is 1.1×10^{-1} *M*, 1.1×10^{-2} *M*, and 1.1×10^{-3} *M*). Unfortunately, the authors selected 0.1 *M* sodium chloride as a standard reference solution for evoking a reaction (Cooper and Hasler 1974, 1976). Sodium chloride is not a potent odorant, and the concentration is above the normal working range of the fish olfactory system (Cooper and Hasler 1974, 1976). The authors claim that the early exposure of the fish to these low concentrations of morpholine induced enhanced responses to the substance in subsequent recordings made up to 18 months after the exposure period using high concentrations of stimuli.

The imprinting process does not require, but might well include, an increased sensitivity to an odor. It might also be that the increased sensitivity establishes itself during a specific period of the salmon life cycle. However, in the imprinting process the specific stimulus information is stored and can be retrieved; to this purpose the brain and not the peripheral olfactory system seems essential, as has been demonstrated for the imprinting process in the domestic duck (for review, see Horn 1981).

Apart from the uncertainties in advocating an increased sensitivity for morpholine in the imprinting process, several studies have expressed doubt about the olfactory relevance of morpholine (Hara 1974, Hara and MacDonald 1976, Hara and Brown 1979). Hara and MacDonald (1976) said that the effect of morpholine on the olfactory epithelium is caused by a mechanism not directly related to the olfactory function and that a 1% solution of morpholine temporarily inhibits the olfactory sensitivity of the rainbow trout. Lowering the pH of the 1% morpholine solution caused a marked reduction of the response amplitude recorded from the bulb (Hara and Brown 1979). The latter authors could not confirm the results of Cooper and Hasler (1976) and found no increase in olfactory responses of the rainbow trout previously exposed to morpholine. Alas, it is not known if Hara and Brown (1979) exposed their fish during the correct period of the trout life cycle.

The atypical responses from the olfactory bulb seen by surface recordings made by Hara and MacDonald (1976) have also been observed in our laboratory (Døving and Thommesen, unpublished results). The behavioral experiments with morpholine related to homing are interesting and very convincing. The electrophysiological studies throw doubt on the belief that morpholine is a potent odorant for the fish. Most probably morpholine would be classified as an uninteresting substance from the olfactory point of view. Its effect on the gustatory system or other chemosensory system of fish does not appear to have been investigated.

5.6 Migration at Sea

Ultrasound tracking experiments have demonstrated that the salmon *Salmo salar*, in the ocean, closely follow the horizontal stratification of the water masses (Westerberg 1982a, b). The records show that the fish, in their migrational route, stay at a certain layer of a given temperature for a considerable time and then suddenly change their preference to a new layer with another temperature.

It was shown that the preference is most probably due not to temperature sensing but to odorants, since a fish deprived of its olfactory sense (anosmic) did not show any temperature preference and, compared with unoperated fish, searched vertically to cover a much larger temperature and depth range (Westerberg 1982b). In the fiord, the anosmic salmon dived under the silt depth, a behavior never seen in the control fish. Døving et al. (1985) studied the activity of single bulbar neurones of the Atlantic salmon when stimulating the olfactory epithelium with water samples taken from different depths of the fiord. Ninety percent of the responding neurones showed differential responses to the water samples, indicating the capacity of the salmon olfactory system to discriminate among stratified layers found in the ocean.

The information obtained from tracking experiments (Westerberg 1982b, Døving et al. 1985), in combination with the results of the electrophysiological study, made the authors propose the following hypothesis for salmon orientation in the coastal zone under the chemical influence of the home-stream odor. In the coastal waters there exist almost homogeneous layers containing home-stream odorants which are considered the layers of interest to the salmon. These layers contain specific odorants and their adjacent layers establish a reference system with which salmon can move in an oriented manner. To maintain its position with respect to the important information-giving layer, the animal moves from the layer of interest to the adjacent one with low amplitude movements up and down through a layer of strong vertical density gradient. If the boundary layer and/or microscale hydrographic featues (Westerberg 1984) become indistinct, the fish will begin large-amplitude up and down movements to locate another layer of interest. Salmon without an olfactory sense will continue to swim up and down with large-amplitude movements in search of a layer of interest. This behavior is observed for anosmic fish (Døving et al. 1985). The microscale hydrographic features found in the gradient layers have been discussed by Westerberg (1984). From these features, salmon may be able to determine an axis and direction for oriented movement.

Orientation with respect to hydrographic features has been demonstrated to be in part mediated by olfaction, but presumably other sensory systems are required for successful homing. As pointed out by Døving et al. (1985) tropotactic orientation by olfaction is less likely, since the fish with unilateral ablation of the olfactory nerve returned to its river of release. In future studies it will be important to design experiments to determine which external directional cues are used, such as current shear, and which sensory systems other than olfactory are required for successful salmon homing.

Pheromones. It has been suggested that the odorants, which are orienting clues for salmonids originate from the environment, i.e., from geological or botanical sources (Hasler 1966). Physiological experiments have been performed to show that each river has odorants that are specific for the population in that river. Thus, Hara (1965) and Hara et al. (1965) recorded EEG activity from the olfactory bulb in the chinook *Oncorhynchus tshawytscha* and coho salmon *O. kisutch* and claimed that the responses were selective. It should be recalled that recordings using their particular EEG techniques do not tell which neuronal pool is activated; they only tell that two unidentified groups of neurones close to the recording site are active. A conclusion of specificity or discrimination is invalid, since it is the spike discharge activity of a specific set of neurones that tells the fish which odor is present; the electrical activity recorded at the bulbar surface does not give the fish such information. Thus, the EEG recordings that have been made suggesting specificity without quantitative measurements of stimulus strength, cannot be used in arguments advocating a specific response to the home-stream water (Oshima et al. 1969a, b, c, Ueda et al. 1967, 1971). A critical study by Bodznick (1975) shows no relationship between the relative magnitude of EEG responses of the fish to the home stream water and the actual behavioral responses of the fish in the same water during migration.

Nordeng (1971, 1977) first suggested that pheromones were used as olfactory cues for migration in salmonid fishes. His proposal was made on the basis of extensive analyses of the migration pattern and experimental work on natural populations of the sea char *Salmo alpinus*, sea trout *Salmo trutta*, and Atlantic salmon *Salmo salar*, in the Salangen River system in northern Norway. Observations on the Atlantic salmon have supported Nordeng's hypothesis (Solomon 1973). Behavior studies on mature Atlantic salmon showed that hatchery-reared fish preferred water from the hatchery to river water. Wild fish did not show any preference for either water (Sutterlin and Gray 1973). Following Nordeng's suggestion, it was shown that olfactory bulb responses could be evoked by stimulation with substances from the fish. The odorous substances emanating from the fish are derivatives of bile salts or substances which have chemical properties similar to bile salts (Selset 1980), and they attract the mature fish in concentrations of 1.5×10^{-9} g/l (Selset and Døving 1980).

It was ascertained from EEG recordings that a population of the salmon *Salmo salar* did not produce substances which were particularly potent as odorants to individuals of their own population (Fisknes and Døving 1982). It is evident from the experiments that the salmon could tell by olfaction that other salmon were present. The odorous substances emanating from fish elicit different kinds of responses in the relay neurones of the olfactory bulb (Døving et al. 1974). Olfac-

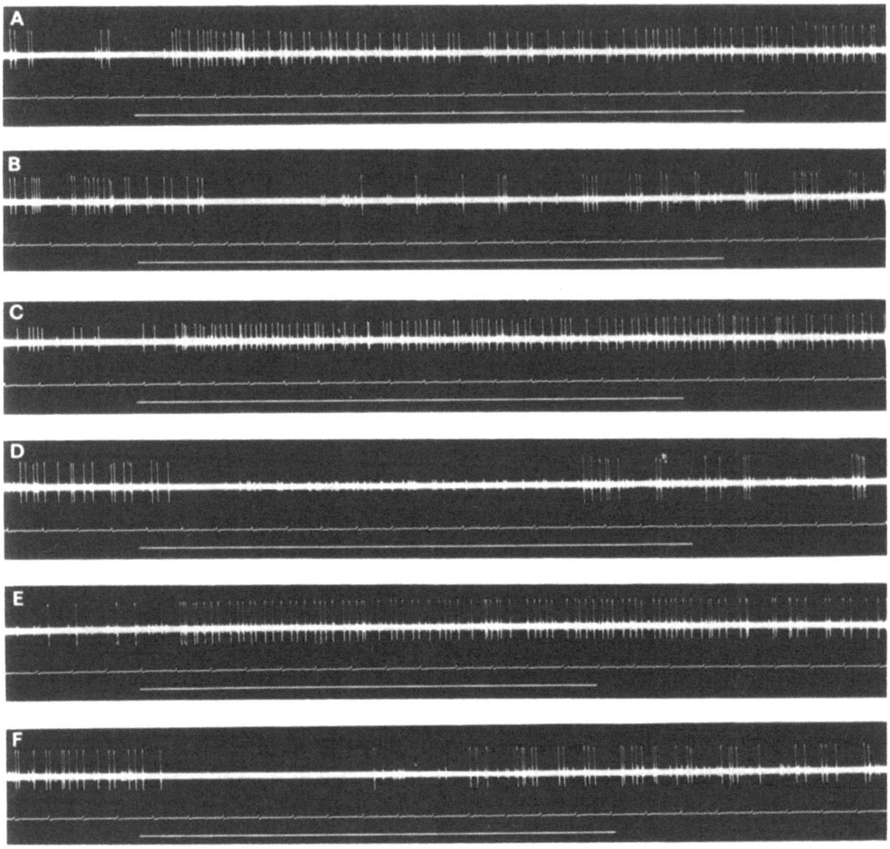

Fig. 24 A – F. Recordings of the activity of a bulbar neurone in char, *Salmo alpinus* (*upper trace*). *Middle trace* monitors the heart rate, the interval between each heart beat is 1 s. The *lower line* indicates stimulation. Tank water from the following char populations was used: **A** young Vangsvatn; **B** adult Vangsvatn; **C** young Løna; **D** adult Løna; **E** young Salangen migratory char; **F** young Salangen stationary char. Note the differential effect on the neural activity evoked by the different char populations. (From Døving, Nordeng and Oakley 1974)

tory bulb units of sockeye salmon react differently to water of various relevant sources (Bodznick 1978b). Thus, the olfactory system in salmon can readily discriminate between different salmonid populations; an example of this feature is shown in Fig. 24.

Experiments by Thommesen (1978) on the distribution of bulbar responses with different qualities of odorants indicated the existence of potent stimuli other than amino acids. In later studies, it has been shown that bile salts are potent olfactory stimuli (Døving et al. 1980, Thommesen 1983). The recordings were made from the olfactory bulb in the char *Salmo alpinus* and grayling *Thymallus thymallus*. They were monopolar and performed using DC amplifiers, which gave responses like those shown in Fig. 15. The mean threshold varied from 10^{-8} to 10^{-6} M for the different bile salts, but were consistently lower than the thresholds for amino acids. Taurine conjugated bile salts were the most potent odorants.

Bile Salts. Bile salts are secretory products of the vertebrate liver. They are synthesized from cholesterol, and are thus steroidal in nature. The biological history of bile salts is an interesting part of vertebrate evolution, with important phylogenic aspects. Bile salts have been the major subject of study for G.A.D. Haslewood, who has reviewed the work on bile salts (Haslewood 1967a, b, 1978). A review on bile salts in fishes is published by Tammar (1974). Recent articles (Ripatti and Sidorov 1973, Denton et al. 1975, Amos et al. 1977, Anderson et al. 1980) describe specific aspects of the bile salts in fishes.

Bile salts of cyclostomes are primitive detergents, but later developments give the efficient C24-5-beta-acids. According to Tammar (1974), the bile salts of the two agnathan families are clearly distinguishable from those of other Pisces, and from each other, and the chondrichthyans are characterized by C24 hydroxylated 5-beta-bile salts. Choanichthyan bile salts show similarities, in that 5-alpha-cyprinol sulphate is found in both superorders (Crossopterygii and Dipnoi), but latimerol sulphate is found only in *Latimeria* and 5-alpha-bufol sulphate only in *Neoceratodus*. Of the actinopterygian superorders, Chondrostei is characterized by about 10% of 5-beta-cyprinol sulphate in otherwise advanced bile salts. As far as is known, the Holostei have lost all relics of their evolutionary history, and those groups of Teleostei that do not have taurine conjugated C24 acids alone (*Anguilliformes, Arapaima,* and *Ostariophysi*) are all separated from the main bulk of the Teleostei in the classification by Greenwood et al. (1966).

The bile salts or similar steroid-based odorants have the potential variation needed to give clues for the multitude of populations of fishes. They are important for the homing behavior of fish and in the future may be shown to be important for speciation in the evolutionary process. Amino acids cannot give the variability necessary for the multitude of fish populations present.

5.7 Chemical Nature of Fish Odorants

Chemical signals in water have certain advantages over visual, auditory, tactile, or mechanical signals. The well-developed sense of smell and the conservative organization of the olfactory system in fishes indicate the importance of olfactory signals in water and the early development of chemical communication routes. We are at the beginning of an exciting era, when the odorants used in aquatic environments will be identified and give new insight into the puzzle of the evolutionary history of fishes.

Studies on the chemical nature of communication signals can be roughly divided into three fields. The extensive use of bait for fishing illustrates the importance of chemical signals in feeding behavior. Surprisingly, the dependence of controlled bait for good fishing has not been reflected in the number of scientific investigations on the subject.

Von Frish (1941a) was the first to describe intraspecific chemical communication when he demonstrated the alarm reactions in the carp fishes. White (1936) suggested that substances from returning salmon attracted the rest of the fish to the spawning site, but the idea seemed to have been forgotten in the enthusiasm over the imprinting hypothesis. The realization that pheromones are a means of communicating in aquatic environments has only been fully appreciated in the last decade. In salmonids, migration pheromones were first proposed by Nordeng (1971) to be clues for correct home-site return. Later, he extended his hypothesis to concern all salmonid species (Nordeng 1977). The results of further physiological and behavioral work confirm the results of Nordeng's hypothesis (Selset 1980, Selset and Døving 1980, Stabell et al. 1982). Time will tell whether pheromones are important for homing in other fishes, as well.

There are several reasons for focusing attention on the identification of the chemical signals important to the life functions of the fish. The substances can be used as *tools* for further studies on the behavior and functional anatomy, not only of the olfactory organ but of the CNS as well. The identification of the substances which evoke specific behavior patterns may be important for *defining the neural substrate* essential for controlling a particular behavior. Knowledge of the nature of the chemical signals will be important for controlling fish behavior in the expanding field of fish cultivation. It will also promote the development of efficient fish baits. Last, but not least, knowledge of the nature of chemical signals will give us a means of forecasting the effects of pollution in the aquatic environment.

6 Olfactory Sense and Pollution

Most studies on the effects of pollution on fish have been concerned with the effect of oil on the survival, or the pollutant's effects on selected organs (Anderson et al. 1974, Moles et al. 1979, Morrow 1974, Hawkes 1977). The effects on the chemosensory systems have been reviewed recently (Evans et al. 1982, Bertmar 1982, Døving 1983). A review on the preference and avoidance of fishes to aquatic contaminants is given by Giathina and Garton (1983).

The olfactory epithelium is necessarily exposed to the environment, a feature that also makes it liable to exposure to pollutants. The pollutants can be effective in different stages of olfactory chemoreception. They may affect the sensory organ directly and destroy the integrity of the sensory epithelium (Zeiske 1983). Crude oil and water-soluble fraction (WSF) of soil induce lesions in the olfactory organ of the common mummichog *Fundulus heteroclitus* and Atlantic riverside *Menidia menidia* (Gardner 1975). Both the sensory and sustentacular cells in the olfactory mucosa of the hogchoker *Trinectes maculatus* exhibited necrosis in the high water-soluble fractions (50% WSF), but no damage in low concentrations (5% WSF) (Solangi and Overstreet 1982). The water-soluble fractions of aviation fuel at a concentration of 5 mg/l sea-water caused cell necrosis in the olfactory rosette of the fathead minnow *Pimephales promelas* (Latendresse and Fisher 1983).

Moles et al. (1979) studied the sensitivity of various species of salmonids to benzene and water-soluble fractions of crude oil. The median tolerance levels in a 96-h test ranged from 2.7 – 14.7 µl/l. Exposure of the pink salmon *Oncorhynchus gorbuscha* to benzene at a concentration of 4.3 mg l^{-1} for 12 days caused the mucus cells of the olfactory lamellae to empty and there were patchy losses of cilia (Babcock 1985).

In behavior tests, the presmolt coho salmon *Oncorhynchus kisutch* avoided a mixture of monocyclic aromatic hydrocarbons at a concentration of 3 – 4 mg/l

(Maynard and Weber 1981). During the smoltification period the fish avoided concentrations of 2 mg/l. Electrophysiological recordings from the olfactory bulb indicated that the olfactory system was stimulated by an aromatic hydrocarbon mixture at concentrations correlating with the observed avoidance behavior. Behavior studies of cod have shown that this fish reacts to water-soluble fractions of diesel oil. The reactions are snapping and restless swimming, types of behavior which are released at remarkably low concentrations, about 1.6×10^{-7} g/l seawater (Hellstrøm and Døving 1983).

Pollutants can mimic the effects evoked by natural odorants. Such an ill effect will induce inappropriate reactions at times when the behavior appears meaningless. Pollutants can also block the effects induced by naturally occurring odorants, either by preventing the natural substances from reaching the receptor sites, or by antagonistic attachments at the receptor sites. It is also conceivable that the pollutants can induce ill effects by interfering with physiological events in the nervous structures connected to the olfactory system.

In the coming years, the nature of substances which induce specific behavior patterns in fish should be identified. This knowledge will make it easier to investigate the possible effects induced by new chemicals in the environment. It is evident that physiological studies along these lines will be essential if we are to be able to predict the effects of pollution.

7 Future Studies

An increasing amount of literature is being published on the chemical senses in fish, in particular, methods have been developed which permit the experimenter to apply and monitor the chemical stimuli in time and space. Techniques have also become available which make it possible to correlate the fish behavior with parameters of the natural environment.

In the future, several areas of olfactory research will be found worthwhile, both from an academic and an applied point of view. Insight into the primary processes of vertebrate olfaction is badly needed. Some species of fish have large olfactory organs and therefore provide an abundant supply of primary receptors. Key stimuli induce specific behavior. The reaction specificity and abundance of material awaits the eager hands of biochemists in their studies of the primary processes. Previous studies on the migration of fishes have been dominated by man's focus on celestial mechanics, an inheritance from the zoologists' interest in bird navigation. Future studies, coming down to earth and out into the water, will correlate the behavior of the fish with their immediate environment, instead of with the position of the sun and the stars. For the neuroethologists the specific behavior patterns evoked by stimulation of the bundlets of the olfactory tract is a new field of interest where combining sensory stimuli can be related to the behavior of the animal. Last, but not least, the uncovering of the chemical nature of pheromones and other key stimuli for fishes will be important, not only for our understanding of fish behavior in general, but, as stated above, also to forecast the effects of pollution. The natural chemicals may become important in fishing industries and aquaculture, and will contribute to a continuous management of our natural resources.

Acknowledgement. I am very grateful to Xenia Toverud for her unfailing help and patient typing of the many versions of this review.

8 References

Adair JS (1964) The olfactory bulb and related nuclei of the sturgeon (*Scaphirhynchus platorhynchus*). Ala J Med Sci 1(4):386–398

Adrian ED (1950) Sensory discrimination. With some recent evidence from the olfactory organ. Br Med Bull 6:330–332

Adrian ED, Ludwig C (1938) Nervous discharges from the olfactory organs of fish. J Physiol (Lond) 94:441–460

Adron JW, Mackie AM (1977) Studies on the chemical nature of feeding stimulants for rainbow trout, *Salmo gairdneri* Richardson. J Fish Biol 12:303–310

Afzelius BA (1976) A human syndrome caused by immotile cilia. Science 193:317–319

Amos B, Anderson IG, Haslewood GAD, Tökes L (1977) Bile salts of the lungfishes *Lepidosiren, Neoceratodus* and *Protopterus* and those of the coelacanth *Latimeria chalumnae* Smith. Biochem J 161:201–204

Anderson IG, Banister KE, Haslewood GAD, Cho D, Tökes L (1980) Bile salts of fishes collected on the Zaire river expedition (1974–75): their chemical nature and its possible significance. Zool J Linnean Soc 68(1):41–51

Anderson JW, Neff JM, Cox BA, Tatem HE, Hightower GM (1974) Characteristics of dispersions and water-soluble extracts of crude and refined oils and their toxicity to estuarine crustaceans and fish. Marine Biol 27:75–88

Andres KH (1970) Anatomy and ultrastructure of the olfactory bulb in fish, amphibia, reptiles, birds and mammals. In: Wolstenholme GEW, Knight J (eds) Taste and smell in vertebrates, a Ciba foundation symposium. Churchill, London, pp 177–196

Appelbaum S (1981) Zum Geruchsorgan der Karpfenlarven (*Cyprinus carpio* L.). Zool Anz Jena 206(3/4):227–233

Atema J (1977) Functional separation of smell and taste in fish and crustacea. In: Le Magnen J, McLeod P (eds) Olfaction and taste VI. Paris IRL, London, pp 165–174

Atema J, Holland K, Ikehara W (1980) Olfactory responses of yellowfin tuna (*Thunnus albacares*) to prey odors: chemical search image. J Chem Ecol 6:457–465

Babcock MM (1985) Morphology of olfactory epithelium of pink salmon, *Oncorhynchus gorbuscha* (Walbaum), and changes following exposure to benzene: a scanning electron microscopy study. In: Gray JS, Christiansen ME (eds) Marine biology of polar regions and affects of stress on marine organisms. 18th European marine biology symposium 1984

Bannister LH (1965) The fine structure of the olfactory surface of teleostean fishes. Q J Microscop Sci 106:333–342

Bardach JE, Villars T (1974) The chemical senses of fishes. In: Grant PI, Mackie AM (eds) Chemoreception in marine organisms. Academic Press, London, pp 49–104

Barlow HB (1972) Single units and sensation: A neuron doctrine for perceptual psychology? Perception 1:371–394

Barnett C (1977) Aspects of chemical communication with special reference to fish. Biosci Commun 3:331–392

Bashor DP, Beuerman RW, Easton DM (1974) Ciliary action and normal movement of odorant wave fronts in garfish nasal capsule of *Lepisosteus asseus*. Experientia 30: 777–779

Bass AH (1981) Organization of the telencephalon in the channel catfish, *Ictalurus punctatus*. J Morphol 169:71–90

Bateson W (1890) The sense organs and perceptions of fishes; with remarks on the supply of bait. J Marin Biol Res 1:225–256
Belghaug R, Døving KB (1977) Odour threshold determined by studies of the induced waves in the olfactory bulb of the char (*Salmo alpinus* L.). Comp Biochem Physiol 57A:327–330
Bell WJ, Tobin TR (1982) Chemo-orientation. Biol Rev 57:219–260
Bertmar G (1972a) Secondary folding of olfactory organ in young and adult sea trout. Acta Zool 53:113–120
Bertmar G (1972b) Scanning electron microscopy of olfactory rosette in sea trout. Z Zellforsch 128:336–346
Bertmar G (1972c) Labyrinth cells, a new cell type in vertebrate olfactory organs. Z Zellforsch 132:245–256
Bertmar G (1972d) Ecostructural studies on olfactory organ in young and adult sea trout (Osteichthyes, salmonidae). Z Morphol Tiere 72:307–330
Bertmar G (1973) Ultrastructure of the olfactory mucosa in the homing Baltic sea trout *Salmo trutta trutta*. Marine Biol 19:74–88
Bertmar G (1978) Electron microscopy of the olfactory mucosa in migrating Baltic sea trout (*Salmo trutta trutta*), with special reference to phagozytic and immunological cells. Beitr Fisch-Ökol – Fisch Umwelt 6:25–42
Bertmar G (1982) Structure and function of the olfactory mucosa of migrating Baltic trout under environmental stresses, with special references to water pollution. In: Hara TS (ed) Chemoreception in fishes. Elsevier, Amsterdam, pp 395–422
Bianchi S (1978) Le cellule mucose e l'istochimia dei mucopolisaccaridi nell'epitelio olfattivo di uegit-glanis zammaranoi gianferrari (claridae, siluriformes), pesce freatico. Riv Istoch Norm Pat 22:101–107
Bloom HD, Perlmutter A (1978) Possible pheromone medial reproductive isolation in two species of cyprinid fishes of the genus, *Brachydanio*. J Fish Biol 13:47–50
Bodznick D (1975) The relationship of the olfactory EEG evoked by naturally occurring stream waters to the homing behavior of sockeye salmon (*Oncorhynchus nerka* Walbaum). Comp Biochem Physiol 52A:487–495
Bodznick D (1978a) Water source preference and lakeward migration of sockeye salmon fry (*Oncorhynchus nerka*). J Comp Physiol 127:139–146
Bodznick D (1978b) Characterization of olfactory bulb units of sockeye salmon with behaviorally relevant stimuli. J Comp Physiol 127:147–156
Bodznick D (1978c) Calcium ion: an odorant for natural water discriminations and the migratory behavior of sockeye salmon. J Comp Physiol 127:157–166
Bodznick D, Northcutt RG (1979) Some connections of the lateral olfactory area of the horn shark (Abstract). In: Proceedings of annual meeting of the Society of Neuroscience, Atlanta, 1979, p 139
Boudreau JC (1962) Electrical activity in the olfactory tract of the catfish. Jpn J Physiol 12:272–278
Bradford MR Jr, Northcutt RG (1974) Olfactory bulb projections in the bichir, *Polypterus*. J Comp Neurol 156:165–178
Brannon EL (1982) Orientation mechanisms of homing salmonids. In: Brannon EL, Salo EO (eds) Proceedings of salmon and trout migratory behaviour symposium. School of Fisheries, University of Washington, Seattle, pp 219–227
Branson BA (1963) The olfactory apparatus of *Hybopsis gelida* (Girard) and *Hybopsis aestivalis* (Girard) (Pisces: Cyprinidae). J Morphol 113:215–229
Brawn VM (1961a) Aggressive behaviour in the cod (*Gadus callarias* L.). Behaviour 18:107–128
Brawn VM (1961b) Reproductive behaviour of the cod (*Gadus callarias* L.). Behaviour 18:177–198

Brawn VM (1969) Feeding behaviour of cod (*Gadus morhua*). J Fish Res Bd Can 26:583–596
Breipohl W (1974) Licht- und elektronenmikroskopische Befunde zur Struktur der Regio olfactoria des Goldfisches. Verh Anat Ges 68:479–485
Breipohl W, Bijvank GJ, Zippel HP (1973) Rastermikroskopische Untersuchungen der olfaktorischen Rezeptoren im Riechepithel. Z Zellforsch 138:439–454
Breipohl W, Laugwitz H, Bijvank GJ (1974) Rastermikroskopische Befunde an olfaktorischen Rezeptoren. Verh Anat Ges 68:793–797
Breucker H, Zeiske E, Melinkat R (1979) Development of the olfactory organ in the rainbow fish, *Nematocentris maccullochi* (Atheriniformes, Melanotaeniidae). Cell Tissue Res 200:53–68
Brown SB, Evans RE, Thompson BE, Hara TJ (1982) Chemoreception and aquatic pollutants. In: Hara TJ (ed) Chemoreception in fishes. Elsevier, Amsterdam, pp 363–393
Brown SB, Hara TJ (1982) Biochemical aspects of amino acid receptors in olfaction and taste. In: Hara TJ (ed) Chemoreception in fishes. Elsevier, Amsterdam, pp 159–180
Burne RH (1909) The antomy of olfactory organ of teleostean fishes. Proc Zool Soc Lond 2:610–663
Cancalon P (1983) Receptor cells of the catfish olfactory mucosa. Chem Senses 8:203–209
Cancalon P, Beidler LM (1975) Distribution along the axon and into various subcellular fractions of molecules labeled with ^{3}H leucine and rapidly transported in the garfish olfactory nerve. Brain Res 89:225–244
Cancalon P, Elam JS (1980a) Rate of movement and composition of rapidly transported proteins in regenerating olfactory nerve. J Neurochem 35:889–897
Cancalon P, Elam JS (1980b) Study of regeneration in the garfish olfactory nerve. J Cell Biol 84:779–794
Caprio J (1977) Electrophysiological distinction between the taste and smell of amino acids in catfish. Nature 266:850–851
Caprio J (1978) Olfaction and taste in the channel catfish: an electrophysiological study of the responses to amino acids and derivatives. J Cell Comp Physiol 123:357–371
Caprio J (1980) Similarity of olfactory receptor responses (EOG) of freswater and marine catfish to amino acids. Can J Zool 58:1778–1784
Caprio J (1982) High sensitivity and specificity of olfactory and gustatory receptors of catfish to amino acids. In: Hara TJ (ed) Chemoreception in fishes. Elsevier, Amsterdam, pp 109–134
Caprio J, Raderman-Little R (1978) Scanning electron microscopy of the channel catfish olfactory lamellae. Tissue Cell 10:1–9
Carr WES (1982) Chemical stimulation of feeding behavior. In: Hara TJ (ed) Chemoreception in fishes. Elsevier, Amsterdam, pp 259–274
Clark ER, Fraser JAL (1983) The survial and growth of six species of freshwater fish, in tap water and diluted and undiluted effluent from sewage percolating filters. J Fish Biol 22:431–445
Colombo L, Belvedere PC, Marconato A, Bentivegna F (1982) Pheromones in teleost fish. In: Richter CJJ, Goos HJT (eds) Proceedings of the international symposium: reproductive physiology of fish. Pudoc, Wageningen, pp 84–94
Cooper JC, Hasler AD (1974) Electroencephalographic evidence for retention of olfactory cues in homing coho salmon. Science 183:336–338
Cooper JC, Hasler AD (1975) Morpholine as olfactory stimulus in fish. Science 187:81–82
Cooper JC, Hasler AD (1976) Electrophysiological studies of morpholine-imprinted coho salmon (*Oncorhynchus kisutch*) and rainbow trout (*Salmo gairdneri*). J Fish Res Bd Can 33:688–694

Cooper JC, Hirsch PJ (1982) The role of chemoreception in salmonid homing. In: Hara TJ (ed) Chemoreception in fishes. Elsevier, Amsterdam, pp 343 – 362
Cooper JC, Scholz AT (1976) Homing of artificially imprinted steelhead (rainbow) trout. *Salmo gairdneri.* J Fish Res Bd Can 33:826 – 829
Cooper JC, Scholz AT, Horrall RM, Hasler AD, Madison DM (1976) Experimental confirmation of the olfactory hypothesis with artificially imprinted homing coho salmon (*Oncorhynchus kisutch*). J Fish Res Bd Can 33:703 – 710
Cragie EH (1926) A preliminary experiment upon the relation of the olfactory sense to the migration of the sockeye salmon (*Oncorhynchus nerka,* Walbaum). Roy Soc Can Trans 5:215 – 224
Crapon de Caprona MD (1980) Olfatory communication in a cichlid fish, *Haplochromis burtoni.* Z Tierpsychol 52:113 – 134
Dahlgren U (1908) The oral opening of the nasal cavity in astroscopus. Science 27:993 – 994
Demski LS (1978) Neuroanatomical substrates of reproductive bahavior in male sunfish (*Genus lepomis*). Ann Biol Anim Biochem Biophys 18:831 – 836
Demski LS, Bauer DH, Gerald JW (1975) Sperm release evoked by electrical stimulation of the fish brain: a functional-anatomical study. J Exp Zool 191:215 – 232
Demski LS, Knigge KM (1971) The telencephalon and hypothalamus of the bluegill (*Lepomis macrochirus*): evoked feeding, aggressive and reproductive behavior with representative frontal sections. J Comp Neurol 143:1 – 16
Denton JE, Yousef MK, Yousef IM, Kuksis A (1975) Bile acid composition of rainbow trout, *Salmo gairdneri.* Lipids 9:945 – 951
Derivot JH, Dupé M, Godet R (1979a) Anatomie fonctionnelle de l'organe olfactif de *Protopterus annectens* Owen (Dipneustes): contribution à la connaissance du mécanisme d'irrigation de l'organe olfactif. Acta Zool (Stockh) 60:251 – 257
Derivot JH, Mattei X, Godet R, Dupé M (1979b) Etude ultrastructurale de la région apicale des cellules de l'épithélium olfactif de *Protopterus annectens* Owen (dipneusters). J Ultrastruct Res 66:22 – 31
Devitsyna GV (1972) Morphology of the organs of olfaction in the gadidae. J Ichthyol 12:994 – 1002
Devitsyna GV (1977) Comparative morphology of the olfactory analyzer in fish. J Ichthyol 17:116 – 125
Devitsyna GV, Malyukina GA (1977) The functional organization of the olfactory organ in macrosmatic and microsmatic fishes. J Ichthyol 17:432 – 440
Donaldson LR, Allen GH (1957) Return of silver salmon, *Oncorhynchus kisutch* (Walbaum) to point of release. Trans Am Fish Soc 87:13 – 22
Doroschenko MA, Pinchuk LE (1974) New data on secretory elements of the olfactory epithelium in certain marine bottom fish. Morphology. Doklady Akademii Nauk SSSR 219:756 – 758
Drongelen W van, Holley A, Døving KB (1978) Convergence in the olfactory system: quantitative aspects of odour sensitivity. J Theor Biol 71:39 – 48
Dubois-Dauphin M, Døving KB, Holley A (1980) Topographical relation between the olfactory bulb and the olfactory tract in tench (*Tinca tinca* L.). Chem Senses 5:159 – 169
Dupé M, Godet R (1969) Caractéristiques et variations de la réponse électrique induite par la stimulation de l'organe olfactif chez un poisson dipneuste. (*Protopterus annectus* Ow). C R Seances Soc Biol 163:267 – 270
Døving KB (1964) Studies of the relation between the frog's electro-olfactogram (EOG) and single unit activity in the olfactory bulb. Acta Physiol Scand 60:150 – 163
Døving KB (1966) The influence of olfactory stimuli upon the activity of secondary neurones in the burbot (*Lota lota* L.). Acta Physiol Scand 66:290 – 299
Døving KB (1967) Comparative electrophysiological studies on the olfactory tract of some teleosts. J Comp Neurol 131:365 – 370

Døving KB (1982) Olfactory induced behaviour in fish and the role of centrifugal fibers in the olfactory pathways. In: Steiner JE, Ganchrow JR (eds) The determination of behaviour by chemical stimuli. IRL Press, Oxford, pp 105–114
Døving KB (1983) Adferd hos torsk. En modell i nevro-etologisk forskning. *Atferd hos marine dyr*, Os, Norge, Feb 1983, pp 15–29
Døving KB, Belghaug R (1977) Frequency of induced waves in the olfactory bulb in char (*Salmo alpinus* L.); a temperature dependence. Comp Biochem Physiol 56A:577–579
Døving KB, Dubois-Dauphin M, Holley A, Jourdan F (1977) Functional anatomy of the olfactory organ of fish and the ciliary mechanism of water transport. Acta Zool (Stockholm) 58:245–255
Døving KB, Enger PS, Nordeng H (1973) Electrophysiological studies on the olfactory sense in char (*Salmo alpinus* L.). Comp Biochem Physiol 47A:1051–1063
Døving KB, Gemne G (1965) Electrophysiological and histological properties of the olfactory tract of the burbot (*Lota lota* L.). J Neurophysiol 28:139–153
Døving KB, Gemne G (1966) An electrophysiological study of the efferent olfactory system in the burbot. J Neurophysiol 29:665–674
Døving KB, Holmberg K (1974) A note on the function of the olfactory organ of the hagfish *Myxine glutinosa*. Acta Physiol Scand 91:430–432
Døving KB, Hyvärinen J (1969) Afferent and efferent influences on the activity pattern of single olfactory neurons. Acta Physiol Scand 75:111–123
Døving KB, Nordeng H, Oakley B (1974) Single unit discrimination of fish odours released by char (*Salmo alpinus* L.) populations. Comp Biochem Physiol 47A:1051–1063
Døving KB, Selset R (1980) Behavior patterns in cod released by electrical stimulation of olfactory tract bundlets. Science 207:559–560
Døving KB, Selset R, Thommesen G (1980) Olfactory sensitivity to bile acids in salmonid fishes. Acta Physiol Scand 108:123–131
Døving KB, Thommesen G (1977) Some properties of the fish olfactory system. Olfaction and taste VI. IRL Paris, pp 175–183
Døving KB, Westerberg H, Johnsen PB (1985) Role of olfaction in the behavioral and neural responses of Atlantic salmon, *Salmo salar*, to hydrographic stratification. Can J Fish Aquat Sci 42:1658–1667
Ebbeson SOE (1972) New insights into the organization of the shark brain. Comp Biochem Physiol 42A:121–129
Ebbeson SOE, Heimer L (1970) Projections of the olfactory tract fibers in the nurse shark (*Ginglymostoma cirratum*). Brain Res 17:47–55
Emanuel ME, Dodson JJ (1979) Modification of the rheotropic behavior of male rainbow trout (*Salmo gairdneri*) by ovarian fluid. J Fish Res Bd Can 36:63–68
Evans RE, Zielinski B, Hara TJ (1982) Development and regeneration of the olfactory organ in rainbow trout. In: Hara TJ (ed) Chemoreception in fishes. Elsevier, Amsterdam, pp 15–37
Faucette JR (1969a) The olfactory bulb and medial hemisphere wall of the rat-fish, *Chimaera*. J Comp Neurol 137:377–406
Faucette JR (1969b) The accessory olfactory bulbs and the lateral telencephalic wall of the rat-fish, *Chimaera*. J Comp Neurol 137:407–432
Faure ML (1944) Contribution à l'étude morphologique de l'appareil olfactif chez quelques téléostéens. Bull Museum d'Histoire Naturelle de Marseille 4:1–17
Finger TE (1975) The distribution of the olfactory tracts in the bullhead catfish. Ictalurus nebulosus. J Comp Neurol 161:125–142
Fisknes B, Døving KB (1982) Olfactory sensitivity to group-specific substances in Atlantic salmon (*Salmo salar* L.). J Chem Ecol 8:1083–1092
Frisch K von (1941a) Die Bedeutung des Geruchsinnes im Leben der Fische. Naturwissenschaften 22/23:321–333

Frisch K von (1941b) Über einen Schreckstoff der Fischhaut und seine biologische Bedeutung. Z Vergl Physiol 29:46–145
Gardner GR (1975) Chemically induced lesions in estuarine or marine teleosts. In: Ribelin WE, Migaki G (eds) Pathology of fishes. University of Wisconsin, Madison, pp 657–693
Garten S (1900) Physiologie der marklosen Nerven. Gustav Fischer, Jena
Gasser HS (1956) Olfactory nerve fibers. J Gen Physiol 39:473–496
Gemne G, Døving KB (1969) Ultrastructural properties of primary olfactory neurons in fish (*Lota lota* L.). Am J Anat 126:457–476
Gerard RW, Young JZ (1937) Electrical activity of the central nervous system of the frog. Proc R Soc Ser B 122:343–352
Gerking SD (1953) Evidence for the concept of home range and territory in stream fishes. Ecology 34:347–365
Giattina JD, Garton RR (1983) A review of the preference-avoidence responses of fishes to aquatic contaminants. Residue Rev 87:43–90
Goel HR (1978a) Functional anatomy of the olfactory organ in the fresh water teleost, *Heteropneustes fossilis* (BL.). Okajimas Folia Anat Jpn 55:289–300
Goel HR (1978b) The structure and functions of the olfactory organs in the fresh water teleost *Notopterus notopterus* (HAM.). Folia Morphol (Praha) 26:392–398
Goel HR (1980a) Functional anatomy of the olfactory organs in the fresh water Indian catfish *Clarias batrachus* (Linn.). Zool Poloniae 28:5–14
Goel HR (1980b) The functional anatomy of the olfactory organs in the fresh water teleost fish, *Channa Marulius* Hamilton. Arch Biol (Bruxelles) 91:1–13
Goh Y, Tamura T (1978) Electrical responses of the olfactory tract to some chemical stimulants in carp. Bull Jpn Soc Sci Fish 44:1289–1294
Goh Y, Tamura T (1979) Olfactory and gustatory responses to amino acids in two marine teleosts – Red Sea bream and mullet. Comp Biochem Physiol 66C:217–224
Goh Y, Tamura T (1980a) The electrical responses of the olfactory tract to amino acids in carp. Bull Jpn Soc Sci Fish 44:341–344
Goh Y, Tamura T (1980b) Effect of amino acids on the feeding behaviour in Red Sea bream. *Chrysophrys. major*. Comp Biochem Physiol 66C:225–230
Goh Y, Tamura T, Kobayashi H (1979) Olfactory responses to amino acids in marine teleosts. Comp Biochem Physiol 62A:863–868
Gooding RM (1963) The olfactory organ of the skipjack *Katsuwonus pelamis*. In: Experience Paper No. 36. World scientific meeting on the biology of tunas and related species. La Jolla, California, USA, 2–14 July 1962. FAO Fish Rep 6:1621–1631
Grant PT, Mackie AM (1974) Chemoreception of marine organisms. Academic Press, London
Graystone P, Low B, Rogers J, McLennan H (1970) The induced wave activity of the olfactory bulbs of toads, iguanas and snakes. Comp Biochem Physiol 37:493–502
Greenwood PH, Rosen DE, Weitzman SH, Myers GS (1966) Phyletic studies of teleostean fishes with a provisional classification of living forms. Bull Am Mus Nat Hist 131:339–456
Grimm RJ (1960) Feeding behaviour and electrical stimulation of the brain of *Carassius auratus*. Science 131:162–163
Groot C (1982) Modification on a theme – a perspective on migratory behaviour of pacific salmon. In: Brannon EL, Salo EO (eds) Proceedings of salmon and trout migratory behavior symposium. School of Fisheries, University of Washington, Seattle, pp 1–21
Gross GW, Beidler LM (1973) Fast axonal transport in the C-fibers of the garfish olfactory nerve. J Neurobiol 4:413–428
Gross GW, Beidler LM (1975) A quantitative analysis of isotope concentration profiles and rapid transport velocities in the C-fibers of the garfish olfactory nerve. J Neurobiol 6:213–232

Gross GW, Kreutzberg GW (1978) Rapid axoplasmic transport in the olfactory nerve of the pike: I. Basic transport parameters for proteins and amino acids. Brain Res 139:65–76
Groves AB, Collins GB, Trefethen PS (1968) Roles of olfaction and vision in choice of spawning site by homing adult chinook salmon (*Oncorhynchus tshawytscha*). J Fish Res Bd Can 25:867–876
Gusel'nikova KG, Tsitolovskii L, Potapov AA (1971) Characteristics of electrical activity of olfactory bulb neurons in fish. Neirofiziologiya 3:505–511
Hahn WE (1976) Electroencephalography of the olfactory bulb in relation to prespawn homing. Experientia 32:1095–1097
Hara A, Hirai H (1982) Structural studies of body fish vitellogenin. In: Richter CII, Goos HJT (eds) Proceedings of the international Symposium Reproductive Physiology of Fish. Pudoe, Wageningen, p 158
Hara TJ (1965) Electroencephalographic studies of homing salmon. Science 149:884–885
Hara TJ (1970) An electrophysiological basis for olfactory discrimination in homing salmon: a review. J Fish Res Bd Can 11:567–586
Hara TJ (1971) Chemoreception. Fish Physiol 5:79–120
Hara TJ (1973) Olfactory responses to amino acids in rainbow trout, *Salmo gairdneri*. Comp Biochem Physiol 44A:407–416
Hara TJ (1974) Is morpholine an effective olfactory stimulant in fish? J Fish Res Bd Can 31:1547–1550
Hara TJ (1975) Olfaction in fish. Prog Neurobiol 5:271–335
Hara TJ (1976a) Structure-activity relationships of amino acids in fish olfaction. Comp Biochem Physiol 54A:31–36
Hara TJ (1976b) Effects of pH on the olfactory responses to amino acids in rainbow trout, *Salmo gairdneri*. Comp Biochem Physiol 54A:37–40
Hara TJ (1976c) Olfactory responses to skin mucous substances in rainbow trout, *Salmo gairdneri*. Comp Biochem Physiol 54A:41–44
Hara TJ (1982a) Chemoreception in fishes. Developments in aquaculture and fisheries science, vol 8. Elsevier, Amsterdam, p 433
Hara TJ (1982b) Structure-activity relationships of amino acids as olfactory stimuli. In: Hara TJ (ed) Chemoreception in fishes. Elsevier, Amsterdam, pp 135–157
Hara TJ, Brown SB (1979) Olfactory bulbar electrical responses of rainbow trout (*Salmo gairdneri*) exposed to morpholine during smoltification. J Fish Res Bd Can 36: 1186–1190
Hara TJ, Freese M, Scott KR (1973a) Spectral analysis of olfactory bulbar responses in rainbow trout. Jpn J Physiol 23:325–333
Hara TJ, Law VMC, Hobden BR (1973b) Comparison of the olfactory response to amino acids in rainbow trout, brook trout and whitefish. Comp Biochem Physiol 45A: 969–977
Hara TJ, MacDonald S (1976) Olfactory responses to skin mucous substances in rainbow trout *Salmo gairdneri*. Comp Biochem Physiol 54A:41–44
Hara TJ, Ueda K, Gorbman A (1965) Electroencephalographic studies on homing salmon. Science 149:884–885
Harden-Jones FR (1968) Fish migration. Arnold, London, p 325
Hasler AD (1966) Underwater guideposts. University of Wisconsin Press, Madison, p 155
Hasler AD, Scholz AT (1983) Olfactory imprinting and homing in salmon. Springer, Berlin Heidelberg New York Tokyo, p 134
Hasler AD, Wisby WJ (1951) Discrimination of stream odors by fishes and its relation to parent stream behavior. American Naturalist 85:223–238
Haslewood GAD (1967a) Bile salt evolution. J Lipid Res 8:535–550
Haslewood GAD (1967b) Bile salts. In: Peters R, Young FG (eds) Methuen's monographs on biochemical subjects. Methuen, London, pp 1–116

Haslewood GAD (1978) The biological importance of bile salts. In: Neuberger A, Tatum EL (eds) North-Holland research monographs frontiers of biology, vol 47. North-Holland, Amsterdam, pp 1–206

Hawkes JW (1977) The effects of petroleum hydrocarbon exposure on the structure of fish tissues. In: Wolfe DA (ed) Fate and effects of petroleum, chap 10. Pergamon, Oxford, pp 115–128

Haynes LJ, Sangster AW, Steven DM, Thomas S (1967) Chemical factors inducing exploratory feeding behaviour in fish – EFB-inducing properties of marine invertebrates. Comp Biochem Physiol 20:755–765

Heimer L (1969) The secondary olfactory connections in mammals, reptiles and sharks. Ann NY Acad Sci 167:129–146

Hellstrøm T, Døving KB (1983) Perception of diesel oil by cod (*Gadus morhua* L.). Aquatic Tox 4:303–315

Henderson R, Strichartz G (1974) Ion fluxes through the sodium channels of garfish olfactory nerve membranes. J Physiol (Lond) 238:329–342

Herberhold C (1968) Vergleichende histochemische Untersuchungen am peripheren Riechorgan von Säugetieren und Fischen. Arch Klin Exp Ohren, Nasen Kehlkopfheilkd 190:166–182

Hidaka I (1982) Taste receptor stimulation and feeding behavior in the puffer. In: Hara TJ (ed) Chemoreception in fishes. Elsevier, Amsterdam, pp 243–257

Høglund LB, Åstrand M (1973) Preferences among juvenile char (*Salvelinus alpinus* L.) to intraspecific odours and water currents studied with the fluviarium technique. Report Inst Freshwater Res (Drottningholm) 53:21–30

Holey DT, Belsare DK (1973) Morphology of the olfactory sac in some teleosts. Z Mikrosk Anat Forsch 87:463–469

Holl A (1965) Vergleichende morphologische und histologische Untersuchungen am Geruchsorgan der Knochenfische. Z Morphol Ökol Tiere 54:707

Holl A (1973a) Feinstruktur des Riechepithels von *Chimaera monstrosa* (Holocephali). Int J Life Oceans Coastal Waters – Marine Biol 23:59–72

Holl A (1973b) Funktionsmorphologie der Nase von *Chimaera monstrosa* (Holocephali). Z Morphol Tiere 74:271–296

Holl A (1974) Aktive, rhythmische Ventilation des Geruchsorgans bei Stachelaalen (Mastacembelidae). Z Naturforsch 29c:787–789

Holl A (1980) Selektivdarstellung der Riechsinneszellen des Zwergwelses *Ictalurus nebulosus* mit Procionfarbstoffen. Z Naturforsch 35c:526–528

Holl A (1981) Marking of olfactory axons of fishes by intravital staining with procion brilliant yellow. Stain Technol 56:67–70

Holl A, Meinel W (1968) Das Geruchsorgan des Tiefseefisches *Aphanopus carbo* (Percomorphi, Trichiuridae). Helgoländer Wiss Meeresunters 18:404–423

Holl A, Schulte E, Meinel W (1970) Funktionelle Morphologie des Geruchsorgans und Histologie der Kopfanhänge der Nasenmuräne *Rhinomuraena ambonensis* (*Teleostei, Anguilliformes*). Helgoländer Wiss Meeresunters 21:103–123

Holland KN, Teeter JH (1981) Behavioral and cardic reflex assays of the chemosensory acuity of channel catfish to amino acids. Physiol Behav 27:699–707

Holley A, MacLeod P (1977) Transduction et codage des informations olfactives chez les vertébrés. J Physiol (Paris) 73:725–828

Hollis KL, Overmier JB (1978) The function of the teleost telencephalon in behavior: a reinforcement mediator. In: Mostofsky DJ (ed) The behavior of fish and other aquatic animals, vol 4. Academic Press, New York, pp 137–195

Holmgren N (1920) Zur Anatomie und Histologie des Vorder- und Zwischenhirns der Knochenfische. Acta Zool (Stockh) 1:137–315

Honda H (1980) Female sex pheromone of rainbow trout, *Salmo gairdneri*, involved in courtship behaviour. Bull Jpn Soc Sci Fish 46:1109–1112

Horn G (1981) Neural mechanisms for learning: an analysis of imprinting in the domestic chick. Proc R Soc Lond [Biol] 213:101 – 137
Howarth JV, Keynes RD, Ritchie JM, Muralt A von (1975) The heat production associated with the passage of a single impulse in pike olfactory nerve fibres. J Physiol (Lond) 249:349 – 368
Hurk R van den, Lambert JDG (1983) Ovarian steroid glucuronides function as sex pheromones for male zebrafish, *Brachydanio rerio*. Can J Zool 61:2381 – 2387
Ichikawa M (1975a) Electron microscopic study of the terminal regions of the olfactory tract fibers in the goldfish, telencephalic hemisphere. J Fac Sci Tokyo. Sec IV 14:127 – 134
Ichikawa M (1975b) The central projections of the olfactory tract in the goldfish, *Carassius auratus*. J Fac Sci Tokyo. Sec IV 13:257 – 263
Ichikawa M (1976) Fine structure of the olfactory bulb in the goldfish, *Carassius auratus*. Brain Res 115:53 – 56
Ichikawa M, Ueda K (1977) Fine structure of the olfactory epithelium. Cell Tissue Res 183:445 – 455
Ichikawa M, Ueda K (1979) Electron microscopic study of the termination of the centrifugal fibers in the goldfish olfactory bulb. Cell Tissue Res 197:256 – 262
Jahn LA (1972) Development of the olfactory apparatus of the cutthroat trout. Trans Am Fish Soc 101:284 – 289
Jakubowski M (1975) Anatomical structure of olfactory organs provided with internal nares in the Antartic fish *Gymnodraco acuticeps* Boul. (*Bathydraconidae*). Bull Acad Polon Sci Cl II 23:115 – 120
Jakubowski M (1981) Ultrastructure (SEM, TEM) of the olfactory epithelium in the wels, *Siluris glanis* L. (silurida, pisces). Z Mikrosk Anat Forsch 95:337 – 352
Jakubowski M, Kunysz E (1979) Anatomy and morphometry of the olfactory organ of the wels *Silurus glanis* L. (siluridae, pisces). Z Mikrosk Anat Forsch 93:728 – 735
Jansen J (1930) The brain of *Myxine glutinosa*. J Comp Neurol 49:359 – 507
Johnsen PB, Hasler AD (1980) The use of chemical cues in the upstream migration of coho salmon, *Oncorhynchus kisutch* Walbaum. J Fish Biol 17:67 – 73
Johnson HE, Brown CJD (1962) Olfactory apparatus in the black rockfish *Sebastodes melanops*. Copeia 4:838 – 840
Johnstone ADF (1980) The detection of dissolved amino acids by the Atlantic cod, *Gadus morhua* L. J Fish Biol 17:219 – 230
Jonsson L (1977) Følsomheten for kjemiske stimuli hos røye (*Salmo alpinus* L.), bestemt ved klassisk kondisjonering. Thesis University of Oslo
Kaji S, Satou M, Kudo Y, Ueda K, Gorbman A (1975) Spectral analysis of olfactory responses of adult spawning chum salmon (*Oncorhynchus keta*) to stream waters. Comp Biochem Physiol 51A:711 – 716
Kandel ER (1964) Electrical properties of hypothalamic neuroendocrine cells. J Gen Physiol 47:691 – 717
Kapoor AS, Ojha PP (1972) Studies on ventilation of the olfactory chambers of fishes with a critical reevaluation of the role of accessory nasal sacs. Arch Biol (Liège) 83:167 – 178
Kapoor AS, Ojha PP (1974) Histology of the olfactory epithelium of the fish, *Channa punctatus* Bloch. Acta Anat 87:534 – 554
Kasumjan AO, Lebedeva NE (1979) Nye data om karpefiskenes alarmpheromon. Akademia Nauk USSR 19:890 – 895
Kim YS, Stumpf WE, Sar M (1978) Topography of estrogen target cells in the forebrain of goldfish, *Carassius auratus*. J Comp Neurol 182:611 – 620
Kjøstolfsen I (1969) Adaptasjon i lukteorganet hos torsk (*Gadus morhua* L.). Thesis, University of Oslo
Kleerekoper H (1969) Olfaction in fishes. Bloomington, Indiana University Press, pp 1 – 222

Kleerekoper H, Erkel GA van (1960) The olfactory apparatus of *Petromyzon marinus* L. Can J Zool 38:209 – 223
Kleerekoper H, Mogensen J (1963) Role of olfaction in the orientation of petromyzon marinus. I. Response to a single amine in prey's body odor. Physiol Zool 36:347 – 360
Kosaka T (1980) Ruffed cell: a new type of neuron with a distinctive initial unmyelinated portion of the axon in the olfactory bulb of the goldfish (*Carassius auratus*). II. Fine structure of the ruffed cell. J Comp Neurol 193:119 – 145
Kosaka T, Hama K (1979) Ruffed cell: a new type of neuron with a distinctive initial unmyelinated portion of the axon in the olfactory bulb of the goldfish (*Carassius auratus*). J Comp Neurol 186:301 – 320
Kosaka T, Hama K (1980) Presence of the ruffed cell in the olfactory bulb of the catfish, *Parasilurus asotus*, and the sea eel, *Conger myriaster*. J Comp Neurol 193:103 – 117
Kosaka T, Hama K (1981) Ruffed cell: a new type of neuron with a distinctive initial unmyelinated portion of the axon in the olfactory bulb of the goldfish (*Carassius auratus*). III. Three-dimensional structure of the ruffed cell dendrite. J Comp Neurol 201:571 – 587
Kosaka T, Hama K (1983) Synaptic organization in the teleost olfactory bulb. J Physiol (Paris) 78:707 – 719
Kracke GR, Chacko GK (1979) Polypeptide components of the (Na + K)-ATPase of garfish olfactory nerve axon plasma membrane. Life Sci 25:2125 – 2129
Kracke GR, O'Neal SG, Chacko GK (1981) Partial purification and characterization of $(Na^+ + K^+)$-ATPase from garfish olfactory nerve axon plasma membrane. J Membrane Biol 63:147 – 156
Kreutzberg GW, Gross GW (1977) General morphology and axonal ultrastructure of the olfactory nerve of the pike, *Esox lucius*. Cell Tissue Res 181:443 – 457
Kudo Y, Satou M, Kaji S, Ueda K (1972) Frequency analysis of olfactory response in fish by bandpass filters. J Fac Sci Tokyo Sec IV 12:365 – 383
Kudo Y, Ueda K (1980) The measurement of the electric activity of the olfactory bulb in free swimming carp (*Cyprinus carpio* L.) by underwater telemetry systems. IEEE Transact Biomed Engineer 27:694 – 698
Kux J, Zeiske E, Melinkat R (1978) Messungen der Strömungsgeschwindigkeit an den Öffnungen des Geruchsorgans von Fischen mit einem Laser-Velocimeter. Laser Elektro Optik 1:21
Kühne W, Steiner J (1879) Beobachtungen über markhaltige und marklose Nervenfasern. Untersuch Physiol Inst Heidelberg 3:149 – 170
Kyle AL, Peter RE (1982) Effects of forebrain lesions on spawning behaviour in the male goldfish. Physiol Behav 28:1103 – 1109
Kyle AL, Stacey NE, Peter RE (1982) Ventral telencephalic lesions: effects on bisexual behavior, activity, and olfaction in the male goldfish. Behav Neural Biol 36:229 – 241
Latendresse II Jr, Fisher JW (1983) Histopathologic effects of JP-4 Aviation fuel on fathead minnows (*Pimephales promelas*). Bull Environ Contam Toxicol 30:536 – 543
Lawry JV Jr (1973) The olfactory epithelium of the lantern fish, *Tarletonbeania crenularis* (Myctophidae). Z Zellforsch 138:31 – 39
Liermann K (1933) Über den Bau des Geruchsorgans der Teleostier. Z Ges Arat 100:1 – 39
Liley NR (1982) Chemical communication in fish. Can J Fish Aquat Sci 39:22 – 35
Lorentzen HH (1983) Luktfølsomhet hos røye (*Salmo alpinus* L.) i ferskvann og saltvann. Thesis, University of Oslo
Losey GS Jr (1969) Sexual pheromone in some fishes of the *Genus hypsoblennius* Gill. Science 163:181 – 183
Lowe GA (1974) The occurrence of ciliary aggregations on the olfactory epithelium of two species of gadoid fish. J Fish Biol 6:537 – 539
Lowe GA, MacLeod NK (1975) The ultrastructural organization of olfactory epithelium of two species of gadoid fish. J Fish Biol 7:529 – 532

Mackie AM (1982) Identification of the gustatory feeding stimulants. In: Hara TJ (ed) Chemoreception in fishes. Elsevier, Amsterdam, pp 275–292
Mackie AM, Adron JW, Grant PT (1980) Chemical nature of feeding stimulants for the juvenile Dover sole, *Solea solea* (L.). J Fish Biol 16:701–708
MacLeod NK (1976a) Spontaneous activity of single neurons in the olfactory bulb of the rainbow trout (*Salmo gairdneri*) and its modulation by olfactory stimulation with amino acids. Exp Brain Res 25:267–278
MacLeod NK (1976b) Field potentials in the olfactory bulb of the codfish (*Gadus morhua*). Comp Biochem Physiol 55A:297–299
MacLeod NK, Lowe GA (1976) Field potentials in the olfactory bulb of the rainbow trout (*Salmo gairdneri*): Evidence for a dendrodendritic inhibitory pathway. Exp Brain Res 25:255–266
Marshall NB (1979) Developments in deep-sea biology. Blandford, Dorset, p 566
Matsumoto Y, Tanner E (1975) Action potential of the garfish non-myelinated olfactory nerve. Life Sci 16:1067–1078
Maynard DJ, Weber DD (1981) Avoidance reactions of juvenile coho salmon (*Oncorhynchus kisutch*) to monocyclic aromatics. Can J Fish Aquat Sci 38:772–778
Mazeaud F (1982) Morpholine, a nonspecific attractant for salmonids. Aquaculture 26:189–191
Melinkat R, Zeiske E (1979) Functional morphology of ventilation of the olfactory organ in *Bedotia geayi* Pellegrin 1909 (Teleostei, Atherinidae). Zool Anz Jena 5/6:354–368
Meredith M (1981) The analysis of response similarity in single neurons of the goldfish olfactory bulb using amino-acids as odor stimuli. Chemical Senses 6:277–293
Meredith M, Moulton DG (1978) Patterned response to odor in single neurones of goldfish olfactory bulb: influence of odor quality and other stimulus parameters. J Gen Physiol 71:615–643
Millot J, Anthony J (1958) Anatomie de Latimeria chalumnae. Paris: CNRS, vol II
Moles A, Rice SD, Korn S (1979) Sensitivity of Alaskan freshwater and anadromous fishes to Prudhoe Bay crude oil and benzene. Trans Am Fish Soc 108:408–414
Morrow JE (1974) Effects of crude oil and some of its components on young coho and sockeye salmon. EPA-660/3-73-018, Washington DC
Moulton DG (1976) Spatial patterning of response to odors in the peripheral olfactory system. Physiol Rev 56:578–593
Muralt A von, Weibel ER, Howarth JV (1976) The optical spike. Structure of the olfactory nerve of pike and rapid birefringence changes during excitation. Pflügers Arch 367:67–76
Mustaparta H (1971) Spatial distribution of receptor responses to stimulation with different odours. Acta Physiol Scand 82:154–166
Nanba R, Djahanparwar B, Baumgarten R von (1966) Erregungsmuster einzelner Fasern des Tractus olfactorius lateralis des Fisches bei Reizung mit verschiedenen Geruchsstoffen. Pflügers Arch 288:134–150
Newcombe C, Hartman G (1973) Some chemical signals in the spawning behavior of rainbow trout (*Salmo gairdneri*). J Fish Res Bd Can 30:995–997
Nieuwenhuys R (1967) Comparative anatomy of olfactory centres and tracts. Prog Brain Res 23:1–64
Nieuwenhuys R (1982) An overview of the organization of the brain of actinopterygian fishes. Am Zool 22:287–310
Nordeng H (1971) Is the local orientation of anadromous fishes determined by pheromones? Nature 233:411–413
Nordeng H (1977) A pheromone hypothesis for homeward migration in anadromous salmonids. Oikos 28:155–159
Nowycky MC, Mori K, Shepherd GM (1981) Blockade of synaptic inhibition reveals long-lasting synaptic excitation in isolated turtle olfactory bulb. J Neurophysiol 46:648–658

Oelschläger HA (1976) Morphologisch-funktionelle Untersuchungen am Geruchsorgan von *Lampris guttatus* (Brünnich 1788) (Teleostei: Allotriognathi). Zoomorphologie 85:89–110
Ojha PP, Kapoor AS (1972) Functional anatomy of nose in the teleost Wallago attu Bl. & Schn. Arch Biol (Liege) 83:105–116
Ojha PP, Kapoor AS (1973) Structure and function of the olfactory apparatus in the freshwater carp, *Labeo rohita* Ham. Buch. J Morphol 140:77–86
Ojha PP, Kapoor AS (1974) Structure and function of the olfactory organs in the fish *Sisor rabdophorus* Ham. Acta Anat (Basel) 87:124–130
Oka Y (1980) The origin of the centrifugal fibers to the olfactory bulb in the goldfish, *Carassius auratus*: and experimental study using the fluorescent dye primuline as a retrograde tracer. Brain Res 185:215–225
Oka Y., Ichikawa M, Ueda K (1982) Synaptic organization of the olfactory bulb and central projection of the olfactory tract. Dev Aquacult Fish Sci 8:61–75
Olla BL, Pearson WH, Miller SE (1980) Chemoreception in the food-searching and feeding behavior of the red hake, *Urophycis chuss* (Walbaum). J Exp Marine Biol Ecol 48:139–150
Oshima K, Gorbman A, Shimada H (1969a) Memory-blocking agents: effects on olfactory discrimination in homing salmon. Science 165:86–88
Oshima K, Hahn WE, Gorbman A (1969b) Olfactory discrimination of natural waters by salmon. J Fish Res Bd Can 26:2111–2121
Oshima K, Hahn WE, Gorbman A (1969c) Electroencephalographic olfactory responses in adult salmon to waters traversed in the homing migration. J Fish Res Bd Can 26:2123–2133
Ottoson D (1956) Analysis of the electrical activity of the olfactory epithelium. Acta Physiol Scand 35:1–83
Ottoson D (1959a) Studies on slow potentials in the rabbit's olfactory bulb and nasal mucosa. Acta Physiol Scand 47:136–148
Ottoson D (1959b) Comparison of slow potentials evoked in the frog's nasal mucosa and olfactory bulb by natural stimulation. Acta Physiol Scand 47:149–159
Ottoson D (1959c) Olfactory bulb potentials induced by electrical stimulation of the nasal mucosa in the frog. Acta Physiol Scand 47:160–172
Ottoson D (1963) Some aspects of the function of the olfactory system. Pharmacol Rev 15:1–41
Ottoson D (1970) Electrical signs of olfactory transducer action. In: Wolstenholme GEW, Knight J (eds) Taste and smell in vertebrates, a CIBA foundation symposium. Churchill, London, pp 343–356
Ottoson D, Shepherd GM (1967) Experiments and concepts in olfactory physiology. Prog Brain Res 23:83–138
Parker GH (1910) Olfactory reactions in fishes. J Exp Zool 8:535–541
Parker GH, Sheldon RE (1912) The sense of smell in fishes. US Bur Fish Bull 32:35–46
Parker RA (1974) Empirical functions relating metabolic processes in aquatic systems to environmental variables. J Fish Res Bd Can 31:1550–1552
Partridge BL, Liley NR, Stacey NE (1976) The role of pheromones in the sexual behaviour of the goldfish. Anim Behav 24:291–299
Pawson MG (1977) Analysis of a natural chemical attractant for withing *Merlangius merlangus* L. and cod *Gadus morhua* L. using a behavioral bioassay. Comp Biochem Physiol 56A:129–135
Pfeiffer W (1962) The fright reaction of fish. Biol Rev 37:495–511
Pfeiffer W (1963a) The morphology of the olfactory organ of the Pacific salmon (*Oncorhynchus*). Can J Zool 41:1233–1236
Pfeiffer W (1963b) Vergleichende Untersuchungen über die Schreckreaktion und den Schreckstoff der Ostariophysen. Z Vergleich Physiol 47:111–147

Pfeiffer W (1964) The morphology of the olfactory organ of Hoplopagrus Guentheri Gill 1862. Can J Zool 42:235 – 237
Pfeiffer W (1967) Schreckreaktion und Schreckstoffzellen bei Ostariophysi und Gonorhynchiformes. Z Vergleich Physiol 56:380 – 396
Pfeiffer W (1968) Das Geruchsorgan der Polypteridae (Pisces, Brachiopterygii). Z Morphol Tiere 63:75 – 110
Pfeiffer W (1982) Chemical signals in communication. In: Hara TJ (ed) Chemoreception in fishes. Elsevier, Amsterdam, pp 307 – 326
Pfeiffer W, Lamour D (1976) Die Wirkung von Schreckstoff auf die Herzfrequenz von *Phoxinus phoxinus* (L.) (Cyprinidae, Ostariophysi, Pisces). Rev Suisse Zool 83:861 – 873
Pfeiffer W, Riegelbauer G (1978) The effect of the alarm substance on the central nervous excitation of the black tetra *Gymnocorymbus ternetzi* (Characidae, Ostariophysi, Pisces) indicated by dorsal light response. J Comp Physiol 123:281 – 288
Pinching AJ, Powell TPS (1971) The neuropil of the glomeruli of the olfactory bulb. J Cell Sci 9:347 – 377
Pipping M (1926) Der Geruchssinn der Fische mit besonderer Berücksichtigung seiner Bedeutung für das Aufsuchen des Futters. Soc Sci Fennica Comm Biol 2:1 – 27
Pyatkina GA (1976) Receptor cells of various types and their proportional interrelation in the olfactory organ of larvae and adults of acipenserid fishes. Citologija 18:1444 – 1449
Quinn TP, Brannon EL, Whitman RP (1983) Pheromones and the water source preferences of adult coho salmon *Oncorhynchus kisutch* Walbaum. J Fish Biol 22:677 – 684
Rahmani AR, Khan SM (1980) Histology of the olfactory epithelium and the accessory nasal sacs of an anabantoid fish, Anabas testudineus (Bloch). Arch Biol (Bruxelles) 91:397 – 411
Rahmani AR, Khan SM (1981) The olfactory organ in a few Indian teleosts. Curr Sci 50:329 – 330
Rao PDP, Dabhade KB (1973) In situ demonstration of changes in the preoptico-neurohypophysial complex after electrical stimulation of the olfactory tract in the catfish *Clarias batrachus* (Linn.). Z Zellforsch 139:95 – 99
Reese TS, Brightman MW (1970) Olfactory surface and central olfactory connexions in some vertebrates. In: Wolstenholme GEW, Knight J (eds) Taste and smell in vertebrates, a CIBA foundation symposium. Churchill, London, pp 115 – 150
Rhein LD, Cagan RH (1980) Biochemical studies of olfaction: isolation, characterization, and odorant binding activity of cilia from rainbow trout olfactory rosettes. Proc Natl Acad Sci USA 77:4412 – 4416
Ripatti PO, Sidorov VS (1973) Quantitative composition of the bile acids of certain vertebrates in connection with the nature of their nutrition. Doklady Akademii Nauk SSSR 212:770 – 773
Ritchie JM, Straub RW (1975) The movement of potassium ions during electrical activity, and the kinetics of the recovery process, in the non-myelinated fibres of the garfish olfactory nerve. J Physiol (Lond) 249:327 – 348
Sangster AW, Thomas SE, Tingling NL (1975) Fish attractants from marine invertebrates. Tetrahedron 31:1135 – 1137
Satou M (1971) Electrophysiological study of the olfactory system in fish. I. Bulbar responses with special reference to adaptation in the carp, *Cyprinus carpio* (L.). J Fac Sci Tokyo Sec IV 12:183 – 218
Satou M, Fujita I, Ichikawa M, Yamaguchi K, Ueda K (1983) Field potential and intracellular potential studies of the olfactory bulb in the carp: evidence for a functional separation of the olfactory bulb into lateral and medial subdivisions. J Comp Physiol 152:319 – 334

Satou M, Ichikawa M, Ueda K, Takagi SF (1979) Topographical relation between olfactory bulb and olfactory tracts in the carp. Brain Res 173:142–146
Satou M, Ueda K (1975) Spectral analysis of olfactory responses to amino acids in rainbow trout, *Salmo gairdneri*. Comp Biochem Physiol 52A:359–365
Satou M, Ueda K (1978) Synchronized rhythmic discharges of the secondary olfactory neurons in carp. Brain Res 158:313–329
Scalia F, Ebbeson SOE (1976) The central projections of the olfactory bulb in a teleost (*Gymnothorax funebris*). Brain Behav Evol 4:376–399
Schnitzlein HN (1966) The olfactory tubercle of the African lungfish, *Protopterus*. Ala J Med Sci 3:39–45
Schnitzlein HN, Crosby EC (1967) The telencephalon of the lungfish, *Protopterus*. J Hirnforsch 9:105–149
Scholz AT, Cooper JC, Madison DM, Horrall RM, Hasler AD, Dizon AE, Poff RJ (1973) Olfactory imprinting in coho salmon: behavioral and electrophysiological evidence. In: Proceeding of the 16th Conference Great Lakes Research, 1973, pp 143–153
Scholz AT, Gosse CK, Cooper JC, Horrall RM, Hasler AD, Daly RI, Poff RJ (1978) Homing of rainbow trout transplanted in Lake Michigan: A comparison of three procedures used for imprinting and stocking. Trans Am Fish Soc 107:439–443
Schulte E (1972) Untersuchungen an der Regio olfactoria des Aals, *Anguilla anguilla* L. I. Feinstruktur des Riechepithels. Z Zellforsch 125:210–228
Schulte E, Holl A (1971) Feinstruktur des Riechepithels von *Calamoichthys calabaricus* J. A. Smith (Pisces, Brachiopterygii). Z Zellforsch 120:261–279
Schulte E, Riehl R (1977) Elektronenmikroskopische Untersuchungen am Geruchsorgan von *Ophicephalus (Channa) obscurus* Guenther, 1861 (Teleostei, Ophicephalidae). Z Naturforsch 32c:307–308
Schutz F (1956) Vergleichende Untersuchungen über die Schreckreaktion bei Fischen und deren Verbreitung. Z Vergl Physiol 38:84–135
Selset R (1980) Chemical methods for fractionation of odorants produced by char smolts and tentative suggestions for pheromone origins. Acta Physiol Scand 108:97–103
Selset R, Døving KB (1980) Behaviour of mature anadromous char (*Salmo alpinus* L.) towards odorants produced by smolts of their own population. Acta Physiol Scand 108:113–122
Shahab M, Rahmani AR, Khan SM (1981) Anatomy of the olfactory apparatus of a siluroid fish, *Rita rita* (Ham.). J Anim Morphol Physiol 28:53–61
Sheldon RE (1911) The sense of smell in selachians. J Exp Zool 10:51–61
Sheldon RE (1912) The olfactory tracts and centers in teleosts. J Comp Neurol 22:177–339
Shepherd GM (1963a) Responses of mitral cells to olfactory nerve volleys in the rabbit. J Physiol (Lond) 168:89–100
Shepherd GM (1963b) Neuronal system controlling mitral cell excitability. J Physiol (Lond) 168:101–117
Shibuya T (1960) The electrical responses of the olfactory epithelium of some fishes. Jpn J Physiol 10:317–326
Silver WL (1979) Olfactory responses from a marine elasmobranch, the Atlantic stingray, *Dasyatis sabina*. Marine Behav Physiol 6:297–305
Silver WL (1982) Electrophysiological responses from the peripheral olfactory system of the American eel, *Anguilla rostrata*. J Comp Physiol 148:379–388
Silver WL, Caprio J, Blackwell JF, Tucker D (1976) The underwater electro-olfactogram: a tool for the study of the sense of smell of marine fishes. Experientia 32:1216–1217
Simontacchi C, Casciotti D, Boiti C (1982) Plasma steroid profile during the spawning season of pike. In: Richter CJJ, Goos HJT (eds) Proceedings of the international symposium reproductive physiology of fish. Pudoc Wageningen, p 110

Singh SP (1978) The functional anatomy of the olfactory organs in a hill-stream fish *Garra mullya* (Ham.). Anat Anz 144:220–224
Smeets WJAJ (1983) The secondary olfactory connections in two chondrichtians, the shark *Scyliorhinus canicula* and the ray *Raja clavata*. J Comp Neurol 218:334–344
Smith RJF (1982) The adaptive significance of the alarm substance-fright reaction system. In: Hara TJ (ed) Chemoreception in fishes. Elsevier, Amsterdam, pp 327–342
Solangi MA, Overstreet RM (1982) Histopathological changes in two estuarine fishes, *Menidia beryllina* (Cope) and *Trinectes maculatus* (Bloch and Schneider), exposed to crude oil and its water-soluble fractions. J Fish Dis 5:13–35
Solemdal P, Tilseth S (1978) Reactions of cod (*Gadus morhua* L.) to smell stimuli from bait. ICES report F 40
Solomon DJ (1973) Evidence for pheromone-influenced homing by migrating Atlantic salmon, *Salmo salar* (L.). Nature 244:231–232
Stabell O (1984) Homing and olfaction in salmonids: a critical review with special reference to the atlantic salmon. Biol Rev 59:333–388
Stabell O, Selset R, Sletten K (1982) A comparative chemical study on population-specific odorants from Atlantic salmon. J Chem Ecol 8:201–217
Stacey NE, Kyle AL (1983) Effects of olfactory tract lesions on sexual and feeding behavior in the goldfish. Physiol Behav 30:621–628
Stasko AB (1971) Review of field studies on fish orientation. Ann NY Acad Sci 188:12–29
Stoddart DM (1979) External nares and olfactory perception. Experientia 35:1456–1457
Story RH (1974) The olfactory bulbar formation and related nuclei of the paddlefish (*Polyodon spathula*). J Comp Neurol 123:285–298
Sutterlin AM (1975) Chemical attraction of some marine fish in their habitat. J Fish Res Bd Can 32:729–738
Sutterlin AM, Gray R (1973) Chemical basis for homing of Atlantic salmon (*Salmo salar*) to a hatchery. J Fish Res Bd Can 30:985–989
Sutterlin AM, Sutterlin N (1971) Electrical responses of the olfactory epithelium of Atlantic salmon (*Salmo salar*). J Fish Res Bd Can 28:565–572
Suzuki N (1977) Intracellular responses of lamprey olfactory receptors to current and chemical stimulation. In: Katsuki Y, Sato M, Tagaki SF, Oomura Y (eds) Food intake and chemical senses. University of Tokyo Press, Tokyo, pp 13–22
Suzuki N (1978) Effects of different ionic environments on the responses of single olfactory receptors in the lamprey. Comp Biochem Physiol 61A:461–467
Suzuki N (1982) Responses of olfactory receptor cells to electrical and chemical stimulation. In: Hara TJ (ed) Chemoreception in fishes, chap 6. Elsevier, Amsterdam, pp 93–108
Suzuki N, Tucker D (1971) Amino acids as olfactory stimuli in freshwater catfish, *Ictalurus catus* (Linn.). Comp Biochem Physiol 40A:399–404
Tammar AR (1974) Bile salts in fishes, chap 15. Academic Press, New York, pp 595–612
Teichmann H (1954) Vergleichende Untersuchungen an der Nase der Fische. Z Morphol Ökol Tiere 43:171–212
Teichmann H (1959) Über die Leistung des Geruchssinnes beim Aal (*Anguilla anguilla* L.) Z Vergleich Physiol 42:206–254
Teichmann H (1962) Die Chemorezeption der Fische. Ergebn Biol 25:177–205
Theisen B (1972) Ultrastructure of the olfactory epithelium in the Australian lungfish *Neoceratodus forsteri*. Acta Zool 53:205–218
Theisen B (1973) The olfactory system in the hagfish. I. Fine structure of the apical part of the olfactory epithelium. Acta Zool 54:271–284
Theisen B (1976) The olfactory system in the Pacific hagfishes *Eptatretus stoutii*, *Eptatretus deani*, and *Myxine circifrons*. Acta Zool (Stockh) 57:167–173
Theisen B (1982) Functional morphology of the olfactory organ in *Spinachia spinachia* (L.) (Teleostei, Gasterosteidae). Acta Zool (Stockh) 63:247–254

Theisen B, Breucker H, Zeiske E, Melinkat R (1980) Structure and development of the olfactory organ in the garfish *Belone belone* (L.) (Teleostei, Atheriniformes). Acta Zool (Stockh) 61:161–170
Thommesen G (1978) The spatial distribution of odour induced potentials in the olfactory bulb of char and trout (Salmonidae). Acta Physiol Scand 102:205–217
Thommesen G (1982) Specificity and distribution of receptor cells in the olfactory mucosa of char (*Salmo alpinus* L.). Acta Physiol Scand 115:47–56
Thommesen G (1983) Morphology, distribution, and specificity of olfactory receptor cells in salmonid fishes. Acta Physiol Scand 117:241–249
Thommesen G, Døving KB (1977) Spatial distribution of the EOG in the rat; a variation with odour quality. Acta Physiol Scand 99:270–280
Thornhill RA (1967) The ultrastructure of the olfactory epithelium of the lamprey *Lampetra fluviatilis*. J Cell Sci 2:591–602
Todd JH, Atema J, Bardach JE (1967) Chemical communication in social behavior of a fish, the yellow bullhead (*Ictalurus natalis*). Science 158:672–673
Toft R (1975) Lukt- och synsinnets roll för lekvandrings-beteendet hos Östersjölax. Swedish Salmon Research Institute, Report no 10, p 40
Trotier D, MacLeod P (1983) Intracellular recordings from salamander olfactory receptor cells. Brain Res 268:225–237
Trujillo-Cenoz O (1961) Electron microscope of observations on chemo- and mechanoreceptor cells in fishes. Z Zellforsch 54:654–676
Tucker D, Suzuki N (1974) Olfactory responses to schreckstoff of catfish. In: Denton A (ed) Proceedings of the international symposium of olfaction and taste. IRL Press, Melbourne, pp 121–127
Uchihashi K (1953) Ecological study of the teleosts in relation to the brain morphology. Jpn Reg Fish Lab Report no 2
Ueda K, Hara TJ, Gorbman A (1967) Electroencephalographic studies on olfactory discrimination in adult spawning salmon. Comp Biochem Physiol 21:133–144
Ueda K, Hara TJ, Satou M, Kaji S (1971) Electrophysiological studies of olfactory discrimination of natural waters by Himé Salmon, a land-locked Pacific salmon, *Oncorhynchus nerka*. J Fac Sci Tokyo Sec IV 12:167–182
Vanegas H, Ebbeson SOE (1976) Telencephalic projections in two teleost species. J Comp Neurol 165:181–196
Vinnikov YA (1982) Evolution of receptor cells. Cytological, membraneous and molcular levels. Springer, Berlin Heidelberg New York, pp 1–141 (Molecular biology, biochemistry and biophysics, vol 34)
Weiss DG (1980) Rapid axonal transport of amino acids and peptides in the olfactory nerve of the pike. Eur J Cell Biol 22:349
Westerberg H (1982a) Ultrasonic tracking of Atlantic salmon (*Salmo salar* L.) – I. Movements in coastal regions. Rep Inst Freshwater Res (Drottningholm) 60:81–101
Westerberg H (1982b) Ultrasonic tracking of Atlantic salmon (*Salmo salar* L.) – II. Swimming depth and temperature stratification. Rep Inst Freshwater Res (Drottningholm) 60:102–120
Westerberg H (1984) The orientation of fish and the vertical stratification at fine- and microstructure scales. In: McCleave JD, Neill WH, Dodson JJ, Arnold GP (eds) Mechanisms of migration in fishes. Plenum, New York, pp 179–204
Westerman RA, Baumgarten R von (1964) Wiederherstellung der Riechfunktionen nach Tractus olfactorius-Durchschneidung beim Karpfen. Pflügers Arch Ges Physiol 279:35
Westerman RA, Wilson JAF (1968) The fine structure of the olfactory tract in the teleost *Carassius carassius* L. Z Zellforsch 91:186–199

Whipple JA, Eldridge MB, Benville P Jr (1981) An ecological perspective of the effects of monocycle aromatic hydrocarbons on fishes. In: Vernberg F (ed) Biological monitoring of marine pollutants. Academic Press, New York, pp 483–551
White HC (1936) The homing of salmon in Apple River, N.S. J Biol Bd Can 2:391–400
Wilson JAF, Westerman RA (1967) The fine structure of the olfactory mucosa and nerve in the teleost *Carassius carassius* (L.). Z Zellforsch 83:196–206
Wisby WJ, Hasler AD (1954) Effect of olfactory occlusion on migrating silver salmon (*O. kisutch*). J Fish Res Bd Can 11:472–478
Yamamoto M, Ueda K (1977) Comparative morphology of fish olfactory epithelium. I. Salmoniformes. Bull Jpn Soc Sci Fish 43:1163–1174
Yamamoto M, Ueda K (1978a) Comparative morphology of fish olfactory epithelium. II. Clupeiformes. Bull Jpn Soc Sci Fish 44:855–859
Yamamoto M, Ueda K (1978b) Comparative morphology of fish olfactory epithelium. III. Cypriniformes. Bull Jpn Soc Sci Fish 44:1201–1206
Yamamoto M, Ueda K (1978c) Comparative morphology of fish olfactory epithelium. IV. Anguilliformes. Bull Jpn Soc Sci Fish 44:1207–1212
Yamamoto M, Ueda K (1978d) Comparative morphology of fish olfactory epithelium. V. Gasterosteiformes, Channiformes and Synbranchiformes. Bull Jpn Soc Sci Fish 44:1309–1314
Yamamoto M, Ueda K (1978e) Comparative morphology of fish olfactory epithelium. VI. Siluriformes. Zool Mag 87:254–261
Yamamoto M, Ueda K (1978f) Comparative morphology of fish olfactory epithelium. VII. Gadiformes, Lophiiformes and Gobiesociformes. J Fac Sci Tokyo Sec IV 14:115–125
Yamamoto M, Ueda K (1979a) Comparative morphology of fish olfactory epithelium. VIII. Atheriniformes. Zool Mag 88:155–164
Yamamoto M, Ueda K (1979b) Comparative morphology of fish olfactory epithelium. IX. Tetraodontiformes. Zool Mag 88:210–218
Yamamoto M, Ueda K (1979c) Comparative morphology of fish olfactory epithelium. X. Perciformes, Beryciformes, Scorpaeniformes, and Pleuronectiformes. J Fac Sci Tokyo sec IV 14:273–297
Yamazaki F, Watanabe K (1979) The role of steroid hormones in sex recognition during spawning behaviour of the goldfish, *Carassius auratus* L. Proc Indian Natl Sci Acad 45:505–511
Zeiske E (1974) Morphologische und morphometrische Untersuchungen am Geruchsorgan oviparer Zahnkarpfen (Pisces). Z Morphol Tiere 77:19–50
Zeiske E (1983) Chemically induced lesions in fish olfactory epithelium. FOH report no 1:40–43
Zeiske E, Breucker H, Melinkat R (1979) Gross morphology and fine structure of the olfactory organ of rainbow fish (Atheriniformes, Melanotaeniidae). Acta Zool (Stockh) 60:173–186
Zeiske E, Kux J, Melinkat R (1976a) Development of the olfactory organ of oviparous and viviparous cyprinodonts (Teleostei). Z Zool Systematik Evolutionsforsch 14:34–40
Zeiske E, Melinkat R, Breucker H, Kux J (1976b) Ultrastructural studies on the epithelia of the olfactory organ of cyprinodonts (teleostei, cyprinodontoidea). Cell Tissue Res 172:245–267
Zippel HP, Voigt R (1982) Neuronal correlates of olfactory behavior in goldfish. In: Hara TJ (ed) Chemoreception in fishes. Elsevier, Amsterdam, pp 181–199

Homeostasis of Extracellular Fluid in Retinas of Invertebrates and Vertebrates

J.A. Coles

Laboratoire d'Ophtalmologie Expérimentale, 22 Rue Alcide-Jentzer, 1211 Genève 4, Switzerland

1 Introduction

When a neurone is active, movement of ions and water across its membrane will tend to modify the composition of the fluid in the adjacent extracellular space. Usually, K^+ is the ion of particular interest because the relative changes in its extracellular concentration (K_o) can be large, up to more than fourfold, and because these changes will modify the excitability both of the neurone itself and of any of its neighbours that are exposed to a raised K_o (e.g., Frankenhaeuser and Hodgkin 1956, Baylor and Nicholls 1969). There is direct evidence that this K^+ mediates signal transmission between giant neurones in the cockroach (Yarom and Spira 1982), and the strongly suggestive evidence that it plays a role in nervous integration in many parts of the central nervous system is the major theme of the reviews by Orkand (1980) and Syková (1983). But in most nervous tissue, the interposition of glial cells between neurones, and the geometry and membrane properties of these glia, have led to the suggestion that one of the roles of the glia is to contribute to the homeostasis of the extracellular fluid, particularly by removing K^+ from those clefts where its concentration is elevated (Orkand et al. 1966, see Connors et al. 1982 for a recent example). General principles have been set out by Gardner-Medwin (1980, 1983b) and there are reviews by Kuffler and Nicholls (1966), Somjen (1979, 1981), Walz and Hertz (1983a) and Newman (1985a).

Extracellular ion changes have been given particular attention in two kinds of retina: that of cold-blooded vertebrates and that of an insect, the honeybee drone. In these two systems what seem to be similar problems have been solved in two different ways, and the core of this review will be a summary of known homeostatic processes in the retina of the drone, where they are unusually accessible to experimental study, and in the neural retina of vertebrates.

2 Techniques

The most useful experimental tool has been the double-barrelled ion-selective microelectrode that simultaneously records an ion activity and the electrical potential. Electrodes of the liquid membrane type (Walker 1971) are the easiest to make, and a variety of sensor solutions are available for making electrodes specific for most of the common biological ions (Ammann et al. 1983). For theory see Morf (1981) and for practical methods see Thomas (1978) and Munoz et al. (1983). The electrode senses the activity of the ion rather than its concentration, and the two quantities are related by the activity coefficient, y, which is in the range $0.7-0.8$ in most Ringer solutions (see Appendix 1 in Coles and Tsacopoulos 1979). In this article free concentrations are usually given, rather than activities, it being assumed that, for ions that are free and not bound, y is approximately the same as in the Ringer solution. Experimental results appear not to contradict this assumption, at least for monovalent ions (see, e.g., Caillé and Hinke 1972, Coles and Rick 1985), except, perhaps, in the case of an extracellular space in leech photoreceptors (Sect. 5.2 below).

A question that has been of particular importance in studying K_o in the vertebrate retina is the extent to which the space created by the electrode tip, of diameter 0.5 – 3 µm, causes the measured changes in [K] to be smaller and slower than those in the undisturbed extracellular clefts whose width is about 0.02 µm (see Oakley 1981, Syková 1983, Newman and Odette 1984). Experiments in a variety of tissues now suggest that in most cases the effect is small. In drone retina, a flash of light can produce an increase in measured K_o that reaches its peak in less than 1 s (Fig. 1 f in Tsacopoulos et al. 1983), which sets an upper limit for the delay introduced by the dead volume in this tissue. C. J. Karwoski (personal communication) measured light-induced changes in K_o in frog retina with electrodes with tip diameters of more than 3 µm and less than 0.5 µm and, with the possible exception described in Sect. 6.5, obtained similar results. C. L. Yamate (personal communication) systematically used a range of tip diameters in rat optic nerve and found that only when the tip diameter was greater than about 2 µm was there a significant fall in signal amplitude.

Experiments of a different kind have been done by Nicholson et al. (1979) and Nicholson and Phillips (1981). They used a micropipette (tip diameter 1 µm) to inject ions to extracellular space in rat brain, creating a step function source with an instantaneous onset and a duration of 50 s. A second micropipette that had a barrel selective for the injected ion and a tip diameter of 2 – 4 µm was located 40 – 200 µm away, and the change in concentration of the ion at this site was recorded. For ions that did not cross the cell membranes, the observed time course of the concentration change could be well described by theoretical equations that made no allowance for possible dead spaces round the electrode tips. In conclusion, in experiments such as these, where the source of extracellular ions either extends over many micrometres in space or lasts for several seconds, extracellular ion-selective microelectrodes appear to record changes that correspond quite closely to those in undisturbed clefts.

3 Comments on Some Basic Principles of K^+ Clearance

3.1 Space-Dependent and Space-Independent Processes

Some of the processes by which excess K^+ can be cleared from extracellular space in one region of tissue depend on there being other regions where K_o has not increased. K^+ can then diffuse along the extracellular clefts, or set in train the spatial buffering by transfer cells which is discussed below (Sect. 3.2). Such processes are called "space-dependent" by Gardner-Medwin (1980). There may, in addition, be processes that would be equally effective if K_o rose uniformly throughout the tissue. These "space-independent" processes involve net movement of ions or water into (or out of) cells. It now appears that usually several processes act together to clear K_o, and results from some earlier papers have subsequently been reinterpreted more plausibly in this light (e.g., Gardner-Medwin 1981a, Coles and Orkand 1983).

3.2 Spatial Buffering

In the conceptually simplest case, K^+ released into extracellular space during neuronal activity within a restricted volume of tissue is prevented from causing an excessive rise in K_o by an elongated "transfer" cell (usually a glial cell) whose membrane is permeable exclusively to K^+. At rest, the cell is confronted by a uniform K_o and the membrane potential is given by the Nernst equation. When K_o increases at one point along its length, this part of the membrane will no longer be polarized to the Nernst potential and there will be an electrochemical imbalance that causes K^+ to enter. The entry of K^+ tends to depolarize the whole cell: for a given K_o, the less the cell depolarizes the greater is the inward driving force. Because the cell does depolarize somewhat, K^+ is no longer in Nernst equilibrium across the membranes facing an unchanged K_o, and here K^+ will leave the cell (Fig. 6A). The return loop for the current is through the narrow, tortuous clefts of the extracellular space, where it is associated with extracellular potential gradients which are typically 0.5 – 3 mV/100 µm. If the electrical resistivity of the cytoplasm is known, the potential gradient within the transfer cell can be estimated from its cross-section relative to that of the extracellular clefts (and taking into account the tortuosity of the latter). For the principal transfer cells of frog retina, the Müller cells (Sect. 6.2 below), the measurements of Newman (1985b) give a value of cytoplasmic resistivity indistinguishable from that of the extracellular medium. Since at their thinnest cross-section they occupy about 2.3% of the volume of the tissue (calculated from Newman et al. 1984 and Newman 1985b), while extracellular space is about 10% (and tortuous) (Karwoski et al. 1985a), the intracellular gradients are probably only about twice those in the extracellular space. Hence the intracellular potential gradients are very small compared to the membrane potential and for many purposes the interior of the cell can be regarded as isopotential. (Consider also the vertebrate rod, which is almost isopotential despite the narrowness of the neck joining the two segments: Bader et al. 1979.) During the first few seconds of K^+ transfer the intracellular current will be carried by all the mobile ions, but an approximately steady state will then be set up and the current carried only by K^+ ions (Appendix in Coles and Orkand 1985).

The optimum specific membrane resistance of the transfer cell will depend on the spatial extent of the K^+ source and its duration: this is discussed by Gardner-Medwin (1983b). Recently, evidence has been obtained for regional variations of membrane resistance in a glial cell (Newman 1984; Sect. 6.2 below). In principle, if the areas of membrane facing the sources and sinks of extracellular K^+ had particularly high K^+ permeability then the spatial buffering would be more effective.

Spatial buffering through glial cells is a major contributor to the sources and sinks of extracellular current that are revealed by current-source density analysis (Nicholson and Freeman 1975, Freeman and Nicholson 1975).

3.3 Net Uptake

The earlier direct evidence for net uptake was for bulk-separated or cultured cells in vitro and came from use of radioactive tracers (e.g., Franck et al. 1978, Hertz 1978). More recently, net uptake has been demonstrated using ion-selective electrodes in identified glial cells in culture (Kettenmann et al. 1983). In intact tissue, it has been argued that net uptake accounts for part of the measured changes in

K_o (Vern et al. 1978, Cordingly and Somjen 1978, Gardner-Medwin 1981a), and recently more direct evidence has been obtained from intracellular measurements in drone retina (Coles and Orkand 1983) and leech neuropil glial cells (Schlue and Wuttke 1983). Net uptake obviously must occur in neurones that cease a period of activity, and may also occur in inactive neurones as well as glia. Where comparisons have been made, the latter have been found to be the more effective (Hertz and Chaban 1982, Coles 1981, Coles and Orkand 1983, Schlue and Wuttke 1983).

Mechanisms for clearing K_o by net uptake include:

1. Passive fluxes of K^+, Cl^- and water through independent membrane channels to meet the requirements of Gibbs-Donnan equilibrium. If K^+ is the principal intracellular cation, then intracellular K^+ concentration, K_i, will be almost unchanged because of dilution by the water uptake (Boyle and Conway 1941).
2. More directly coupled transport of K^+ and Cl^- that may also involve Na^+ and is blocked by furosemide and bumetanide (for references see Russell 1983, Walz and Hertz 1984b, Ellory and Hall 1984). The energetics of this system are not yet certain: both the electrochemical gradient of Na^+ and ATP may provide energy to pump Cl^- into the cell.
3. Na/K exchange by the ouabain-sensitive Na-K ATPase or Na pump found in virtually all cells. The pumping rate, which is usually limited by Na_i, is increased in neurones after entry of Na during activity (e.g., Ritchie and Straub 1957, Tsacopoulos et al. 1983). The pump can also be stimulated by an increase in K_o so that inactive cells, such as glial cells, can make a contribution to K_o clearance by Na/K exchange. But except when the cells were artificially loaded with Na, the contributions that have been observed in tissues have been small (Tang et al. 1980, Walz et al. 1983). This may be either because most of the external K sites are saturated at physiological K_o, as is the case in muscle and nerve (e.g., Baker and Connelly 1966, Gadsby 1980), or simply because the pumping rate is low, as is the case for glial cells in drone retina (Coles and Orkand 1985). In cultured astrocytes, Walz and Hertz (1982, 1984a) describe what appears to be a Na pump with unusual properties.

4 Drone Retina

4.1 Anatomy and Function

Behind each corneal facet of the compound eye of the drone is an optical system that focuses light on to the end of a light guide called the rhabdom. This extends for about 400 μm to a basal membrane and is composed of densely packed microvilli contributed, essentially, by six large photoreceptor cells (see Perrelet 1970 for details). The photoreceptors are electrically coupled in a rosette-like cluster called the retinula (Fig. 1; Shaw 1969). The microvilli contain visual pigment and the photoreceptors respond directly to light. Nearly all the space between the retinulae is filled by outer pigment cells. These have the histological properties of glial cells (Perrelet 1970, Coles and Tsacopoulos 1981) and are electrically coupled (Coles and Orkand 1983). Coles and Tsacopoulos (1979) esti-

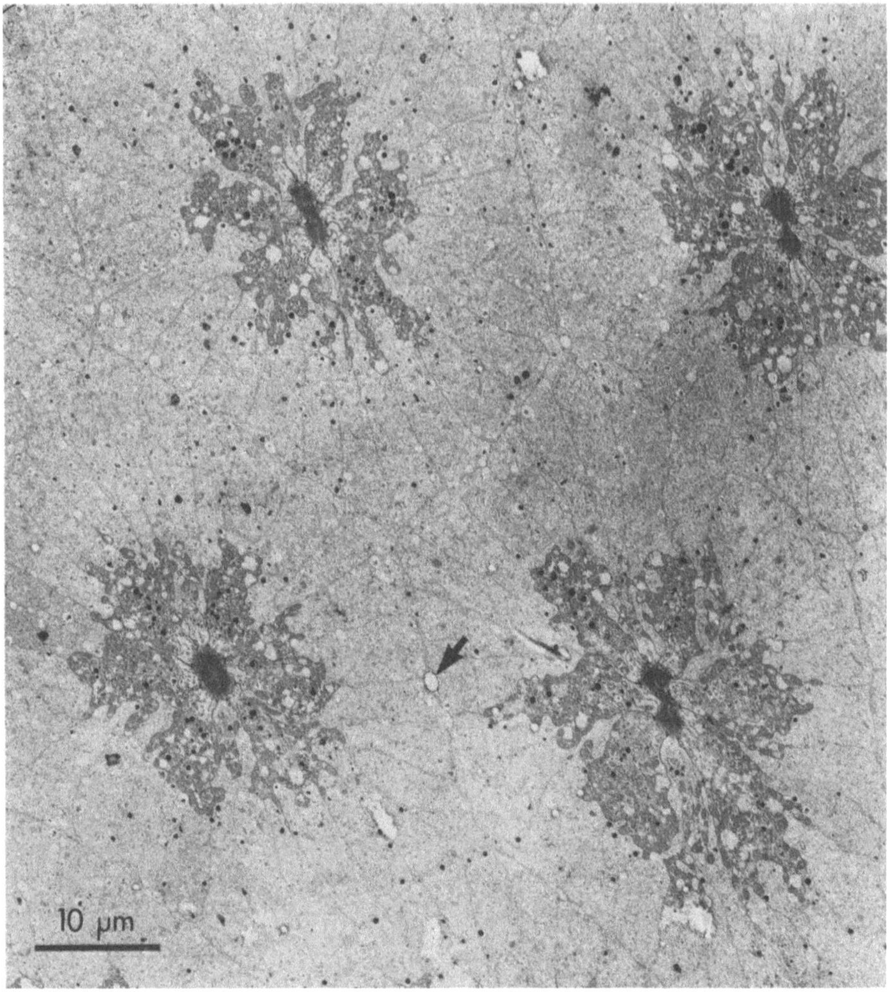

Fig. 1. Electron micrograph of a tangential section of retina of honeybee drone (*Apis mellifera* ♂) showing four retinulae cut towards their distal ends. The retinulae, which are clusters of photoreceptor cells, are seen embedded in the syncytium of outer pigment cells. Each photoreceptor contributes microvilli to the rhabdom, the *dark structure* at the centre of each retinula. The photoreceptors are packed with mitochondria: granules of screening pigment are scattered through both types of cell. In vivo, the tissue is oxygenated by tracheoles (*arrow*). (Coles and Rick 1985)

mated that the glia occupied 57% of the total volume, extracellular space 5% and photoreceptors 38%. The basal membrane in a similar retina, that of the locust, is a barrier to diffusion: "Tracer equilibration measurements imply that the extracellular cation exchange pool must lie totally within the eye, for events on a time scale even exceeding an hour" (Shaw 1978). This is very different from the vertebrate retina with its blood supply and contact with the vitreous. Because of the high electrical resistance of the basal membrane, stimulation of the photoreceptors set up current loops that accentuate differences in the signals reaching the afferent synapses of the different photoreceptors (Shaw 1975). In the drone, the main function of the eye is presumably to seek out the small black dot that is the queen on her nuptial flight, and to do this, each retinula must respond to very small fractional changes in light flux. At the same time there will always be at least one facet directed at the sun and intensely stimulated.

4.2 Experimental Preparation

In order to oxygenate the drone retina without the respiratory movements of the abdomen, and to permit experiments in which the extracellular fluid can be changed artificially, most experiments have been done on isolated slices of retina: two cuts are made parallel to the axes of the retinulae and 300 – 800 μm apart at the cornea (Fig. 2). Superfusion with oxygenated Ringer solution provides adequate oxygenation to a distance of 400 μm from the superfused surface in the dark and to about 200 μm during repetitive stimulation with 20-ms flashes, one per second (Tsacopoulos et al. 1981). The experimental illumination is perpendicular to the axes of the retinulae, instead of through the cornea as in life.

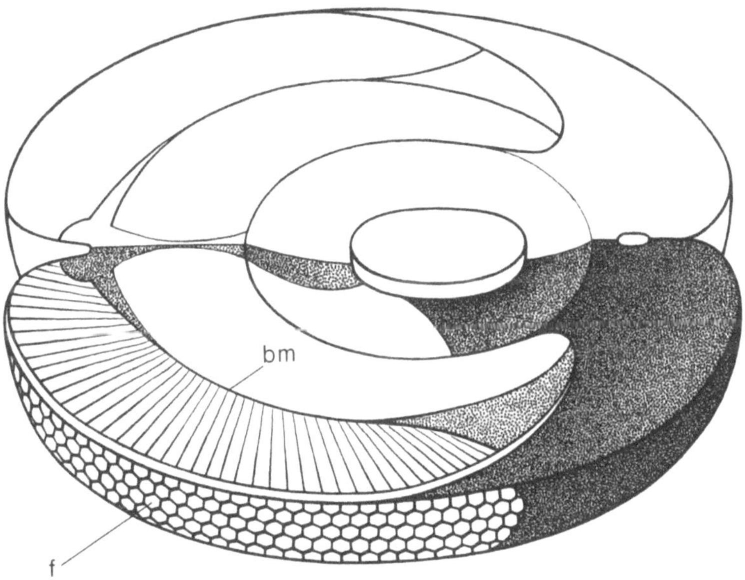

Fig. 2. Schematic view of a slice of drone retina as used for electrophysiological experiments. The posterior side is *upwards* and dorsal is *to the left*. Behind each facet, *f*, is an ommatidium about 400 μm long that ends at the electrically resistant basal membrane (*bm*). In vivo, light enters through the facets of the cornea, but the experimental preparation is stimulated by light incident perpendicular to the axes of the ommatidia. The facets are drawn larger than to scale. (Courtesy of F. von Kaenel)

4.3 Stimulus-Induced Ion Movements in Drone Retina

The drone retina, being composed essentially of homogeneous populations of one kind of neurone and one kind of glial cell, is exceptional among nervous tissues in that it is not unreasonable to attempt to quantify the stimulus-induced ion movements and check the results by seeing if the book-keeping of the gains and losses in each compartment is correct. Measurements have been made of activities (with microelectrodes) and concentrations (with electron microprobe X-

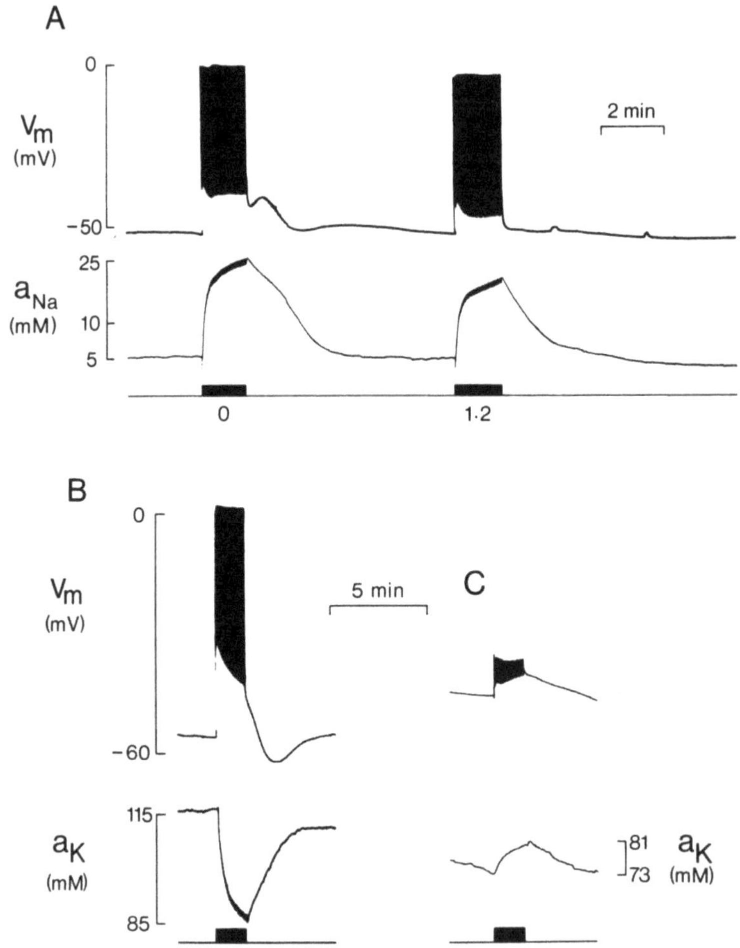

Fig. 3A,B. Light-induced changes in intracellular ion activities in drone retina measured with ion-selective microelectrodes. **A** Na in a photoreceptor. The *top trace*, recorded by the reference barrel of the double-barrelled microelectrode, is the membrane potential (V_m). During the period indicated by the *solid bar* (*bottom line*), the retina was stimulated diffusely with 20-ms flashes of light, one per second for 90 s. For the second stimulation period, the light intensity was reduced by 1.2 log units. The *second trace* is the difference in potential between the two barrels and gives sodium activity, a_{Na}. (Record by R. K. Orkand and J. A. Coles) **B** Potassium activity, a_K, in a photoreceptor. **C** a_K in a glial cell. (*B* and *C*, Coles and Orkand 1983)

ray analysis) of Na, K and Cl. Comparison of the two kinds of measurement suggests that as a working hypothesis these ions can be considered to be essentially freely dissolved in the cell water, with activity coefficients close to those in the extracellular fluid (Coles and Rick 1985). Data exist for changes in $[Na^+]$, $[K^+]$ and $[Cl^-]$ in each of the three compartments: photoreceptors, glia and extracellular space. In addition, stimulus-induced changes in the volume of the extracellular space have been monitored (Orkand et al. 1984). The requirements of electroneutrality and conservation of matter impose strong tests on the validity of the measurements and their interpretation. In contrast, the preparation is probably less well suited to a detailed study of the molecular mechanism of glial membrane transport than are preparations of organs composed only of glial cells (Orkand et al. 1973, Ritchie et al. 1981, Ransom and Yamate 1984) or cultured glial cells (Hertz et al. 1982, Kettenmann et al. 1983, 1984).

The surface-to-volume ratio of the photoreceptors is about 1.8 μm^{-1}. This is the same as a cylinder with a diameter of 2.3 μm and, compared with, say, a squid giant axon or a muscle cell, the changes in intracellular ion composition during activity are very large. As in most invertebrate photoreceptors, the effect of light is to increase the permeability of the membrane to Na (Fulpius and Baumann 1969). The mean value of intracellular $[Na^+]$ (Na_i) in the dark measured by Coles and Orkand (1985) was 14 m*M*, but any errors in the measurements, such as damage to the cells or leaks round the electrode (Taylor and Thomas 1984), would tend to increase Na_i, so this may be an overestimate. Tsacopoulos et al. (1983) reported that after a single 40-ms light flash, the mean increase in Na_i was 3.6 m*M* (see also Fig. 3A). The charge associated with an increase in Na_i of only 0.01 m*M* would be sufficient to depolarize the capacitance of the surface membrane, so there must be movements of other ions to maintain electroneutrality. Those measured so far include an entry of Cl (Coles and Orkand 1984) and an efflux of K (Coles and Tsacopoulos 1979, Coles and Orkand 1983).

4.4 Routes of Clearance of Extracellular Potassium

As shown in Fig. 3B, repetitive stimulation causes photoreceptor K_i to fall by about 25% in 90 s (Coles and Tsacopoulos 1979). Because of the regularity of the structure of the drone retina, it is straightforward to make approximate calculations of how the K^+ that leaves the photoreceptors might change K_o in various hypothetical situations. Curve *a* in Fig. 4 is an estimate of what would happen during stimulation with a 90-s series of light flashes if the K^+ simply accumulated in the extracellular clefts: K_o rapidly reaches unphysiological levels. Curve *b* shows that diffusion through the clefts to the bath would not reduce K_o to the levels actually measured with ion-selective microelectrodes (Fig. 5 and solid line *e* in Fig. 4). Curve *c* suggests what would happen if diffusion were quite unrestricted and the photoreceptors confronted a very large (semi-infinite) reservoir of extracellular space. Excursions of K_o are greatly reduced and, indeed, during the first 14 s of the stimulation K_o is lower than what is observed (line *e*). At longer times, as $[K^+]$ builds up in the reservoir this mechanism becomes less effective. Curve *d* illustrates another possibility, spatial buffering. Instead of a large reservoir, the photoreceptor membrane is now supposed to be confronted

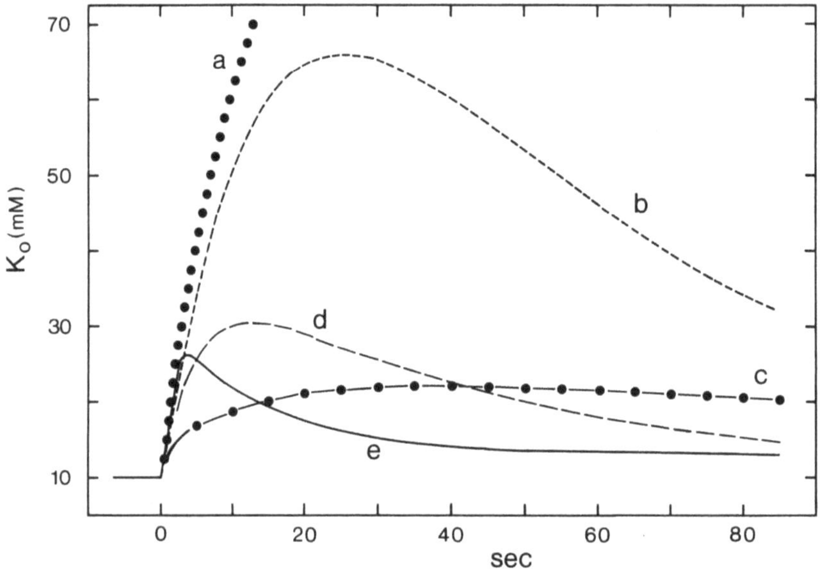

Fig. 4. Calculated changes in K_o in drone retina for four hypothetical cases. It is assumed that at the onset of repetitive light stimulation the fall in $[K^+]$ in the photoreceptors is exponential:

$$\Delta[K^+] = \Delta K_p(\infty)\{1 - \exp(-t/\tau\} \qquad (1)$$

where t is the time after the onset of the stimulation, the final change $\Delta K_p(\infty) = -30.7$ m*M* and $\tau = 45$ s, as in Coles and Tsacopoulos (1981). The *line a* shows K_o if the K^+ simply stayed in an extracellular space with a volume 5/38 that of the photoreceptors. *Line b* shows K_o if K^+ diffused along tortuous extracellular clefts to the bath, 100 μm away (for details see legend to Fig. 3 in Coles and Tsacopoulos 1981). *Line c* is an indication of the consequences of diffusion into a large volume of extracellular space. A surface-to-volume ratio was calculated that ignored the great increase in area introduced by the microvilli: a value of 0.6 μm^{-1} was obtained (using figures from Coles and Tsacopoulos 1979). The K^+ flux density across this surface was calculated and it was supposed that this flux came out of an infinite plane membrane into a semi-infinite reservoir of stagnant Ringer solution. K_o at the surface of this membrane is given by an equation analogous to 2.9.9 in Carslaw and Jaeger 1959, which was evaluated numerically by Mr. S. Poitry. *Line d* was calculated from a very simple model for glial cell spatial buffering. It is supposed that the K^+ is released into extracellular clefts confronting a glial syncytium whose membrane is specifically permeable to K^+. At rest, the K^+ is passively distributed. When local K_o rises, the rest of the syncytium prevents the membrane at this point from depolarizing more than 0.3 of the way to the new local Nernst equilibrium. The curve was obtained from equation A8 in Appendix A, with $k_2 = 2k_1$. The *solid line e* is a typical measured response

by a narrow cleft bounded by a membrane that is electrically resistant but selectively permeable to K^+, as described in Appendix A. This is at all times more effective than diffusion through extracellular clefts (curve *b*), and after 42 s it is more effective than unrestricted diffusion into a semi-infinite space (curve *c*; see also Gardner-Medwin and Coles 1980). This effectiveness at longer times is due to the fact that the K^+ is carried across the glial membrane mainly by the elec-

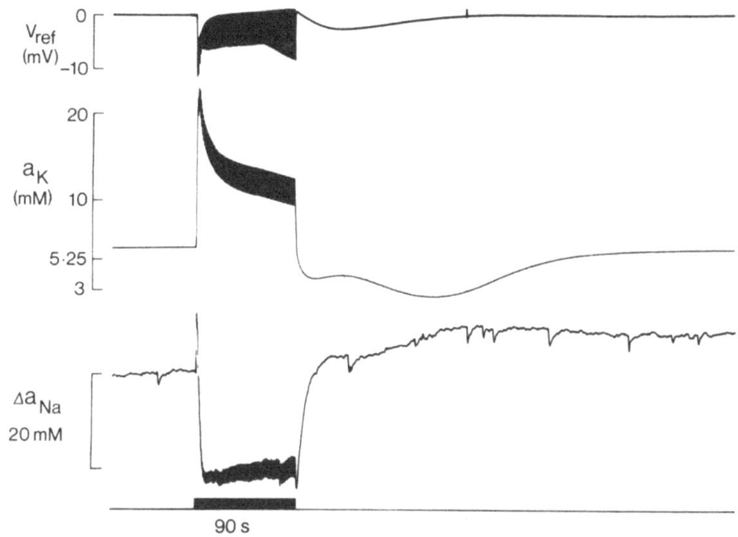

Fig. 5. Light-induced changes in ion activity in extracellular space in drone retina. Potassium activity, a_K, change in sodium activity, Δa_{Na}, and field potential, V_{ref}, were measured simultaneously with a triple-barrelled ion-selective microelectrode. During the time indicated by the *solid bar* the retina was stimulated with a series of 20-ms light flashes, one per second for 90 s; each flash evoked a negative-going field potential, which was recorded by the reference barrel. In the *third trace*, the baseline a_{Na} was close to that in the Ringer solution, 147 m*M*. (Coles and Orkand 1985)

trical gradient and is affected not so much by the absolute increase in local glial $[K^+]$ as by the fractional increase, which is small.

Figure 3C shows the increase in glial K_i that is observed when the photoreceptors are stimulated by a train of light flashes (Fig. 3B). In their first work on the drone retina, using not a slice but half a head (the "cut head" preparation), Coles and Tsacopoulos (1979) estimated that the increase in glial $[K^+]$ was sufficient to account for all the K^+ that left the photoreceptors, apart from the very small quantity that remained in the extracellular clefts. Subsequent work, including use of O_2 electrodes (Tsacopoulos et al. 1981), has shown that light stimulation increases the amount of deep tissue that is hypoxic in this preparation, and this factor probably contributed to increasing glial $[K^+]$. The present estimate, from measurements in fully oxygenated slices, is that if the glial cell volume did not change significantly during stimulation, then net uptake of K^+ would account for just over half of the K^+ lost by the photoreceptors (Table 2 in Coles and Orkand 1985).

The quantity of extra K^+ that remains in the extracellular clefts is very small because not only is the volume small, but it shrinks during light stimulation by some 30% (Orkand et al. 1984). Most of the remainder of the K^+ is presumably cleared by spatial buffering, as indicated in Fig. 6A. Given the circumstances, spatial buffering must occur: it is facilitated by the electrical coupling between the glial cells (Coles and Orkand 1983). The extracellular electrical gradients associated with it have been measured in the cut head by Gardner-Medwin et al. (1981), who concluded that the currents might account for all the K^+ clearance.

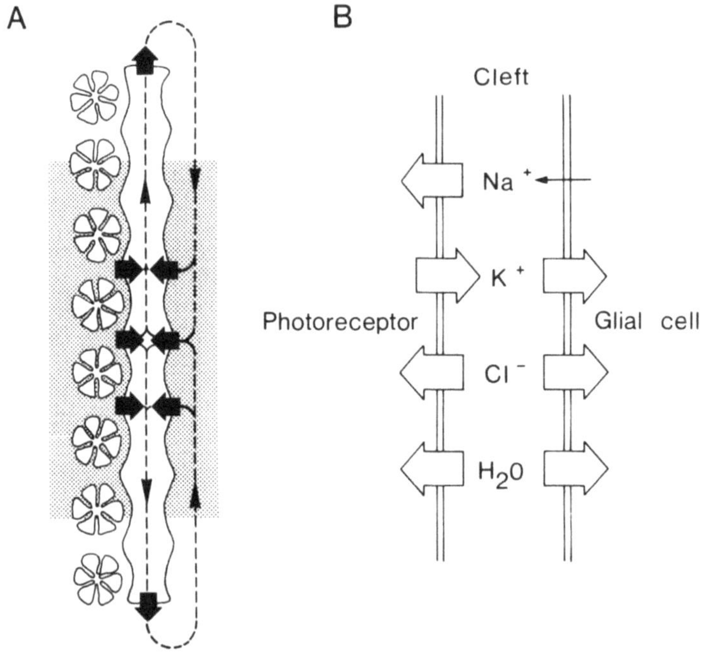

Fig. 6A, B. Scheme of light-induced ion movements in drone retina. **A** Spatial buffering currents. The retinulae of a slice of retina are represented by the *rosettes* and the glial cell syncytium by the *column* traversing the slice. During light stimulation K_o increases most at the centre of the slice (*dotted area*). K^+ enters the glia in this region and other K^+ ions are discharged into the superfusate at the two surfaces of the slice (*fat arrows*). The *dashed lines* indicate the intracellular and extracellular limbs of the current loops. (Coles and Orkand 1983). **B** In addition to the spatial buffering currents there are net movements of ions and water across the cell membranes of the photoreceptors and the glial cells. In the glial cells in addition to the entry of K^+, Cl^- and H_2O, there may perhaps be a small efflux of Na^+

Extracellular potential gradients in slices appear to be smaller (J. A. Coles, unpublished results), but this does not necessarily mean that the currents are smaller: the high K_o in the cut head would tend to shrink the extracellular space to a permanently small volume (see Coles et al. 1981).

4.5 Mechanisms of Net Uptake of K^+ into Glia in Drone Retina

When bath $[K^+]$ is increased from its normal value in bee Ringer of 7.5 m*M* to 18 m*M*, K_i in the photoreceptors hardly changes, but in the glia it increases. Cl_i increases in both glia and photoreceptors when K_o is increased (Coles and Orkand 1984), so there is presumably an entry of KCl and therefore of water into the cells. The more recent measurements of K_i with ion-selective electrodes have given similar values, about 130 m*M*, in the two cell types. The measured osmolarity of the Ringer solution was 530 mosmol, so the measured K_i accounts for only about half the predicted intracellular cation concentration. In this case, entry of isotonic KCl should increase K_i. The difficulty is to explain the absence of change in the photoreceptors, rather than the increase in the glia. Another classical mechanism is the Na/K exchange or Na pump mediated by the ouabain-sensitive Na/K ATPase. In drone retina, raising K_o causes glial Na_i to fall, albeit slowly. The initial fall is probably due in part to swelling of the cells, resulting in dilution of Na_i, but the fall continues for tens of minutes, which is difficult to explain except by a Na pump. However, it is not known whether Na_i falls simply because the driving force for the Na leak is reduced when the membrane is depolarized, or whether the pump is stimulated. During photostimulation for 90 s, the decrease in Na_i is only 2 m*M*, which sets an upper limit for the contribution of Na/K exchange to the increase in K_i (Coles and Orkand 1985). At present, although the data are not conclusive, some additional mechanism of K^+ uptake would seem to be necessary.

4.6 Homeostasis of Na_o

If the volume of the extracellular space remained constant at 10% of the volume of the photoreceptors, then the quantity of Na^+ involved in an increase in photoreceptor Na_i as large as that shown in Fig. 3A would correspond to a fall in Na_o of over 200 m*M*. The observed fall in Na_o is only about a tenth of this (Fig. 5). Efflux of Na from the glia makes at most a small contribution (Coles and Orkand 1985). Two other factors are more important. One is the arrival of Na^+ ions through the extracellular clefts, carrying about half the charge in the extracellular limb of the spatial buffer current loops (Fig. 6A; see Dietzel et al. 1980, 1982). The other is the fact that the extracellular space shrinks during activity (Orkand et al. 1984) and the remaining Na^+ ions are confined in a smaller volume. This shrinking can provide a large quantity of ions if they are present in a high concentration, as for Na^+. It has an adverse effect on ΔK_o, but since K_o is much less than Na_o, the quantity of ions involved is much smaller. Shrinkage of extracellular space during activity has also been observed in other nervous tissues (Dietzel et al. 1982, Ransom et al. 1985).

5 Other Invertebrates

5.1 Insects with Open Rhabdoms

In *Musca* and some other insect species the microvilli of the photoreceptor cells in an ommatidium do not abut to form a single fused rhabdom as in the drone (Fig. 1) but are separated by a large volume of extracellular space as illustrated in Fig. 7. Curves *c* and *d* of Fig. 4 would suggest that this arrangement should be most effective for brief changes in K^+ release.

5.2 Leech

The eye of the leech (*Hirudo*) is composed of some 30 similar photoreceptor cells arranged within a small eyecup (Fig. 8A). Half of the volume of each photo-

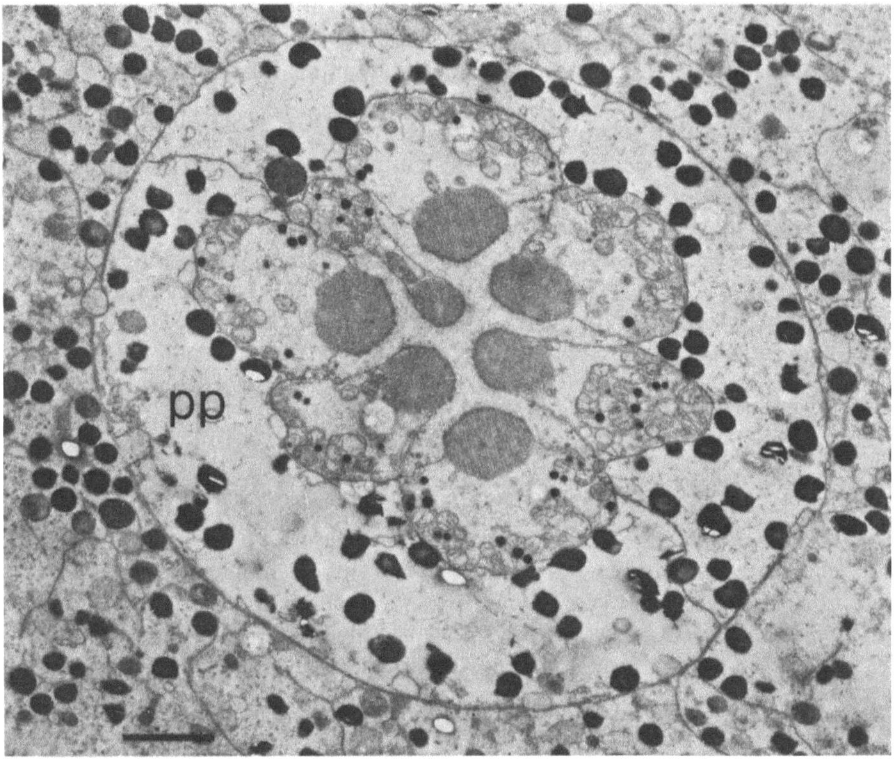

Fig. 7. Cross-section of an ommatidium of *Musca domestica*. The picture shows seven photoreceptor cells whose rhabdomeres are separated by a large extracellular space. They are surrounded by two primary pigment cells (*pp*). *Scale bar*, 2 μm. For details see Boschek 1971. (Courtesy of Dr. C. B. Boschek)

receptor is comprised by a "vacuole", which is connected by several narrow clefts to the external extracellular space (Fig. 8B). The microvilli that bear the photopigment project into this vacuole. The outer surface of the cell confronts a wide extracellular cleft except where it is invested by slender glial processes from a sheath of glial cells. Hence the two parts of the cell are in situations like those corresponding to curves *c* and *d* in Fig. 4. B. Walz (1985) measured [K^+] in the two spaces and found that the light-induced changes had the qualitative features of the theoretical ones in Fig. 4. Walz and Somlyo (1984) measured the concentrations of elements in the vacuole by electron microprobe X-ray analysis and found that the concentration of K^+ was double that in the extracellular space, although there was good evidence that the activities were equal. It is probable that the high density of negative charges on the matrix substance of the vacuole bind some of the K^+. If the equilibrium constant were in the millimolar range then the matrix would be a chemical buffer, albeit far less effective than the more familiar buffers that keep free concentrations of H^+ or Ca^{2+} below micromolar. The ratio of Na^+ concentrations was smaller (1.3), suggesting that the effect might be greater for K^+ than for Na^+. If this is correct, a similar phenomenon might be expected in the extracellular space in the rhabdom of *Musca*.

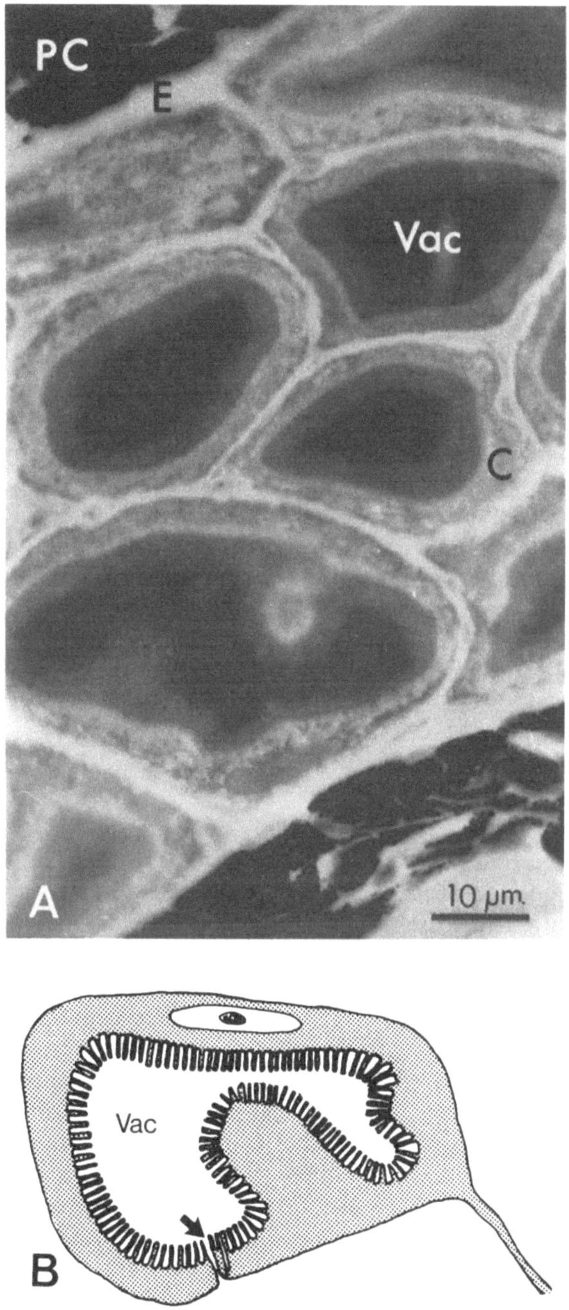

Fig. 8. A Light micrograph of part of an eyecup of *Hirudo medicinalis*. The photoreceptor cells are separated from each other by wide extracellular spaces (*E*). Each cell contains a large "vacuole" (*Vac*) surrounded by a thin layer of cytoplasm (*C*). The vacuole is lined by microvilli. *PC*, pigment cells. (Courtesy of Dr. B. Walz) **B** A single photoreceptor. The *arrow* indicates one of the clefts that connect the vacuole to the external extracellular space. (After Walz 1982)

6 Vertebrate Retina

6.1 The Electroretinogram

Light-induced changes in K_o in the vertebrate retina have been of great interest since it became apparent that they contributed to the light-evoked potentials that constitute the electroretinogram (ERG, Fig. 11 A). The history and clinical usefulness of the ERG are outlined by Armington (1974), and the origins of its various component waves have been discussed by Tomita and Yanagida (1981). When K_o is changed experimentally, transmission of information through the neural retina is affected: in *Raja* (skate), doubling K_o reduces sensitivity as measured by the b-wave of the ERG by 0.3 log units (Fig. 2 in Dowling and Ripps 1976). But it seems unlikely that this is a major mechanism of light adaptation since, as will be described, the measured changes in K_o are relatively small, while light adaptation can cause sensitivity to change by over 6 log units, and a considerable part of the volume of the retina is given over to a cell that seems to be specially well suited to clearing K_o, the Müller cell. The common postulate that K_o tends to be stabilized in the neural retina will therefore be accepted. The effect of localized illumination, which sets up currents in the tangential plane (Proenza and Freeman 1975, Karwoski and Proenza 1981), will be given by only passing notice, and discussion will be further limited to only two components of the ERG, slow P-III and the b-wave. Both of these involve spatial buffering by the Müller cell.

6.2 The Müller Cell

The Müller cell is the only one that extends from the vitreous to the level of the photoreceptors (Fig. 9 B). At the sclerad end, the surface area is increased by microvilli that extend beyond the outer limiting "membrane" and, at least in *Necturus* and *Rana*, by deep infoldings in the outer nuclear layer (Lasansky 1965, Miller and Dowling 1970). In the nuclear layers there are lateral branches that stain readily with the Golgi method and embrace the neuronal cell bodies. In the plexiform layers there are finer and more extensive processes. Those in the outer plexiform layer are not visible in many published figures obtained by the Golgi method (e.g., Cajal 1933, Sigelman and Ozanics 1982, Fig. 35), nor are they conspicuous on the isolated cells of Sarthy and Bunt (1982). But they are described by electron microscopists (Pedler 1963, Miller and Dowling 1970, Uga and Smelser 1973). At the vitread extremity the cell widens to an endfoot and the membrane area is increased by lamellar infoldings (e.g., Lasansky 1965, Rasmussen 1974). At the sclerad end, neighbouring cells are joined by what are probably gap junctions (Miller and Dowling 1970). But at both sclerad and vitread ends substances can diffuse freely through the extracellular spaces between the Müller cells (Lasansky and Wald 1962, Lasansky 1965). The Müller cells constitute 8% of the volume of the retina in the grass snake and 21% in a nocturnal gecko (Rasmussen 1975). Their frequency in the tangential plane is illustrated (for mouse retina) in Fig. 3 of Dräger 1983, and for *Rana pipiens* is 4.8×10^5 cells cm^{-2} (Newman et al. 1984). Mitochondria tend to be located close to the sclerad

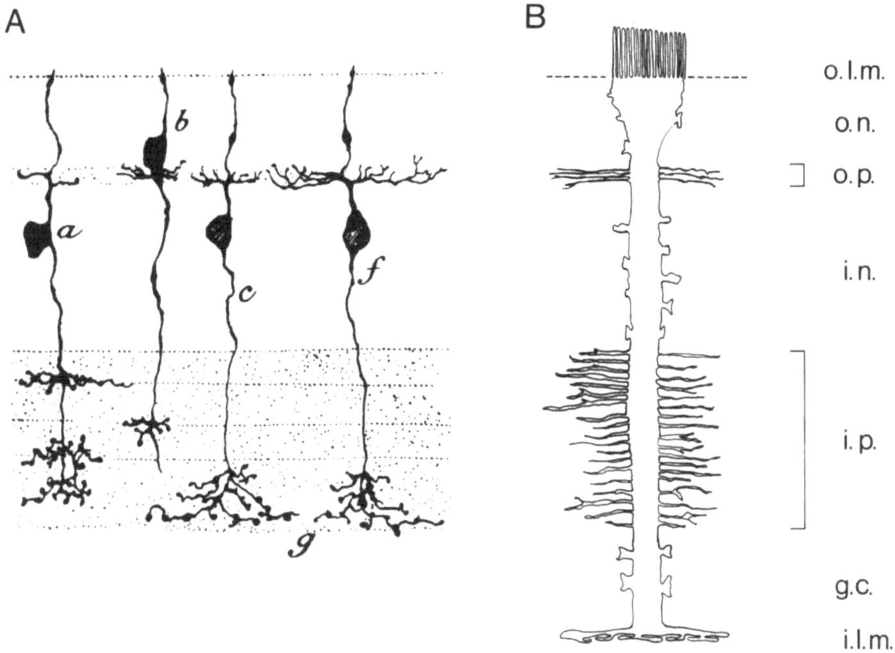

Fig. 9A, B. The principal radically oriented cells in the frog retina, sclerad upwards. **A** Some of the different types of bipolar cells, stained by the Golgi method and classified by Cajal. (Cajal 1933) **B** Tentative schematic representation of a Müller cell. *o.l.m.*, outer limiting membrane; *o.n.*, outer nuclear layer; *o.p.*, outer plexiform layer; *i.n.*, inner nuclear layer; *i.p.*, inner plexiform layer; *g.c.*, ganglion cell layer; *i.l.m.*, inner limiting membrane. (Based mainly on Cajal 1933, Miller and Dowling 1970, Uga and Smelser 1973, Lasansky 1965 and Sarthy and Bunt 1982)

and vitread membranes, which might suggest the existence of active membrane transport here, but their distribution can vary and, for example, in *Gekko* there are none near the vitread membrane, while in *Bufo* they are found throughout the length (Lasansky 1965). Attempts have been made to relate differences in Müller cell structure to differences in the patterns of vascularization of different retinas (Uga and Smelser 1973, Rasmussen 1975).

The surface membranes of the Müller cells of *Rana* and *Ambystoma* are almost exclusively permeable to K^+ (Newman 1985b). When K^+ is ejected from a micropipette at the surface of a slice of retina, above a Müller cell, then the cell depolarizes and the amplitude of the depolarization is at least four times greater if the ejection is opposite the optic fibre layer than if it is in the inner nuclear layer or further sclerad. The uneven distribution of K^+ conductance indicated by this experiment is revealed as even more extreme in isolated cells (from *Ambystoma*), where some 95% of the membrane conductance is localized to the endfoot. In addition to these K^+ ejection experiments (Fig. 10), this was elegantly and convincingly demonstrated in three other ways:

1. The input resistance of the cell was lower when the penetration was near the endfoot.

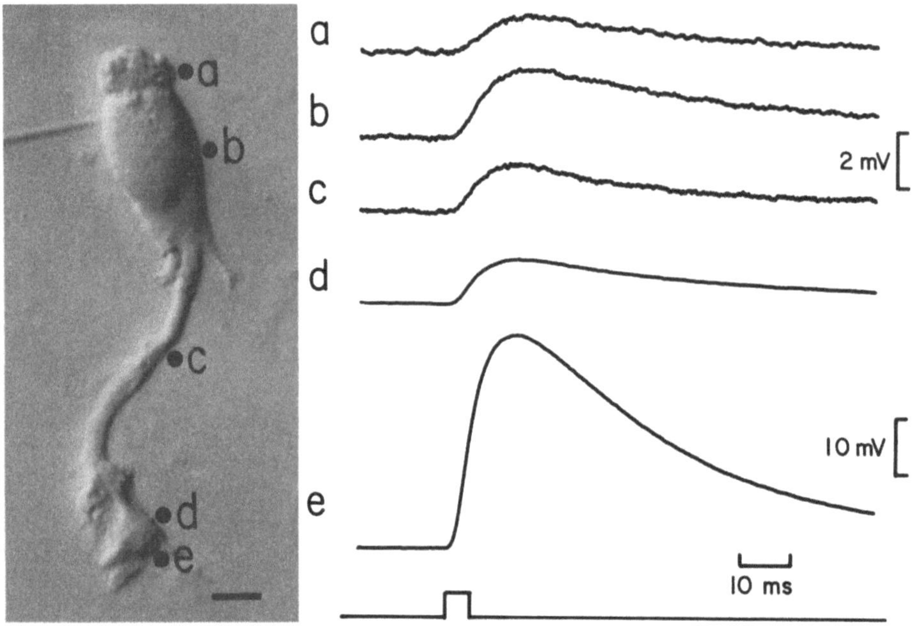

Fig. 10. *Left:* A dissociated *Ambystoma* Müller cell viewed with Nomarski optics. It has been impaled with a micropipette near the nucleus. *Scale bar,* 10 μm. *Right:* Depolarizations produced by ejection of pulses of K^+ from a second extracellular micropipette at the positions indicated in *A*. (Newman 1984)

2. When the cell was cut in two near the endfoot, the small vitread piece had a much lower input resistance than did the larger sclerad piece (Newman 1984, 1985b).
3. When K^+ was ejected from a micropipette near the sclerad end, a K^+-selective microelectrode close outside the cell near the endfoot gave a larger and more rapid response than one near other parts of the cell (Newman et al. 1984).

This high, localized permeability was destroyed by mechanical disturbance, a finding that raises questions about how closely the properties of these isolated cells resemble those of cells in situ. The input resistance appears to be about the same in the two situations (Newman 1985b), but although the isolated cell would maintain its physiological elongated form, within minutes of isolation the lateral processes were resorbed.

By being elongated and having a membrane selectively permeable to K^+, the Müller cell has the properties of an effective transfer cell for spatial buffering. In addition, the asymmetrical concentration of the permeability at the endfoot will tend to direct the efflux of K^+ into the huge extracellular reservoir that is the vitreous, rather than into the extracellular clefts confronting the rest of the cell membrane.

Possession of an endfoot is not unique to retinal Müller cells, and end feet are found also on astrocytes, where they embrace capillaries, and on tanycyte ependymal cells, which have endfeet facing the third cerebral ventricle (e.g., Bruni et al. 1983). At present, there is no direct electrophysiological evidence for a localized conductance in these endfeet (see e.g., Ransom and Connors 1985), although measurements of extracellular potentials in frog brain can be most readily interpreted if there is a glial conductance localized near the surface (Gardner-Medwin 1985). Some, but not all, of these cells have dense arrays of particles in the endfoot membrane (Landis and Reese 1981).

When K^+ is injected extracellularly in the retina, the extracellular potentials created are further strong evidence for such asymmetrical spatial buffering. Fujimoto and Tomita (1981) found that injection of K^+ at any depth in the retina more than 20 µm from the vitreous produced a potential difference across the retina that was vitreous positive: a symmetrical Müller cell would have produced a vitreous negative potential when the injection site was less than half-way along the Müller cell (35% depth).

Intracellular recordings from Müller cells show, as for glia elsewhere in the nervous system, depolarization when the neurones are activated, in this case by light (Miller and Dowling 1970, Karwoski and Proenza 1977). They continue to do this even when their histological appearance has been severely modified by treatment of the retina with L-alpha-aminoadipic acid (Zimmerman and Corfman 1984).

6.3 Slow PIII

Light-induced activity in the retina can be reduced to its simplest form by blocking synaptic transmission from the photoreceptors: this can be done conveniently by application of aspartate (Furukawa and Hanawa 1955, Shimazaki et al. 1984) or cobalt (see Dick and Miller 1985). At light onset, a negative potential still appears across the layers of the retina proximal to the photoreceptors (Witkovsky et al. 1973). Light causes vertebrate photoreceptors to hyperpolarize, there is net entry of K^+ (Somlyo and Walz 1985) and K_o at the level of the photoreceptors falls (Matsuura et al. 1978, Oakley et al. 1979). This sets up spatial buffer currents through the Müller cells: K^+ ions leave at the sclerad end, other K^+ ions enter at the vitread end and the return current loop through the extracellular space, in the vitread direction, accounts for the vitreous negative potential (Faber 1969, Zuckerman 1973, Hanitzsch 1973, Witkovsky et al. 1975, Fujimoto and Tomita 1979). A satisfactory imitation of the response has been generated from a mathematical model of these processes by Newman and Odette (1984). The recovery of K_o during prolonged illumination has been studied by Shimazaki and Oakley (1984) and Hanitzsch (1985).

6.4 The Current Sources of the b-Wave

The onset of uniform illumination of the retina, stimulating either rods or cones, produces a potential that at all depths in the retina is negative with respect to the vitreous (Fig. 12A). The most detailed experimental studies have been carried out on Amphibia, and in these, as illustrated in Fig. 11A, the voltage peaks about 300 ms after the onset of light and recovers in 2 – 3 s. The maximum amplitude is in the distal one-third of the thickness, the exact location depending somewhat

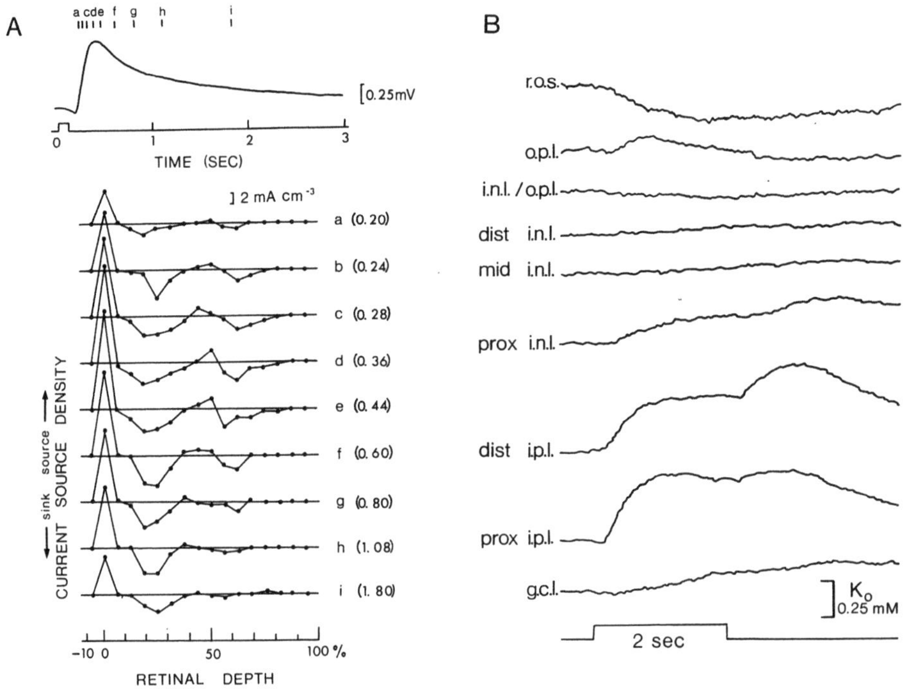

Fig. 11. A Light-induced current sources and sinks in *Rana* retina (Newman 1980). **B** Changes in K_o (Karwoski et al. 1985b). Note that in *A* responses at different times are plotted as a function of depth, while in *B* the converse is true. *r.o.s.*, rod outer segment layer; *o.p.l.*, outer plexiform layer; *i.n.l./o.p.l.*, boundary between inner nuclear layer and *o.p.l.*; *dist*, mid, and prox *i.n.l.*, distal, middle and proximal inner nuclear layers; *i.p.l.*, inner plexiform layer; *g.c.l.*, ganglion cell layer

on the shunt resistance between the vitreous and the back of the retina (see Newman and Odette 1984). Such a profile of extracellular potentials implies that currents flow sclerad through the extracellular clefts and vitread within one or more classes of cells: cells with a radial orientation would be the most effective. When the resistance of the extracellular pathways is taken into account, the distribution of the sources and sinks can be calculated in more detail (Byzov 1959, Faber 1969, Vogel and Green 1979). Newman (1979, 1980) was the first to extend a careful examination of potentials into the vitreous and he showed clearly that the major, and sometimes the only, net source is found within the first 10 µm from the vitreal surface. The locations of the sinks were much more variable but usually there were two, a proximal one centred at 26% of the depth through the retina from the vitreous, and a distal one centred at about 58% (Fig. 11 A). Other workers have referred to deformation of the tissue by the electrode (Karwoski et al. 1985a), and perhaps this could account for some of the variability. The return loops for the current must be intracellular, and the predominant radially oriented cells are the bipolar cells and the Müller cells (Fig. 9). Of these, only the latter are in a position to generate the source near the vitreous.

6.5 Light-Induced Changes in K_o

If the b-wave does result from spatial buffering, then the current sinks where charge leaves the extracellular space should correspond in location and amplitude to increases in K_o. The first measurements of K_o did indeed describe light-induced changes, but the only increases were in the proximal layers of the retina (Oakley and Green 1976, Tomita 1976, Karwoski and Proenza 1977, 1978). These results appeared to be in conflict with the measured distribution of current sinks (Fig. 11 A). A distal, as well as a proximal, increase was required. Several groups were prompted to look harder for a distal increase, and this was duly found, although it was small (about 0.05 m*M*) and not always present (Fig. 11 B; Kline et al. 1978, Dick and Miller 1978, 1985, Shimazaki et al. 1984, Karwoski et al. 1985b).

From the beginning, these workers drew attention to the likelihood that local tissue damage round the electrode tip (dead space) or leaks along the barrel would have caused the measured changes in K_o to be smaller and slower than those in undisturbed tissue (see Oakley 1981). The effect of a dead space depends on: (a) the diameter of the dead space; (b) the volume fraction, α, of the extracellular space; (c) the spatial extent of the region of tissue in which K_o increases; and (d) the rapidity of the K_o changes. Newman and Odette (1984) have computed the effects of the first two factors on the measured K_o to be expected from the distal current sink measured by Newman (1980). The spatial extent of the distal K^+ source appears to be small (Dick and Miller 1978, 1985, Kline et al. 1978, Shimazaki et al. 1984, Karwoski et al. 1985b), as is to be expected if it coincides with the outer plexiform layer, which is 5 μm thick in frog. α in the distal neural retina also appears to be small compared to its value in most nervous tissues (Karwoski et al. 1985a; Sect. 6.7 below).

In response to the onset of white light, all the photoreceptors hyperpolarize and take up K^+. Of the cells with which they synapse in the outer plexiform layer the horizontal cells also hyperpolarize (with white light), while bipolar cells are of different types and some depolarize while others hyperpolarize (see Kaneko 1979). The former presumably lose K^+ while the latter take it up, so the arrangement tends to minimize changes in K_o. Apparently, the loss is greater than the uptake at both light on and light off, a behaviour that could readily be accounted for by a non-linear dependence of K^+ release on membrane potential.

6.6 Depolarization of the Müller Cell

Of the various cells in the retina, the Müller cell is the one whose light-induced depolarization most closely parallels the time course of the b-wave (Miller and Dowling 1970). It is, however, distinctly slower (Karwoski and Proenza 1977, 1978). This fits with even the simplest spatial buffering theory because buffer currents will be set up immediately when K_o increases in the plexiform layers: their amplitude will depend largely on the gradient of K_o between these layers and the vitreous. In contrast, the degree of membrane depolarization will depend mainly on the ΔK_o confronting the membrane area of greatest permeability, i.e., the endfoot, and here K_o will slowly build up as K^+ flows out of the cell into the vitreous. The amplitude of the stimulus-induced Müller cell depolarization can be over 1 mV. If this were due to increases in K_o over half the area of the membrane that was not endfoot, then the measurements on isolated Müller cells would suggest that K_o must have increased by some 2 m*M*, which is large compared to the observed values.

An experimental manipulation that should modify the Müller cell depolarization is to drain the vitreous and thereby reduce the sink available to the endfoot. Karwoski and Proenza (1980) studied the effect of this and found that the depolarizations were greater, as was to be expected since the buildup of $[K^+]$ outside the endfoot would have increased. Since spatial buffer currents would be reduced by this buildup, the b-wave should fall more rapidly from its peak: it appears that this remains to be demonstrated.

6.7 Does Spatial Buffering by the Müller Cell Account for the b-Wave?

In any explanation of the b-wave that is based on spatial buffering, the locations of the K_o increases should roughly correspond to the locations of the current sinks. The experimental results are perhaps compatible with this. However, the theory should also meet other requirements. Notably, it should account quantitatively for the relations between the K^+ sources and the current sinks.

The most baffling question posed by the early results was why current that entered the Müller cell 60% of the way towards its sclerad end should flow towards the remote vitread end, instead of towards the near-by sclerad end. The demonstration of the high permeability confined to the endfoot (Fujimoto and Tomita 1981, Newman 1984) showed qualitatively how this could arise, since the currents would tend to flow as indicated in Fig. 12B. Newman and Odette (1984) examined the question of quantitative agreement between the theory and observation by developing a mathematical model. They did achieve a good simulation of the potentials, but they used values that have subsequently been found to differ from those observed. They imposed a distal K^+ source that matched the requirements of the measured current sink, rather than the measured K_o. This had an amplitude 0.62 times that of the proximal source, whereas the observed ratio is not greater than 0.11 (Karwoski et al. 1985b). Further, they used too small an endfoot permeability. There are several factors that might contribute to an explanation of the discrepancy. First, as discussed above (Sect. 6.5), measurements of K_o may be underestimates, although there seems to be no strong experimental evidence to support this hypothesis (Sect. 2). Second, the source amplitude may have been overestimated on account of the way the resistance to current flow was measured. This was done by passing current pulses through the tissue. Some of this current will have flowed through the cells, particularly the Müller cells (Gardner-Medwin 1983a). But the currents that give rise to the b-wave flow exclusively through the extracellular clefts, whose resistance will be higher than that measured for the tissue (Gardner-Medwin 1980). However, the Müller cells occupy about 2.3% of the cross-section of the retina (in the frog, calculated from data in Newman et al. 1984 and Newman 1985b), and α is about 10% (Karwoski et al. 1985a). If the tortuosity factor, λ, of the extracellular clefts has its usual value of about 1.6 then, depending on the membrane resistances of the different parts of the Müller cell, the difference could not be more than 30% (see Gardner-Medwin 1983b). This is probably within the errors of the model. For example, the conductivity, σ, of the 70% of the retinal thickness occupied by the Müller cells is 0.19 mS cm^{-1} in Fig. 1 B of Newman 1980 and is taken as 0.32 mS cm^{-1} in the model. It may be remarked that these values for σ are very low compared to those in brain tissue, which are usually in the range 1 – 10 mS cm^{-1} (see Gardner-Medwin 1980, p. 223). Despite this, it seems unlikely that spatial buffering by Müller cells can, alone, account for the b-wave.

6.8 Other Possible Contributions to the b-Wave

There is also the possibility of other contributions to transretinal potential that are intimately linked to synaptic activity in the outer plexiform layer. One, which was raised by Brindley (1956) but which appears not to have been reinvestigated with more recent insights, is that there is a standing potential across the neural retina, as there is across the pigment epithelium (Miller and Steinberg 1977), and that light stimulation causes a change in extracellular resistance and hence a change in the potential. In the Amphibian eyecup preparation, a standing gradient of K_o, falling from sclerad to vitread, has been reported (Oakley and Green 1976, Dick and Miller 1985). If there is spatial buffering by the Müller cell, then standing current loops must be set up. In a number of tissues, the extracellular space has been shown to shrink during activity (e.g., Dietzel et al. 1982, Orkand et al. 1984). If this occurred in the retina, in the presence of a standing current, then a potential change with the sign of the b-wave would be produced. Fatehchand (1964) found only very small, slow light-induced changes in impedance measured across the back of the eye, but he used a measuring frequency

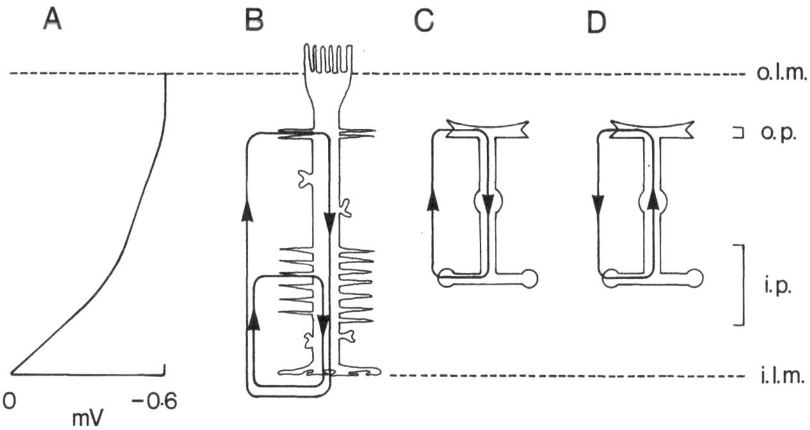

Fig. 12. **A** Scheme of b-wave amplitude as a function of depth. (After Newman 1980) **B** Spatial buffer current loops associated with an asymmetrical Müller cell. **C** Current loop set up at light on in a depolarizing bipolar cell. **D** Current loop set up at light on in a hyperpolarizing bipolar cell. *o.l.m.*, outer limiting membrane; *o.p.*, outer plexiform layer; *i.p.*, inner plexiform layer; *i.l.m.*, inner limiting membrane

of 1 kHz, which may have been too high to restrict the current to the extracellular space. Another possibility is stressed by Yanagida and Tomita (1982). They attempted to mimic the potential distribution of the b-wave by making simultaneous injections of K^+ in the inner and outer plexiform layers. They found that the amplitude of the potential was greatest between the two injection pipettes, whereas the b-wave amplitude continues to increase further sclerad, and concluded that spatial buffering by the Müller cell cannot be the only source of the b-wave. This is in agreement with Vogel and Green (1979), but not with Newman and Odette (1984), who successfully simulated the K^+ injection results with their spatial buffering model. Yanagida and Tomita support the idea that the depolarization of bipolar cells at their sclerad ends will also contribute a sink in the outer plexiform layer and a diffuse source in the inner nuclear layer (Fig. 12C). However, hyperpolarizing (off-centre) bipolar cells will make a contribution in the wrong direction (Fig. 12D). It is also to be noted that all bipolar cells by their orientation, even if their membrane properties are not ideal, will make some contribution to spatial buffering (Fujimoto and Tomita 1981). A complete and approximately quantitative description of the origin of the b-wave is not yet available.

6.9 Effects of Local and Diffuse Illumination

The b-wave arises because light stimulation sets up an electromotive force across the retina in the stimulated area; the measured potential difference is smaller than the electromotive force because current flows back through the retina. Obviously, unless the electromotive force is dramatically greater with local illumination than with diffuse illumination, the former will produce a smaller transretinal potential, as is in fact observed. In contrast, Müller cell depolarization and K_o increase are local phenomena and with different stimulus patterns they should change in parallel. Müller cell depolarization is greater with a small stimulus or, in other words, there is surround inhibition, and this may reduce the amplitude by a factor of more than 2 (Karwoski and Proenza 1980). For K_o there is certainly surround inhibition in the inner plexiform layer (Karwoski and Proenza 1978, Kline et al. 1978). In the outer plexiform layer, surround inhibition has been observed by Karwoski et al. (1985b) in *Necturus*, but not by Kline et al. (1978) in *Raja.*

6.10 K_o Homeostasis in the Vertebrate Neural Retina

Light signals reaching the retina are converted to electrical signals in the photoreceptors and these are transmitted through synapses in the outer plexiform layer that have exceptionally low noise (e.g. Ashmore and Falk 1981). Vitread the signals are amplified and digitized and noise will be less of a problem. By its presence, its orientation, its fine processes in the outer plexiform layer and its endfoot facing a reservoir, the Muller cell seems to have some properties that would make it an effective spatial buffer for the outer plexiform layer. But it is then surprising that its membrane resistance away from the endfoot seems to be

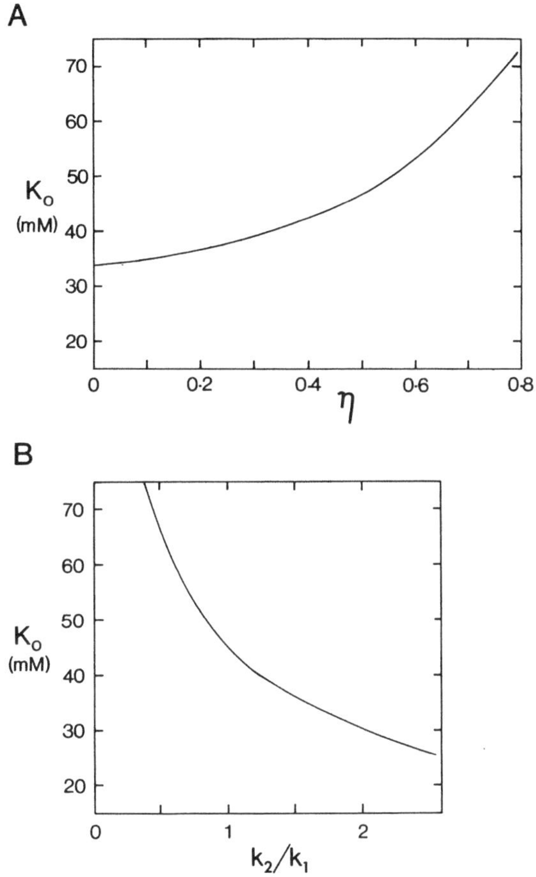

Fig. 13. A Peak K_o plotted as a function of $\eta = \Delta V_m/\Delta E_K$ for the model in Appendix A, with $k_2/k_1 = 1.5$. η is small if most of the K^+ conductance of the glial cell or syncytium is in parts of the membrane that confront a K_o that changes little during the stimulation. **B** Peak K_o plotted as a function of k_2/k_1 with $\eta = 0.3$. Increasing conductance of the glial membrane at the region of the K^+ source (larger values of the abscissa) increasingly limits peak K_o

higher than that necessary to give the cell a space constant adequate for electrotonic spread of depolarization as far as the endfoot (Brew and Attwell 1985). This high resistance will greatly impede the entry of K^+, as illustrated by Fig. 13B. Perhaps it is the design of the neuronal network itself, with some cells hyperpolarizing and others depolarizing at the onset of light stimulation, that is the main factor in maintaining a sufficiently stable K_o under normal conditions, and the membrane properties of the Müller cell are a compromise between this and some other function.

7 Comparative Physiology of K_o Clearance in Retinas

7.1 The Relation Between Neurones and Extracellular Space

Three arrangements have been described in this article. In insect eyes with open rhabdoms and in leech, the microvilli of the photoreceptors face into large volumes of extracellular space. This device appears to be effective at short times (Fig. 4, curve *c*) and may perhaps be made somewhat more effective by chemical buffering properties of the extracellular matrix. In retinas of both drone and

vertebrates the extracellular volume fraction seems to be small, less than a half of that in mammalian brain. The vertebrate retina nevertheless makes use of a large extracellular reservoir, the vitreous, which serves as a K^+ dump for spatial buffering by the Müller cells. On present evidence it is not clear whether an extreme localization of membrane conductance is a feature unique to the Amphibian Müller cell, or whether there are several types of glial cell that transfer K^+ by spatial buffering to extracellular reservoirs. A highly selective permeability to K^+ is also found in the glia of *Necturus* optic nerve (Kuffler et al. 1966).
The third homeostatic mechanism is net uptake, which has now been demonstrated in drone outer pigment cells (Coles and Orkand 1983), leech neuropil glial cells (Schlue and Wuttke 1983) and cultured mammalian oligodendrocytes and astrocytes (Kettenmann et al. 1983, Walz et al. 1984). Net uptake of K^+ requires movements of other ions, and entry of Cl^- is the principal one that has been demonstrated (Walz and Hertz 1983b, Coles and Orkand 1984, see also Boyle and Conway 1941). In astrocytes the Cl^- uptake appears to be principally through a coupled transport system (Walz and Hertz 1984a, b); in drone retina the mechanism is not known.

7.2 *What Is the Homeostasis for?*

If we accept that in retinas changes in K_o do not serve as signals between neurones (Sects. 1 and 6.1), then the question remains whether the homeostatic mechanisms are there because a very stable extracellular milieu is necessary for the retina to respond to small changes in the visual field, or whether they are necessary simply in order to avoid rises in K_o so great that the system could no longer respond to light at all. The latter function would seem to be essential in diurnal animals with compound eyes, which often have one facet directed at the sun and intensely stimulated. One can also argue that a network of neurones without glia would be unstable, and that glia are necessary to prevent catastrophic increases in K_o. Indeed, even with the presence of glia, the phenomenon of spreading depression, in which raised K_o plays an important part (Nicholson and Kraig 1981), is readily induced in isolated retinas (Mori et al. 1876), and is perhaps a fairly widespread pathology in vivo (Gardner-Medwin 1981b). However, some of the capabilities of visual systems, such as that of *Carcinas*, which can detect a rotation of its visual environment of 1° after 15 min in the dark (Horridge 1966), are difficult to imagine without a very stable extracellular milieu. As Gardner-Medwin (1981a) says: "We are somewhat in the position of an engineer trying to assess whether the stabilizing fins on a ship are any use, when he knows neither how rough a typical sea is nor what kinds of ship movement would be disturbing to the passengers."

Acknowledgements. I am indebted to Drs. F. Baumann, C. J. Karwoski, E. A. Newman, R. K. Orkand and B. Walz for correcting mistakes in earlier versions of the text, and thank J.-M. Yersin for photographic work. Supported by USPHS grant EY 03504 and the Cloëtta Foundation.

8 Appendix A (In collaboration with J. Estreicher)

8.1 Model for Spatial Buffering in Drone Retina

In order to illustrate how K_o would change in drone retina at the onset of light stimulation in the presence of spatial buffering, a calculation is made that gives curve *d* in Fig. 4. It is assumed that the glial membrane potential is initially at the K^+ equilibrium potential and that the rise in K_o causes the membrane at the point considered to depolarize a fraction η of the way to the new local equilibrium potential. η will depend on the geometry of the glia, their membrane conductance, and the distribution of K_o. At $t = 0$, K^+ starts to leave a photoreceptor with an exponential time course into an extracellular cleft facing a glial syncytium.

Symbols

y	$K_o(t)$, $[K^+]$ in the extracellular cleft
y_B	$K_o(0)$, baseline value of y
ξ	Volume of cleft between photoreceptor and glial cell per unit area of photoreceptor (or glial cell) membrane, taken as constant
β	(Volume of extracellular space)/(volume of photoreceptors)
$V_m(t)$	Glial membrane potential at the region considered
$E_K(t)$	K^+ equilibrium potential of glial membrane
P	Permeability of glial membrane, assumed to be K^+ specific
$K_p(t)$, K_g:	$[K^+]$ in photoreceptors and glia. Changes in K_g are neglected. $\Delta K_p(\infty)$: steady state change in K_p during prolonged photostimulation
τ	Time constant of light-induced change in K_p
η	$\Delta V_m/\Delta E_K$
ψ	RT/F

$$dy(t)/dt = dy/dt|_p + dy/dt|_g$$

$$y' = y'_p + y'_g \qquad \text{A1}$$

where subscripts p and g indicate contributions from fluxes out of the photoreceptors and into the glia.
During photostimulation K_p declines exponentially:

$$\Delta K_p(t) = \Delta K_p(\infty)\{1 - \exp(-t/\tau)\} \qquad 0 < t < 90\text{ s} \qquad \text{A2}$$

Hence

$$dy/dt|_p = -\Delta K_p(\infty)\exp(-t/\tau)/\beta\tau \qquad 0 < t < 90\text{ s} \qquad \text{A3}$$

Flux into the glia:
Assuming that K^+ flux obeys the constant field equation of Goldman (1943),

$$y'_g = +FPV_m/\psi\xi \cdot \{K_g \exp(V_m/\psi) - y\}/(\exp(V_m/\psi) - 1) \qquad \text{A4}$$

$$V_m(t) = V_m(0) + \Delta V_m$$

and

$$V_m(0) = E_K(0) = RT/F \ln(y_B/K_g) \qquad t \leqslant 0 \qquad \text{A5}$$

so

$$V_m(t) = E_K(0) + \eta\{E_K(t) - E_K(0)\}$$

$$= -\psi \ln(K_g/y_B) + \eta \ln(y/y_B)$$

$$V_m/\psi = \ln\{y^\eta y_B^{(1-\eta)}/K_g\} \qquad \text{A6}$$

$$\exp(V_m/\psi) = y^{\eta} y_B^{(1-\eta)}/K_g \tag{A7}$$

$$y' = k_1 \exp(-t/\tau) + k_2 \frac{(y^{\eta} y_B^{(1-\eta)} - y)}{(y^{\eta} y_B^{(1-\eta)} - K_g)} \ln\left\{\frac{y^{\eta} y_B^{(1-\eta)}}{K_g}\right\} \tag{A8}$$

where

$k_1 = \Delta K_p(\infty)/\beta\tau$

$k_2 = FPK_g/\xi$

8.2 Numerical Values

Approximately Measured Quantities

y_B	10 mmol/l (Coles and Orkand 1983)
K_g	130 mmol/l (Coles and Orkand 1985)
τ	45 s (Coles and Tsacopoulos 1979)
$\Delta K_p(\infty)$	-30.7 mmol/l (Coles and Tsacopsoulos 1979)

Estimated Quantities. (From Appendix 2 in Coles and Tsacopoulos 1979)

ξ	A 15-nm cleft, i.e., 1.5×10^{-10} m^3/m^2
β	5/38

Hence $k_1 = 5.18$ mmol l^{-1} s^{-1}. Equation A8 was integrated numerically using a program[1] based on an explicit Runge-Kutta-Dormand-Prince method of order 5. Values of k_2 were in the range $0.5\,k_1 - 2\,k_1$ and η was in the range $0-0.8$. The effect of changing η is shown in Fig. 13 A; it is seen that for $\eta < 0.3$ its value is not critical. Figure 13 B, in contrast, shows that increasing k_2 for a given k_1 has a comparatively strong effect. The curve *d* in Fig. 4 is for $k_2 = 2\,k_1$ and $\eta = 0.3$.

9 References

Amman D, Morf WE, Anker P, Meier PC, Pretsch E, Simon W (1983) Neutral carrier based ion-selective electrodes. Ion-Selective Electrode Rev 5:3–92

Armington JC (1974) The electroretinogram. Academic, New York

Ashmore JF, Falk G (1981) An analysis of voltage noise in rod bipolar cells of the dogfish retina. J Physiol (Lond) 332:273–297

Bader CR, Macleish PR, Schwartz EA (1979) A voltage-clamp study of the light response in solitary rods of the tiger salamander. J Physiol (Lond) 296:1–26

Baker PF, Connelly CM (1966) Some properties of the external activation site of the sodium pump in crab nerve. J Physiol (Lond) 185:270–297

Baylor DA, Nicholls JG (1969) After-effects of nerve impulses on signalling in the central nervous system of the leech. J Physiol (Lond) 203:571–589

Boschek B (1971) On the fine structure of the peripheral retina and lamina ganglionaris of the fly *Musca.* Z Zellforsch (Abt Histochemie) 8:369–409

Boyle PJ, Conway EJ (1941) Potassium accumulation in muscle and associated changes. J Physiol (Lond) 100:1–63

1 Developed with support from SNSF grant no. 3.920.0.83

Brew H, Attwell D (1985) Is the potassium channel distribution in glial cells optimal for spatial buffering of potassium? Biophys J 45:(Nov. 1985)
Brindley GS (1956) The passive electrical properties of the frog's retina, choroid and sclera for radial fields and currents. J Physiol (Lond) 134:339–352
Bruni JE, Clattenburg RE, Millar E (1983) Tanycyte ependymal cells in the third ventricle of young and adult rats: a Golgi study. Anat Anz 153:53–68
Byzov AL (1959) Analysis of the distribution of potentials and currents within the retina on light stimulation. I. Activity of the bipolars of the two types. Biophysics London 4:46–59
Caillé JP, Hinke JAM (1972) Evidence for Na sequestration in muscle from Na diffusion measurements. Can J Physiol Pharmacol 50:228–237
Cajal SR (1933) La rétine des vertébrés. Travaux du Laboratoire de Recherches Biologiques de l'Université de Madrid. 28(Appendix):1–141
Carslaw HS, Jaeger JC (1959) Conduction of heat in solids, 2nd edn. Oxford University Press, London
Coles JA (1981) Changes in K^+ activity in photoreceptors and glial cells of the drone retina when $[K^+]$ in the superfusate is changed. J Physiol (Lond) 318:16–17P
Coles JA, Orkand RK (1983) Modification of potassium movement through the retina of the drone (*Apis mellifera* ♂) by glial uptake. J Physiol (Lond) 340:157–174
Coles JA, Orkand RK (1984) Chloride enters receptors and glia in response to light or raised external potassium in the retina of the bee drone, *Apis mellifera.* J Physiol (Lond) 357:11P
Coles JA, Orkand RK (1985) Changes in sodium activity during light stimulation in photoreceptors, glia and extracellular space in drone retina. J Physiol (Lond) 362:415–435
Coles JA, Rick R (1985) An electron microprobe analysis of photoreceptors and outer pigment cells in the retina of the honeybee drone. J Comp Physiol A 156:213–222
Coles JA, Tsacopoulos M (1979) K^+ activity in photoreceptors, glial cells and extracellular space in the drone retina: changes during photostimulation. J Physiol (Lond) 290:525–549
Coles JA, Tsacopoulos M (1981) Ionic and possible metabolic interactions between sensory neurones and glial cells in the retina of the honeybee drone. J Exp Biol 95:75–92
Coles JA, Tsacopoulos M, Rabineau P, Gardner-Medwin AR (1981) Movement of potassium into glial cells in the retina of the drone, *Apis mellifera*, during photostimulation. In: Syková E, Hník P, Vyklický L (ed) Ion-selective microelectrodes and their use in excitable tissues. Plenum, New York, pp 345–349
Connors BW, Ransom BR, Kunis DM, Gutnick MJ (1982) Activity-dependent K^+ accumulation in the developing rat optic nerve. Science 218:1341–1343
Cordingley GE, Somjen GG (1978) The clearing of excess potassium from extracellular space in spinal cord and cerebral cortex. Brain Res 157:291–306
Dick E, Miller RF (1978) Light-evoked potassium activity in mudpuppy retina: its relationship to the b-wave of the electroretinogram. Brain Res 154:388–394
Dick E, Miller RF (1985) Extracellular K^+ activity changes related to electroretinogram components. I. Amphibian (I-type) retinas. J Gen Physiol 85:885–909
Dietzel I, Heinemann U, Hofmeier C, Lux HD (1980) Transient changes in the size of the extracellular space in the sensorimotor cortex of cats in relation to stimulus-induced changes in potassium concentration. Exp Brain Res 40:432–439
Dietzel I, Heinemann U, Hofmeier G, Lux HD (1982) Stimulus-induced changes in extracellular Na^+ and Cl^- concentration in relation to changes in the size of the extracellular space. Exp Brain Res 46:73–84
Dowling JE, Ripps H (1976) Potassium and retinal sensitivity. Brain Res 107:617–622
Dräger UC (1983) Coexistence of neurofilaments and vimentin in a neurone of adult mouse retina. Nature 303:169–172

Ellory JC, Hall AC (1984) Na, K, Cl-cotransport stoichiometry; measurement of bumetanide-sensitive fluxes in ferret red cells. J Physiol (Lond) 357:63P
Faber DS (1969) Analysis of the slow transretinal potentials in response to light. Diss Abst Int B 30:2878 (Order no. 69-20.590)
Fatehchand R (1964) Impedance changes in the toad eye following illumination. Vision Res 4:605–608
Franck G, Grisar Th, Moonen G, Schoffeniels E (1978) Potassium transport in mammalian astroglia. In: Schoffeniels E, Franck G, Hertz L, Tower BD (eds) Dynamic properties of glia cells. Pergamon, Oxford, pp 315–325
Frankenhaeuser B, Hodgkin AL (1956) After-effects of impulses in the giant nerve fibres of *Loligo*. J Physiol (Lond) 131:341–376
Freeman JA, Nicholson C (1975) Experimental optimization of current source-density technique for anuran cerebellum. J Neurophysiol 38:369–382
Fujimoto M, Tomita T (1979) Reconstruction of the slow PIII from the rod potential. Invest Ophthal Visual Sci 18:1090–1093
Fujimoto M, Tomita T (1981) Field potentials induced by injection of potassium ions into the frog retina: a test of current interpretations of the electroretinographic (ERG) b-wave. Brain Res 204:51–64
Fulpius B, Baumann F (1969) Effects of sodium, potassium, and calcium ions on slow and spike potentials in single photoreceptor cells. J Gen Physiol 53:541–561
Furukawa T, Hanawa I (1955) Effects of some common cations on electroretinogram of the toad. Jpn J Physiol 5:289–300
Gadsby DC (1980) Activation of electrogenic Na^+/K^+ exchange by extracellular K in canine cardiac Purkinje fibers. Proc Natl Acad Sci USA 77:4035–4039
Gardner-Medwin AR (1980) Membrane transport and solute migration affecting the brain cell microenvironment. In: Nicholson C: Dynamics of the brain cell microenvironment. Neurosci Res Program Bull 18:208–226
Gardner-Medwin AR (1981a) The role of cells in the dispersal of brain extracellular potassium. In: Syková E, Hník P, Vyklický L (eds) Ion-selective electrodes and their use in excitable tissues. Plenum, New York, pp 339–343
Gardner-Medwin AR (1981b) Possible roles of vertebrate neuroglia in potassium dynamics, spreading depression and migraine. J Exptl Biol 95:111–127
Gardner-Medwin AR (1983a) A study of the mechanisms by which potassium moves through brain tissue in the rat. J Physiol (Lond) 335:353–374
Gardner-Medwin AR (1983b) Analysis of potassium dynamics in brain tissue. J Physiol (Lond) 335:393–426
Gardner-Medwin AR (1985) Contrasting glial potassium dynamics in rat brain, frog brain and retina. In: Roitbak AI (ed) Functions of neuroglia. Metzniereba, Tbilisi (in press)
Gardner-Medwin AR, Coles JA (1980) Dispersal of potassium through glia. Proc Int Union Physiol Sci 14
Gardner-Medwin AR, Coles JA, Tsacopoulos M (1981) Clearance of extracellular potassium: evidence for spatial buffering by glial cells in the retina of the drone. Brain Res 209:452–457
Goldman DE (1943) Potential, impedance, and rectification in membranes. J Gen Physiol 27:37–60
Hanitzsch R (1973) Intraretinal isolation of PIII subcomponents in the isolated rabbit retina after treatment with sodium aspartate. Vision Res 13:2093–2102
Hanitzsch R (1985) Properties of mammalian slow PIII – a retinal glial potential. In: Roitbak AI (ed) Functions of neuroglia. Metzniereba, Tbilisi (in press)
Hertz L (1978) An intense potassium uptake into astrocytes, its further enhancement by high concentrations of potassium, and its possible involvement in potassium homeostasis at the cellular level. Brain Res 145:202–208

Hertz L, Chaban G (1982) Indications for an active role of astrocytes in potassium homeostasis at the cellular level: potassium uptake and metabolic effects of potassium. In: Pfeiffer SE (ed) Neuroscience approached through cell culture, vol 1. CRC Press, Boca Raton, pp 157 – 174

Hertz L, Juurlink BHJ, Fosmak H, Schousboe A (1982) Astrocytes in primary cultures. In: Pfeiffer SE (ed) Neuroscience approached through cell culture, vol 1. CRC Press, Boca Rotan, pp 175 – 186

Horridge GA (1966) Optokinetic memory in the crab, *Carcinus*. J Exptl Biol 44:233 – 245

Kaneko A (1979) Physiology of the retina. Ann Rev Neurosci 2:169 – 191

Karwoski CJ, Proenza LM (1977) Relationship between Müller cell responses, a local transretinal potential, and potassium flux. J Neurophysiol 40:244 – 259

Karwoski CJ, Proenza LM (1978) Light-evoked changes in extracellular potassium concentration in mudpuppy retina. Brain Res 142:515 – 530

Karwoski CJ, Frambach DA, Proenza LM (1985a) Laminar profile of resistivity in frog retina. J Neurophysiol (in press)

Karwoski CJ, Newman EA, Shimazaki H, Proenza LM (1985b) Light-evoked increases in extracellular K^+ in the plexiform layers of amphibian retinas. J Gen Physiol 86:189 – 213

Karwoski CJ, Proenza LM (1980) Neurons, potassium, and glia in proximal retina of *Necturus*. J Gen Physiol 75:141 – 162

Karwoski CJ, Proenza LM (1981) Spatio-temporal variables in the relationship of neuronal activity to potassium and glial responses. Vision Res 21:1713 – 1718

Kettenmann H, Orkand RK, Lux HD (1984) Some properties of single potassium channels in cultured oligodendrocytes. Pflügers Arch 400:215 – 221

Kettenmann H, Sonnhof U, Schachner M (1983) Exclusive potassium dependence of the membrane potential in cultured mouse oligodendrocytes. J Neurosci 3:500 – 505

Kline RP, Ripps H, Dowling JE (1978) Generation of *b*-wave currents in the skate retina. Proc Natl Acad Sci USA 75:5727 – 5731

Kuffler SW, Nicholls JG (1966) The physiology of neuroglial cells. Ergebnisse Physiol 57:1 – 90

Kuffler SW, Nicholls JG, Orkand RK (1966) Physiological properties of glial cells in the central nervous system of Amphibia. J Neurophysiol 29:768 – 787

Landis DMD, Reese TS (1981) Membrane structure in mammalian astrocytes: a review of freeze-fracture studies on adult, developing, reactive and cultured astrocytes. J Exptl Biol 95:35 – 48

Lasansky A (1965) Functional implications of structural findings in retinal glial cells. Progr Brain Res 15:48 – 72

Lasansky A, Wald F (1962) The extracellular space in toad retina as defined by the distribution of ferrocyanide. A light and electron microscope study. J Cell Biol 15:463 – 479

Matsuura T, Miller WH, Tomita T (1978) Cone-specific c-wave in the turtle retina. Vision Res 18:767 – 775

Miller RF, Dowling JE (1970) Intracellular responses of the Müller (glial) cells of the mudpuppy retina: their relation to the b-wave of the electroretinogram. J Neurophysiol 33:323 – 341

Miller SS, Steinberg RH (1977) Passive ionic properties of frog retinal pigment epithelium. J Membrane Biol 36:337 – 372

Morf WE (1981) The principles of ion-selective electrodes and of membrane transport. Akadémiai Kiadó, Budapest; Elsevier, New York

Mori S, Miller WH, Tomita T (1976) Microelectrode study of spreading depression (SD) in frog retina. General observations of field potential associated with SD. Jpn J Physiol 26:203 – 217

Munoz JL, Deyhimi F, Coles JA (1983) Silanization of glass in the making of ion-sensitive microelectrodes. J Neurosci Methods 8:231 – 247

Newman EA (1979) *b*-wave currents in the frog retina. Vision Res 19:227–234
Newman EA (1980) Current source-density analysis of the *b*-wave of frog retina. J Neurophysiol 43:1355–1366
Newman EA (1984) Regional specialization of retinal glial cell membrane. Nature 309:155–157
Newman EA (1985a) Regulation of potassium levels by glial cells in the retina. Trends in Neurosciences 8:156–159
Newman EA (1985b) Membrane physiology of retinal glial (Müller) cells. J Neurosci 5:2225–2239
Newman EA, Frambach DA, Odette LL (1984) Control of extracellular potassium levels by retinal glial cell K^+ siphoning. Science 225:1174–1175
Newman EA, Odette LL (1984) Model of electroretinogram *b*-wave generation: a test of the K^+ hypothesis. J Neurophysiol 51:164–182
Nicholson C, Freeman JA (1975) Theory of current source-density analysis and determination of conductivity tensor for anuran cerebellum. J Neurophysiol 38:356–368
Nicholson C, Kraig RP (1981) The behaviour of extracellular ions during spreading depression. In: Zeuthen T (ed) The application of ion-selective microelectrodes. Elsevier/North Holland, Amsterdam, pp 217–238
Nicholson C, Phillips JM (1981) Ion diffusion modified by tortuosity and volume fraction in the extracellular microenvironment of the rat cerebellum. J Physiol (Lond) 321:225–257
Nicholson C, Phillips JM, Gardner-Medwin AR (1979) Diffusion from an iontophoretic point source in the brain: role of tortuosity and volume fraction. Brain Res 169:580–584
Oakley B II (1981) Light-evoked changes in retinal extracellular potassium. In: Zeuthen T (ed) The application of ion-selective microelectrodes. Elsevier, Amsterdam, pp 109–128
Oakley B II, Flaming G, Brown KT (1979) Effects of the rod receptor potential upon retinal extracellular potassium concentration. J Gen Physiol 74:713–739
Oakley B II, Green DG (1976) Correlation of light-induced changes in retinal extracellular potassium concentration with c-wave of the electroretinogram. J Neurophysiol 39:1117–1133
Orkand PM, Bracho H, Orkand RK (1973) Glial metabolism: alteration by potassium levels comparable to those during neural activity. Brain Res 55:467–471
Orkand RK (1980) Extracellular potassium accumulation in the nervous system. Federation Proceedings 39:1515–1518
Orkand RK, Dietzel I, Coles JA (1984) Light induced changes in extracellular volume in drone retina. Neurosci Lett 45:273–278
Orkand RK, Nicholls JG, Kuffler SW (1966) Effect of nerve impulses on the membrane potential of glial cells in the central nervous system of Amphibia. J Neurophysiol 29:788–806
Pedler C (1963) The fine structure of the radial fibers in the reptile retina. Exptl Eye Res 2:296–303
Perrelet A (1970) The fine structure of the retina of the honeybee drone. Z Mikrosk-anat Forsch 108:530–562
Proenza LM, Freeman JA (1975) Contributions of extracellular currents, resistivity, and $[K^+]_o$ flux to retinal potentials. ARVO Abstracts, p 76
Ransom BR, Connors BW (1985) Electrophysiological studies of ependymal cells in turtle cortex. In: Roitbak AI (ed) Functions of neuroglia. Metzniereba, Tbilisi (in press)
Ransom BR, Yamate CL (1984) The rat optic nerve following enucleation: a pure preparation of mammalian glia. Neurosci Abstracts 10
Ransom BR, Yamate CL, Connors BW (1985) Developmental studies on brain extracellular space: activity-dependent K^+ accumulation and shrinkage. In: Kessler M,

Harrison DK, Höper J (eds) Recent advances in the theory and application of ion selective electrodes in physiology and medicine. Springer, Berlin Heidelberg New York Tokyo, pp 206–213

Rasmussen KE (1974) The Müller cell: a comparative study of rod and cone retinas with and without retinal vessels. Expl Eye Res 19:243–257

Rasmussen KE (1975) A morphometric study of the Müller cell in rod and cone retinas with and without retinal vessels. Expl Eye Res 20:151–166

Ritchie JM, Straub RW (1957) The hyperpolarization which follows activity in mammalian non-medullated fibres. J Physiol (Lond) 136:80–97

Ritchie T, Packey DJ, Trachtenberg MC, Haber B (1981) K^+-induced ion and water movements in the frog spinal cord and filum terminale. Expl Neurol 71:356–369

Russell JM (1983) Cation-coupled chloride influx in squid axon. Role of potassium and stoichiometry of the transport process. J Gen Physiol 81:909–925

Sarthy PV, Bunt AH (1982) The ultrastructure of isolated glial (Müller) cells from the turtle retina. Anat Rec 202:275–283

Schlue WR, Wuttke W (1983) Potassium activity in leech neuropile glial cells changes with external potassium concentration. Brain Res 270:368–372

Shaw SR (1969) Interreceptor coupling in ommatidia of drone honeybee and locust compound eyes. Vision Res 9:999–1029

Shaw SR (1975) Retinal resistance barriers and electrical lateral inhibition. Nature 255:480–483

Shaw SR (1978) The extracellular space and blood-eye barrier in an insect retina: an ultrastructural study. Cell Tiss Res 188:35–61

Shimazaki H, Karwoski CJ, Proenza LM (1984) Aspartate-induced dissociation of proximal from distal retinal activity in the mudpuppy. Vision Res 24:587–595

Shimazaki H, Oakley B II (1984) Reaccumulation of $[K^+]_0$ in the toad retina during maintained illumination. J Gen Physiol 84:475–504

Sigelman J, Ozanics V (1982) Retina. In: Jakobiec FA (ed) Ocular anatomy, embryology and teratology. Harper and Row, Philadelphia, pp 441–506

Somjen GG (1979) Extracellular potassium in the mammalian central nervous system. Ann Rev Physiol 41:159–177

Somjen GG (1981) Physiology of glial cells. In: Salánki J (ed) Adv Physiol Sci, vol 3. Physiology of non-excitable cells. Akademiai Kiado, Budapest, pp 23–43

Somlyo AP, Walz B (1985) Elemental distribution in *Rana pipiens* retinal rods: quantitative electron probe analysis. J Physiol (Lond) 258:188–195

Syková E (1983) Extracellular K^+ accumulation in the central nervous system. Prog Biophys Molecul Biol 42:135–189

Tang CM, Cohen MW, Orkand RK (1980) Electrogenic pumps in axons and neuroglia and extracellular potassium homeostasis. Brain Res 194:283–286

Taylor PS, Thomas RC (1984) The effect of leakage on microelectrode measurements of intracellular sodium activity in crab muscle fibres. J Physiol (Lond) 352:539–550

Thomas RC (1978) How to make and use ion-sensitive intracellular microelectrodes. Academic, London

Tomita T (1976) Electrophysiological studies of retinal cell function. Invest Ophthalmol Vis Sci 15:171–187

Tomita T, Yanagida T (1981) Origins of the ERG waves. Vision Res 21:1703–1707

Tsacopoulos M, Orkand RK, Coles JA, Levy S, Poitry S (1983) O_2 uptake occurs faster than Na pumping in bee retina after a light flash. Nature 301:604–606

Tsacopoulos M, Poitry S, Borsellino A (1981) Diffusion and consumption of oxygen in the superfused retina of the drone (*Apis mellifera*) in darkness. J Gen Physiol 77:601–628

Uga S, Smelser GK (1973) Comparative study of the fine structure of retinal Müller cells in various vertebrates. Invest Opthal Vis Sci 12:434–448

Vern BA, Schuette WH, Thibault LE (1979) $[K^+]_0$ clearance in cortex: a new analytical model. J Neurophysiol 40:1015 – 1023
Vogel DA, Green DC (1979) Potassium release and current flow in the frog retina: a test of the Müller cell hypothesis of electroretinogram b-wave generation. Neuroscience Abstracts 5:813
Walker JL (1971) Ion specific liquid ion exchanger microelectrodes. Anal Chem 43:89A – 92A
Walz B (1982) Ca^{2+}-sequestering smooth endoplasmic reticulum in an invertebrate photoreceptor. I. Intracellular topography as revealed by OsFeCN staining and in situ Ca accumulation. J Cell Biol 93:839 – 848
Walz B (1985) Light-induced changes of extra- and intracellular potassium concentration in photoreceptors of the leech *Hirudo medicinalis*. J Comp Physiol 157:199 – 210
Walz B, Somlyo AP (1984) Quantitative electron probe microanalysis of leech photoreceptors. J Comp Physiol A 154:81 – 87
Walz W, Hertz L (1982) Ouabain-sensitive and ouabain-resistant net uptake of potassium into astrocytes and neurons in primary cultures. J Neurochem 39:70 – 77
Walz W, Hertz L (1983a) Functional interactions between neurons and astrocytes. II. Potassium homeostasis at the cellular level. Prog Neurobiol 20:133 – 183
Walz W, Hertz L (1983b) Intracellular ion changes of astrocytes in response to extracellular potassium. J Neurosci Res 10:411 – 423
Walz W, Hertz L (1984a) Sodium transport in astrocytes. J Neuroscience Research 11:231 – 239
Walz W, Hertz L (1984b) Intense furosemide-sensitive potassium accumulation in astrocytes in the presence of pathologically high extracellular potassium levels. J Cerebral Blood Flow and Metabolism 4:301 – 304
Walz W, Wuttke W, Hertz L (1984) Astrocytes in primary cultures: membrane potential characteristics reveal exclusive potassium conductance and potassium accumulator properties. Brain Res 292:367 – 374
Walz W, Wuttke W, Schlue WR (1983) The Na^+-K^+ pump in neuropile glial cells of the medicinal leech. Brain Res 267:93 – 100
Witkovsky P, Nelson PJ, Ripps H (1973) Action spectra and adaptation properties of carp photoreceptors. J Gen Physiol 61:401 – 423
Witkovsky P, Dudek FE, Ripps H (1975) Slow PIII component of the carp electroretinogram. J Gen Physiol 65:119 – 134
Yanagida T, Tomita T (1982) Local potassium concentration changes in the retina and the electroretinographic (ERG) b-wave. Brain Res 237:479 – 483
Yarom Y, Spira ME (1982) Extracellular potassium ions mediate specific neuronal interaction. Science 216:80 – 82
Zimmerman RP, Corfman TP (1984) A comparison of the effects of isomers of alpha-aminoadipic acid and 2-amino-4-phosphonobutyric acid on the light response of the Müller glial cell and the electroretinogram. Neurosci 12:77 – 84
Zuckerman R (1973) Ionic analysis of photoreceptor membrane currents. J Physiol (Lond) 235:333 – 354

Slowly Conducting Afferent Fibers from Deep Tissues: Neurobiological Properties and Central Nervous Actions *

S. Mense

Anatomisches Institut III, Universität Heidelberg, Im Neuenheimer Feld 307, 6900 Heidelberg, FRG

---

* To Inge

1 Introduction

Subjective sensations from deep tissues, which are probably mediated by slowly conducting afferent fibers, play an important role in everyday life. Thus, deep pain is a much commoner reason for consulting a physician than is cutaneous pain. Similarly, reflexes elicited by these fibers (e.g., cardiopulmonary (CP) adjustments during exercise) are essential for the normal functioning of a living organism. Yet most studies dealing with the function of afferent fiber systems have concentrated either on cutaneous fibers or on deep receptors with rapidly conducting afferent fibers. For example, in two of the most elaborate reviews on muscle receptors published to date (Matthews 1972, Barker 1974), only a few pages are devoted to receptive endings with fibers of small diameter. The present report gives an overview of the current knowledge of the physiology and morphology of these afferent units.

1.1 Definitions

Deep tissues are defined here as comprising all subcutaneous structures which do not belong to the viscera, i.e., muscles, fascia, tendons, ligaments, joints, subcutaneous fat, and periosteum. This article will concentrate on tissues for which experimental data from systematic studies are available, namely, skeletal muscle (including tendon) and joints.

"Slowly conducting" or "small-diameter" afferent fibers denote units corresponding to the Aδ- and C-fibers of Erlanger and Gasser (1924, 1930), or to the group III and IV fibers of Lloyd (1943), respectively. The nomenclature of Lloyd will be used throughout this review. Group III fibers denote thin myelinated units (fiber diameter, 1 – 6 μm) with a conduction velocity between 2.5 and 30 m/s; group IV fibers are nonmyelinated units (fiber diameter, less than 1 μm) conducting at less than 2.5 m/s (Gasser 1950, Brock et al. 1951, Paintal 1967). For brevity, "fine deep afferents" will be used for group III and IV fibers supplying receptors in deep tissues.

1.2 Earlier Views and Concepts

In the older literature, several different functions were ascribed to fine deep afferent units, one being the mediation of pain during noxious stimulation or pathological alterations of the tissue, e.g., during ischemic contractions of a limb (Lewis 1942, Bessou and Laporte 1958). Electrical stimulation of a muscle nerve, at a strength sufficient to activate small-diameter afferent fibers, was shown to evoke pain-equivalent reactions (vocalization, struggling) in experimental animals (Bishop and Heinbecker 1935). Mechanical stimulation of muscle receptors supplied by group III fibers proved to facilitate the flexion reflex (Paintal 1961). At the single-fiber level, evidence supporting the assumption that group III (Paintal 1960) and group IV (Iggo 1961) muscle afferents might fulfill nociceptive functions, was likewise obtained. A prerequisite for these experiments was the technique of single-fiber recordings from a dissected nerve which was introduced by Adrian and Zotterman (1926) and Matthews (1931) for thick afferents and by Zotterman (1936) for fine fibers.

The second function ascribed to the small-diameter muscle afferents was to induce CP adjustments during muscular exercise. Paterson (1928) suggested that these adjustments are reflex in nature "rather than due to psychic causes". In later studies, it was assumed that the accumulation of, as yet unidentified, metabolites in a working muscle produces discharges in muscle afferents which elicit the reflexes (Alam and Smirk 1937, 1938). The use of electrical stimulation of muscle nerves later yielded evidence that the smaller (electrically high-threshold) fibers from skeletal muscle are particularly important as the afferent arc of the reflex (Senapati 1966). The muscular receptors responsible for eliciting the exercise hyperpnea or cardiovascular (CV) adjustments were labeled ergoreceptors (Iaria et al. 1959, Kao 1963), metabolic receptors (Stegemann and Kenner 1971), or exercise receptors (Pérez-González and Coote 1972). In the present paper, "ergoreceptor" will be used in a more general way for all receptors, in or near the contracting muscle, which are able to measure the amount of work performed and the activation of which leads to CP reflexes.

In 1951, Brock et al. studied the effects of afferent volleys in muscle nerves on motoneurones with respect to locomotion, rather than to nociception. They found that the actions of group II and III muscle afferents were very similar, in that both produce the pattern of a flexion reflex by activation of flexor motoneurones and inhibition of extensor motoneurones. Later, it was suggested that the group III muscle receptors involved in this reflex are different from nociceptors (Eccles and Lundberg 1959b); therefore, the concept of the existence of "flexion reflex afferents" does not imply that the reflex is nocifensive and due to activation of high-threshold receptors by a noxious stimulus (Eccles and Lundberg 1959b, Holmqvist et al. 1960, Holmqvist and Lundberg 1961). Thus, the third function proposed for small-diameter muscle afferents was related to locomotive reflexes at the spinal level.

2 Anatomy and Morphology of Afferent Units

2.1 Receptive Endings

2.1.1 Skeletal Muscles and Tendons

The great number of afferent fibers in a muscle nerve was already mentioned by Sherrington in 1894, but at that time the various fiber types could not be assigned to particular types of nerve endings. According to the review by Barker (1974), the following types of receptors can be distinguished morphologically in a muscle and its associated tissues (tendons, ligaments, fascia, aponeuroses): muscle spindle, tendon organ, Pacinian corpuscle, paciniform corpuscle, Golgi ending, Dogiel ending, Ruffini ending, and free nerve endings. Most of these receptors are supplied by thick afferent fibers and will not be dealt with in this article.

Concerning the small-diameter afferent units, it was already shown by Hinsey (1927, 1928) that the thin myelinated and nonmyelinated fibers give rise to fine, unencapsulated terminations ("free fibrils") which end in the adventitia of blood vessels (excluding capillaries) or in the connective tissue between the muscle fibers (cf. also Barker 1962, 1974). Hinsey stated that thin myelinated stem fibers can produce nonmyelinated branches with fine endings. Owing to the lack of association with any specialized muscle receptors, the small-diameter muscle afferents were considered to be "pain fibers" (Tiegs 1953). Yet, some of the group III fibers were found to supply paciniform corpuscles and, thus, appeared not to be involved in the production of pain (Barker et al. 1962). In tendons and aponeuroses, free nerve endings have been described as occurring in two subtypes, one arborizing in the loose connective tissue and the other coursing in a clasping fashion. The critical fiber diameter below which the associated receptive endings were found to be invariably "naked axoplasmic threads" was 2 μm (Stilwell 1957a, b, c).

A study dealing specifically with free nerve endings and their stem fibers in cat skeletal muscle was performed by Stacey (1969). One important finding of this investigation was that all nonmyelinated afferent fibers terminate in free nerve endings (Fig. 1), whereas, the thin myelinated afferents can supply a variety of

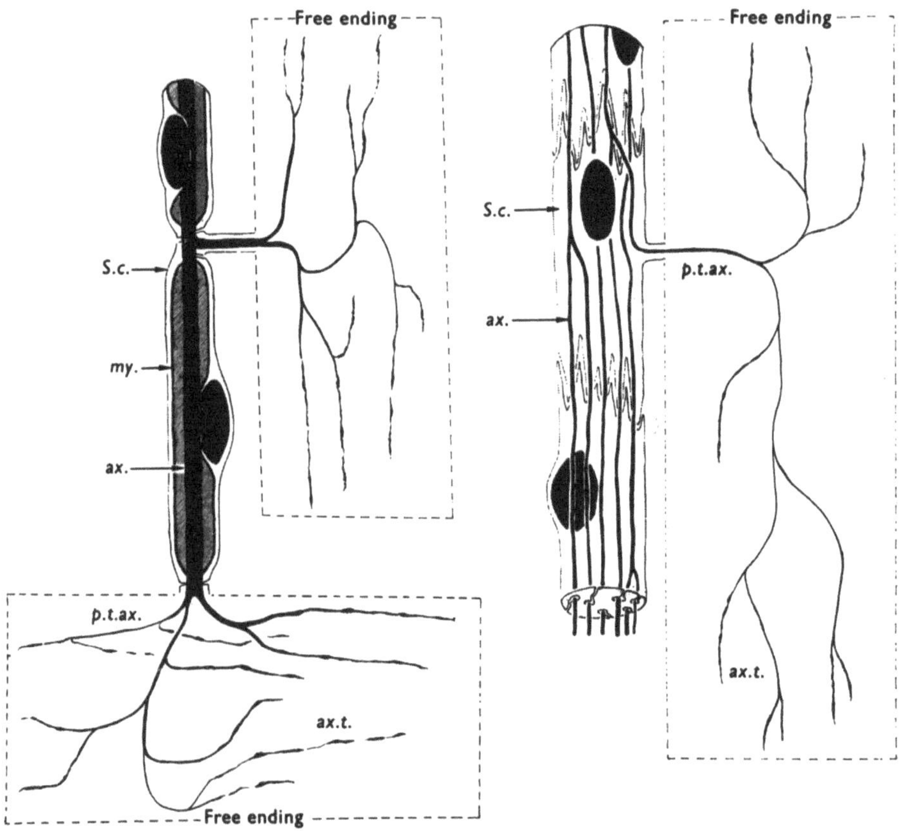

Fig. 1. Schematic diagram of free nerve endings in the skeletal muscle of the cat with thin myelinated (*left*) and nonmyelinated (*right*) stem fiber. *S.c.*, cytoplasm of Schwann cell; *my.*, myelin; *ax.*, axon; *p.t.ax.*, preterminal axon; *ax.t.*, axon terminal. (From Stacey 1969)

receptive endings besides free nerve endings, namely, muscle spindle secondary endings, Pacinian corpuscles, and paciniform corpuscles. Of these, the paciniform corpuscles are the least numerous (the highest number per muscle found by Barker et al. (1962) was five), so that only few of the group III fibers in a muscle nerve are used up for this receptor.

Afferent endings in the diaphragm were studied by Dogiel (1902), who described mainly muscle spindles, tendon organs, and Pacinian corpuscles. In addition, he found some nonencapsulated endings in the peritoneum (the afferent nature of which was not clear) and pericapillar meshworks of fine endings, which were numerous in the tendinous center and probably originated from sympathetic fibers. These sympathetic fibers were supposed to play an important role in diaphragmatic sensibility, particularly for the production of pain (Felix 1922). Dogiel's findings were confirmed by Masumoto (1934a, b), who studied the nervous supply of the diaphragm in a large number of mammals, including man.

2.1.2 Joint Capsules and Ligaments

The types, distribution, and even the denomination of joint receptors in the literature are extremely variable, depending on the group performing the investigation and the joint(s) studied. The main receptor type with myelinated afferents in the knee joint capsule of the cat was found to be the Ruffini ending (Gardner 1944); in addition Golgi-Mazzoni endings are present (Grigg et al. 1982). In the ligaments of the joint, Golgi endings were described; whereas, Pacinian or paciniform corpuscles appear to be rare (Skoglund 1956, O'Connor and Gonzales 1979) or even absent (Gardner 1944) in or near the joint. Recently, Golgi endings and paciniform corpuscles have been reported to exist in the posterior attachements of the lateral meniscus of the dog knee joint (O'Connor 1984).

Free nerve endings are numerous in the joint capsule (Gardner 1944, Skoglund 1973). In sympathectomized joints, the majority of nonmyelinated fibers were found to form a nerve plexus in the inner zones of the fibrous capsule; myelinated fibers contributed nonmyelinated branches to this plexus (Samuel 1952). Many of the fine joint afferents showed a close association with blood vessels and often ended as "fine fibrils".

In order to avoid the existing problems of nomenclature, Freeman and Wyke (1967) used Roman numerals to desginate the different joint receptors. In their system, the types I through III were endings showing a specialized structure (e.g., Ruffini ending, Golgi ending), while type IV consisted of two subtypes of nonmyelinated endings, one forming plexuses and the other free nerve endings. Small-diameter afferent fibers supplied mainly the type IV, group III fibers giving rise to the unmyelinated plexuses and group IV fibers to free nerve endings. No type IV receptors have been found in the synovialis, but they were abundant in other joint regions.

2.1.3 Ultrastructure of Free Endings

In a study using a combination of physiological and anatomical methods, Kruger et al. (1981) showed that mechanonociceptive units of the cat skin with group III

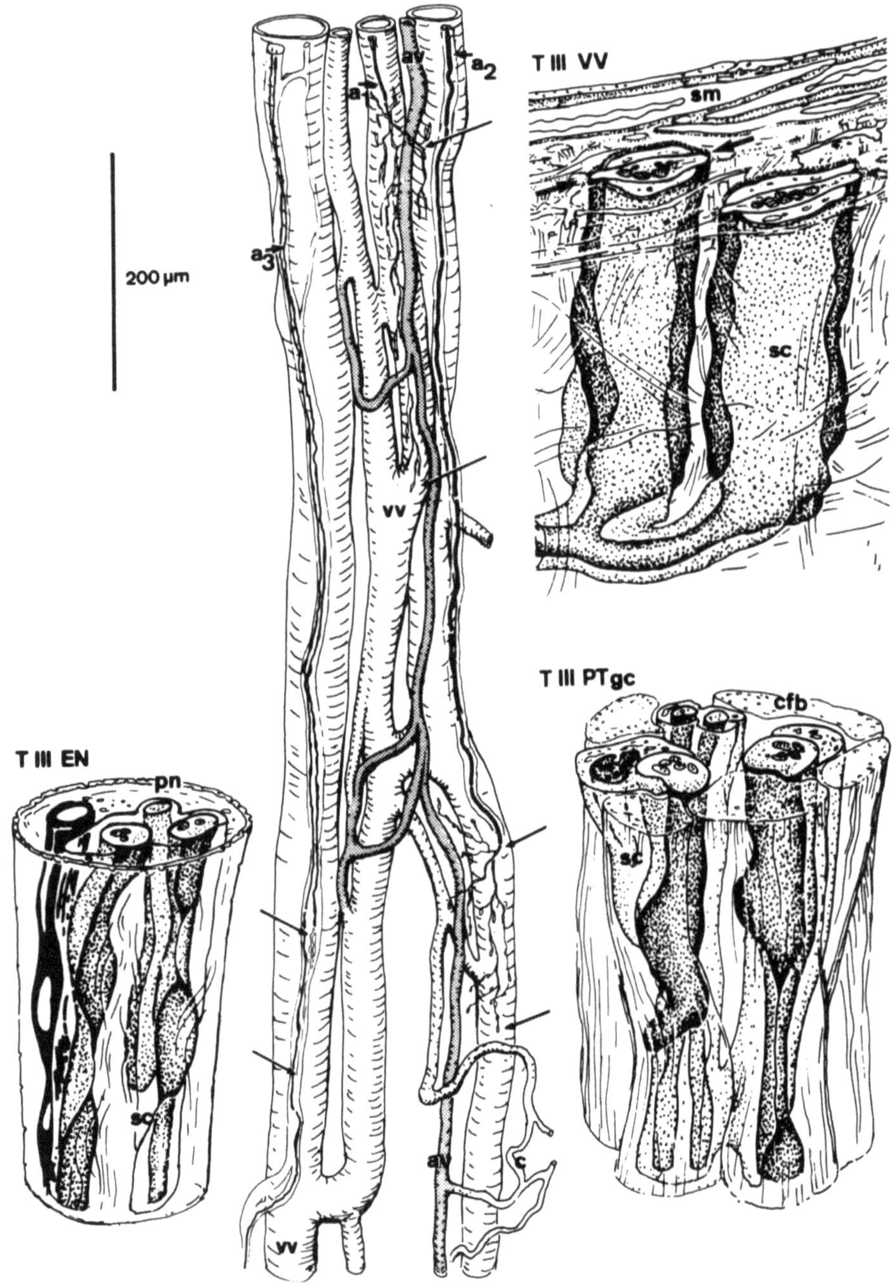

Fig. 2. Reconstruction of a complete series of sections over a total length of 1 mm showing three different terminals of group III afferents. A fascicle of arterial (*av*) and venous (*vv*) vessels with some arteriovenous anastomoses are accompanied by small nerve fiber bundles of which the myelinated axons (a_1, a_2, a_3) are reconstructed: down to their receptive areas the schematic drawings show segments of three different terminals at a higher magnification (*T III VV, T III PTgc, T III EN*) with their typical relations to the surrounding tissue structures. Axon membrane segments without any Schwann-cell (*sc*)

afferent fibers possess nonmyelinated terminals which are located in the basal epidermis. The putative receptive apparatus contained vesicles and was still surrounded by processes of a Schwann cell; therefore, the term "free nerve ending" does not mean that it consists of bare branches of the afferent axon (cf. also Cauna 1966). The terminal fiber-Schwann cell complex with its vesicles was supposed to constitute the morphological substrate of the functional specialization of a receptive ending. In some aspects, the appearance of the mechanonociceptive ending was similar to that of cutaneous cold receptors in the cat, which are likewise supplied by small-diameter afferent fibers (Hensel et al. 1974).

Detailed information on the structure of deep receptors connected to group III and IV afferent fibers was obtained in an electron microscopic study on endings in the gastrocnemius muscle and calcaneal tendon of the cat (Andres et al. 1985). The authors were able to distinguish six different types of terminals with regard to morphology or localization of the ending: group III fibers become nonmyelinated and form five different endings which end in the adventitia of small vessels, within the perineural sheath or between collagen fiber bundles. Three of the five group III types found by the authors are shown in Fig. 2. Group IV afferents end as penicillate endings in the adventitia of blood and lymphatic vessels and contain large numbers of granulated vesicles. Some terminals showed signs of a structural specialization (e.g., close contact with collagen fibers) which are thought to be suggestive of a mechanoreceptive function.

Recently, the anterograde transport of horseradish peroxidase (HRP) has been introduced to visualize the morphology of peripheral receptive terminals, including free nerve endings, using light microscopic (Marfurt and Turner 1983) and electron microscopic techniques (Turner and Marfurt 1983).

2.2 *Afferent Fibers*

In general, the afferent fiber spectrum of a muscle nerve is similar to that of a skin nerve (Boyd and Davey 1968), with the exception that the fastest conducting fibers in a skin nerve belong to group II, i.e., group I fibers are missing in a cutaneous nerve (Lloyd 1943). The borderline between large- and small-diameter afferent fibers (or group I, II on the one side and group III, IV on the other) is usually drawn at a fiber diameter of 4 – 6 µm, corresponding to a conduction ve-

Fig. 2 (*continued*) covering are marked by *darker stippling*. The specific terminal of the a_1 axon is the *T III VV* (tendon group III venous vessel) type with two lanceolate terminals in the adventitia of the vessel. The *arrows* indicate the free edges of the terminal. *Sm*, Smooth muscle cell. The specific terminal of the a_2 axon is type *T III PTgc* (tendon group III peritenonium ground substance-collagen) in the connective tissue of the vessel-nerve-fiber fascicle. The terminals are orientated to collagen fiber bundles (*cfb*). The specific terminal of the a_3 axon is the *T III EN* (tendon group III endoneural) type. The terminals ramify within the endoneural space. Another small myelinated axon passes the terminals. *Pn*, perineural sheath. The length of the receptive areas of each myelinated axon is indicated in the reconstruction by *arrows*. The close association of the *T III VV* ending with the adventitia and of the *T III PTgc* terminal with collagen fiber bundles might be indicative of a mechanosensitive function. (From Andres et al. 1985)

locity of approximately 30 m/s (Gasser and Grundfest 1939, Lloyd 1943, Brock et al. 1951, Stacey 1969). Histograms of fiber diameter and conduction velocity of fine afferent fibers from feline muscles are shown in Figs. 3 and 4.
Sherrington (1894) reported that out of 946 fibers in the medial gastrocnemius (MG) nerve of the cat, 375 (39.6%) were afferent. In this light microscopic study, nonmyelinated axons could not be correctly counted. Later, it was shown that a considerable number of nonmyelinated afferent fibers were present in a muscle nerve. Originally, the ratio between nonmyelinated and myelinated afferents was estimated to be less than one (Ranson and Davenport 1931), but exact counts at the electron microscopic level demonstrated that nonmyelinated muscle afferents outnumber the myelinated ones by a factor of about two. In such an experiment, the nerve to the posterior tibial muscle has been found to contain 130 myelinated and 261 nonmyelinated afferent fibers (Stacey 1969). In Table 1, the data from fiber counts of the nerve to the lateral gastrocnemius-soleus (LGS) muscle of the

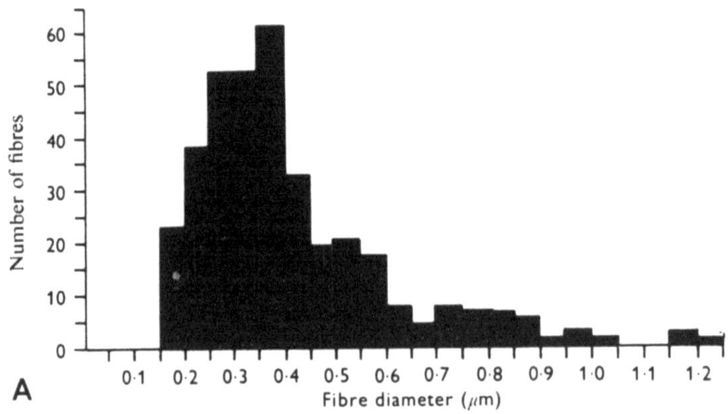

A

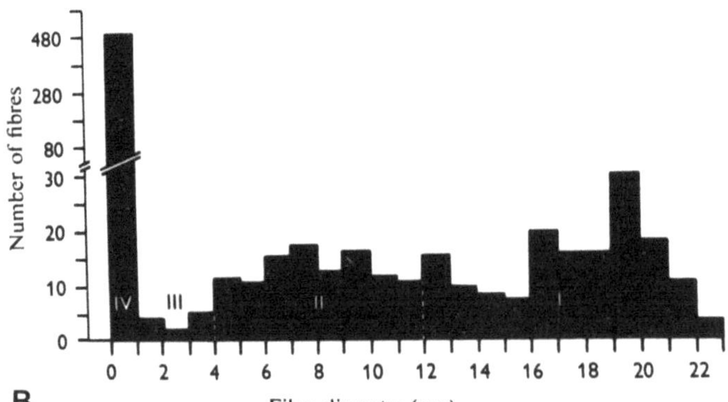

B

Fig. 3. The diameter (**A**) and extent (**B**) of nonmyelinated sensory nerve fibers in the de-efferentated and sympathectomized tibialis posterior (TP) muscle nerve of the cat. **A** was constructed from diameter measurements, made from electron micrographs, of 357 non-myelinated fibers from two TP muscle nerves. **B** is a composite histogram of the total sensory nerve fiber component compiled from both EM and LM observations of two TP muscle nerves. (From Stacey 1969)

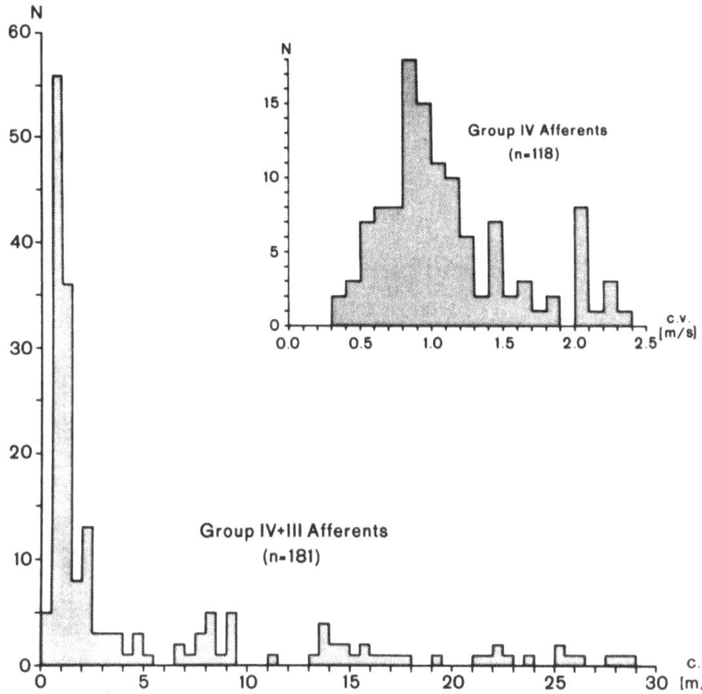

Fig. 4. Histogram of conduction velocities of group III and IV afferent fibers from the gastrocnemius soleus (GS) muscle of the cat. All the fibers shown (118 group IV and 63 group III) were isolated from dorsal rootlets; the conduction velocity was calculated from conduction distance and latency after electrical stimulation of the GS muscle nerves in the popliteal fossa. In the *inset,* all fibers conducting at less than 2.5 m/s are shown at a smaller bin width. *N*, number of fibers, *c.v.*, conduction velocity. (From Mense 1978)

cat are shown. From these numbers it can be concluded that afferent fibers constitute almost half (in this case, 46%) of all the fibers in a muscle nerve. The great majority (in this case, 75%) of the afferent fibers have a small diameter, i.e., belong to group III and IV. In the MG nerve of the cat, 58% of the nonmyelinated fibers were found to be afferent (Langford 1983). There may be differences in the fiber spectrum between muscles of different function. Thus, the nerves to the postural neck muscles in the cat have been shown to contain unusually high numbers of group II and III fibers (Richmond et al. 1976, Abrahams et al. 1984).
A recent electron microscopic study of the rat phrenic nerve demonstrated that the proportion of nonmyelinated fibers in this nerve is relatively small (43%). Of these, 59% were found to be of dorsal root ganglion origin (Langford and Schmidt 1983b). An interesting observation was reported by Landau et al. (1962) in their study on myelinated fibers of the phrenic nerve in the dog. According to their results, the phrenic nerve does not possess a γ-fiber system, and the afferent group II fibers are likewise missing. This would mean that the diaphragmatic spindles differ from those of other muscles.
In comparison with the data of Table 1, the application of HRP to the proximal end of the severed nerve to the GS muscle and subsequent counting of labeled

Table 1. Fiber composition of the lateral gastrocnemius-soleus nerve in the cat

Myelinated 1200								Unmyelinated 2000	
Motor		720	60%	Sensory		480	40%	Motor	Sensory
Aα	Skeletomotor	382	53%	I	Spindle primary sens. endings (Ia)	144	30%		
					Tendon organs (Ib)	72	15%		
Aβ	Skeleto- and fusimotor	14	2%	II	Spindle second. sens. endings	144	30%		
Aγ	Fusimotor	324	45%		Spray (Ruffini) endings	5	<1%		
					Lamellated (Pacini-form) endings	5	<1%		
				III	"Free nerve endings"	110	23%	C vasomotor 1000 50%	IV Sensory 1000 50%

The combined nerve to the lateral gastrocnemius and soleus muscles of the cat contains 1200 myelinated and 2000 unmyelinated nerve fibers. Of the latter, 50% are afferent; the other 50% are postganglionic sympathetic vasomotor nerve fibers as revealed by nerve fiber counts from chronically sympathectomized cats (M. von Düring and R. F. Schmidt, unpublished). The percentages of the various afferent and efferent myelinated fiber types are those reported by Boyd and Davey (1968). Note that the majority of afferent fibers (1110 from 1480) belong to groups III and IV. (From Mitchell and Schmidt 1983) cells in the lumbosacral dorsal root ganglia yielded similar (Craig and Mense 1983) or somewhat lower numbers (McLachlan and Jänig 1983) for the afferent fiber population. The nerve of a flexor muscle (anterior tibial) has been found to contain more myelinated afferent fibers than the nerve of an extensor muscle (GS), the difference being due to a greater portion of small afferents in the flexor nerve (Rexed and Therman 1948).

The great number of afferent fibers in a muscle nerve suggest that a skeletal muscle possesses, besides its contractile properties, the functions of a sense organ. Owing to the abundance of small-diameter afferent fibers, a great deal of the sensory functions appear to be subserved by group III and IV units.

This conclusion seems to be valid also for other deep tissues. In electron microscopic studies on the posterior and medial articular nerve of the knee joint, the proportion of nonmyelinated fibers was reported to be around 80% in the monkey (Gardner and Lenn 1977), dog (O'Connor et al. 1982), and cat. In the latter species about 50% of these fibers were found to be afferent (Langford and Schmidt 1983a).

It is generally accepted that afferent fibers reach the spinal cord via the dorsal roots, However, a small portion of these fibers (consisting mainly of fine afferents) have been shown to course through the ventral roots (Coggeshall and Ito 1977, Light and Metz 1978, Coggeshall et al. 1980). Some of the fine ventral root afferents possessed receptive fields in the deep tissues (Coggeshall and Ito 1977), so that the ventral roots have to be regarded as a possible pathway conveying sensory information from deep tissues to the spinal cord.

When considering the composition of muscle or joint nerves, it has to be kept in mind that probably not all afferent fibers contained within such a nerve are of the same origin. It has been shown that a muscle nerve, for example, may consist partly of afferent fibers from the joint moved by that muscle (Skoglund 1956, Freeman and Wyke 1967). The same is true for cutaneous nerves which often have branches supplying deep tissues such as skeletal muscles and tendons (Stilwell 1957c, Chang et al. 1983). Thus, the differentiation between cutaneous, muscular, and articular nerves does not mean that the respective nerves are "pure" with respect to the origin or destination of the fibers making up the nerve.

2.3 Spinal Termination

2.3.1 Projections of Whole Nerves

The projection of afferent fibers to the spinal cord or to the brain stem, respectively, has been examined with the use of degeneration methods (Cassirer 1899, Sprague and Ha 1964, Westrum et al. 1976, Gobel and Binck 1977, Grant and Ygge 1981, Culberson and Brown 1984) or with application of a neuronal marker (e.g., cobalt, radioactive amino acids) to the dorsal root ganglion or dorsal root filaments (Iles 1976, Ralston and Ralston 1979, Réthelyi et al. 1979). For several years now, HRP has been extensively used for this purpose (cf. Proshansky and Egger 1977). The enzyme has the added advantage that, after being applied to a peripheral nerve, it is transported transganglionically (across the dorsal root ganglion) to the spinal cord (Grant et al. 1979, Mesulam and Brushart 1979).

After a suitable survival period (usually 2–3 days), the terminal branches of the primary afferent fiber can be visualized using the histochemical protocol of Mesulam (1978), which employs tetramethyl benzidine (TMB) as a chromogen and appears to be the most sensitive method for the detection of the enzyme in neuronal structures (Mesulam and Rosene 1979). As far as peripheral nerves are concerned, this technique has mainly been used to study the projections of cutaneous nerves to the dorsal horn (Koerber and Brown 1980, 1982).

The available data on the spinal termination of muscle nerves are somewhat controversial, particularly with regard to the small-diameter afferent fibers. It has been shown that, in general, group III and/or group IV fibers project to the superficial layers (laminae I and II; Rexed 1952) and to the neck of the dorsal horn (lamina V) (Réthelyi and Szentágothai 1973, Light and Perl 1979a, Gobel et al. 1981, Snyder 1982). In the first study on the spinal projection of muscle nerves, in which the transganglionic transport of HRP was employed, no significant labeling was found in the superficial dorsal horn of rats (Mesulam and Brushart 1979), but application of HRP bound to wheat germ agglutinin produced intensely stained terminal fields, mainly in the substantia gelatinosa (Brushart and Mesulam 1980). In contrast, Kalia et al. (1981) described projections of muscle afferents to all laminae of the dorsal horn. As i.m. injections of the enzyme were used in these studies, it was not clear whether the labeling in the dorsal horn exclusively represented muscle afferents or whether a contamination with cutaneous afferents existed, owing to the spread of HRP from the injected muscle to the overlying skin.

A study in which the central terminations of the feline GS nerves were visualized with restricted application of HRP to the muscle nerves, yielded different results (Craig and Mense 1983). In the dorsal horn, the GS afferents were found to terminate in lamina I and lamina V, but not to a significant extent in lamina II, or laminae III, and IV (Fig. 5). In the latter laminae, only traces of reaction prod-

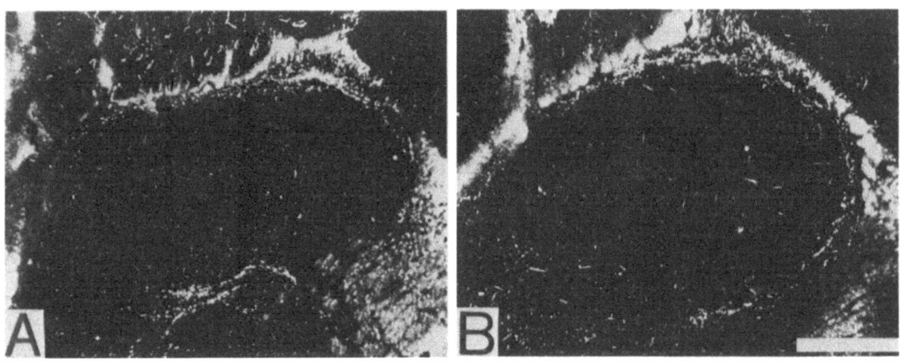

Fig. 5. Labeling pattern of afferent fiber from cat GS muscle in the dorsal horn at rostral S_1 (**A**) and mid-S_2 levels (**B**). Dark-field photomicrographs. Horseradish peroxidase (HRP) solution was applied to the severed muscle nerve; for the detection of the enzyme the TMB reaction was used. In **A**, the labeling is focused in Lissauer's tract, the apex of lamina I, and the lateral portion of lamina V. In **B**, it extends throughout the mediolateral extent of lamina I, and into lateral lamina V. Transverse, 48-μm sections, dorsal up, lateral right; scale bar, 310 μm in **A** and 200 μm in **B**. (From Craig and Mense 1983)

uct were present, the possible source of which is discussed below (cf. Sect. 2.3.2). A further finding was that i.m. injections of HRP inevitably resulted in leakage of the enzyme to neighboring tissues which led to artifactitious labeling in the dorsal horn. From this study it can be concluded that laminae II and III, which receive strong projections from skin nerves (Koerber and Brown 1982), are not a major terminus of GS muscle afferents. Thus, it appears that the pattern of spinal termination of muscular afferents is distinctly different from that of cutaneous fibers and resembles more the central projection of visceral afferent fibers (De Groat et al. 1978, Kuo et al. 1983, Cervero and Connell 1984).

2.3.2 Termination Patterns of Identified Single Fibers

Recently, a combination of neurophysiological and histochemical methods has been introduced for the visualization of functionally identified neuronal structures. The technique employs micropipettes filled with HRP solution and connected to an iontophoretic apparatus. First, the functional properties of a given fiber or soma are determined by recording its discharges in response to natural stimulation, then HRP is injected iontophoretically into the neuron.

This technique has been used extensively for studying the intraspinal course and terminations of thick afferent fibers from skin and deep tissues (for a review, see Brown 1981). The method is not (yet) applicable to group IV fibers, but group III units from the cat skin have been stained using this method (Light and Perl 1979b). The only available study dealing with the spinal terminations of identified small-diameter afferent units from deep tissues is a preliminary one by Mense et al. (1981). The results showed that nociceptive afferent units from the deep tissues of the cat tail possess spinal terminations similar to those of cutaneous nociceptive units. Both terminate mainly in laminae I and V and possess additional synaptic connections in the gray matter dorsal to the central canal (Fig. 6). The few low-threshold mechanosensitive units stained in this study projected to more dorsally located layers, mainly lamina IV.

In a systematic survey (Lehmann-Willenbrock et al. 1984, Mense et al. 1985), stained axons conducting in the group II – III range (9 – 40 m/s), from deep receptors of the cat hindlimb, were reconstructed from histological sections. The axons showed clear differences in their spinal course and site of termination, depending on the type of receptors they supplied. Owing to the mechanical sensitivity of the experimental set-up, the units were only roughly characterized as either nociceptive (requiring noxious local pressure or forceful movement of a joint to an extreme position for activation) or low-threshold mechanosensitive (responding to innocuous deformation of the deep tissues by deep pressure or gentle movement of joints). All stimuli were applied through the intact skin. A more detailed description of the receptive properties of fine deep afferents is given in later sections.

Nociceptive Units. Most of the axons from deep nociceptors ascended laterally in the dorsal columns or within Lissauer's tract. They produced a small number of collaterals which, in about half the units, were restricted to lamina I (Fig. 7 A). The collaterals ran medially within lamina I and possessed bouton-like enlarge-

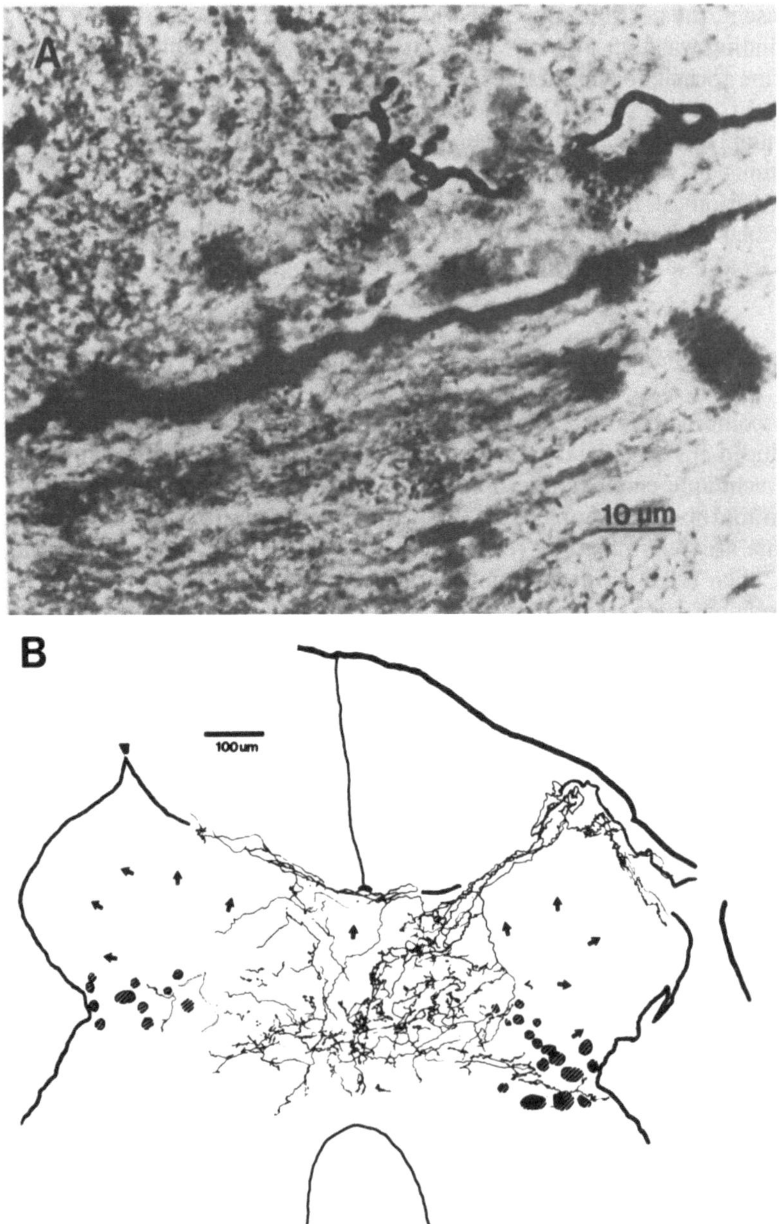

Fig. 6 A, B. High-threshold mechanosensitive afferent unit from a muscle of the cat tail stained with intraaxonal iontophoresis of HRP, conduction velocity 46 m/s. **A** En passant and terminal boutons in the region just dorsal to the central canal. *Top* of the figure is dorsal. The collateral running through the *lower half* of the figure terminated in the contralateral lamina V (toward the *left* of the figure). **B** Same unit as in **A**. Camera lucida drawing of all collaterals of the unit rostral to its point of entry into the spinal cord. *Arrows* indicate approximate boundary between laminae II and III. (From Mense et al. 1981)

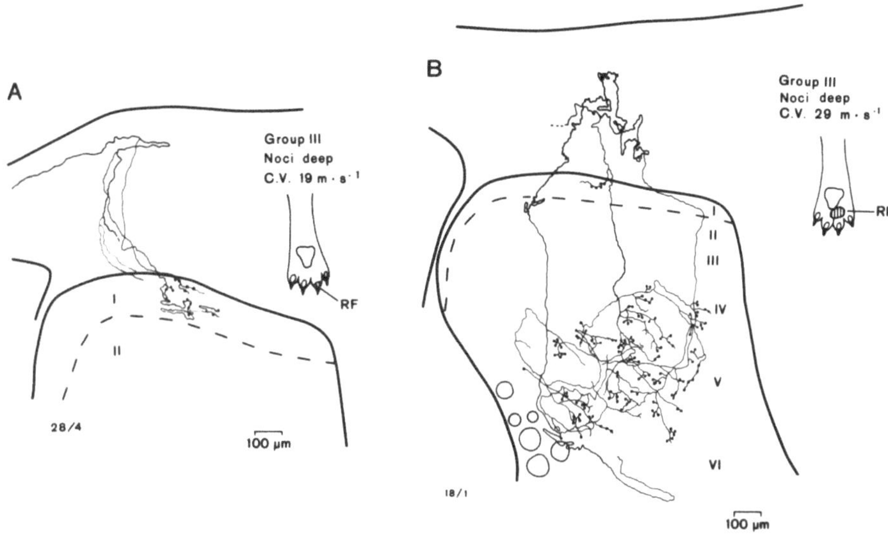

Fig. 7 A, B. Spinal termination patterns of nociceptive group III afferent units from deep tissues of the cat lower hindlimb. **A** Unit with terminal arborizations confined to lamina I. **B** Unit terminating in laminae IV/V. The approximate border between lamina I and II is marked by the *dashed line*. *C. V.*, conduction velocity; *RF*, subcutaneous receptive field which was stimulated through the intact skin. (From Mense et al. 1985)

ments, in the form of terminal and en passant boutons. The other half of the nociceptive units had collaterals terminating in laminae I and IV/V or predominantly in laminae IV – V (Fig. 7 B). It has to be pointed out that some of the nociceptive afferents had conduction velocities exceeding 30 m/s and, thus, belonged to group II units. These units likewise possessed terminals in laminae I and/or IV/V.

In a combined light and electron microscopic study, the boutons of high-threshold mechanosensitive afferents from the cat skin were found to have smaller dimensions than boutons of D-hair afferents. Both types of afferents contained clear, spherical vesicles in their presynaptic terminals (Réthelyi et al. 1982).

Low-Threshold Mechanosensitive Units. The spinal course of low-threshold mechanosensitive afferents differed from that of nociceptors in that the parent axon did not directly approach the gray matter after entering the spinal cord, but followed an oblique path across the dorsal columns from the dorsal root entry zone to the medial corner of the dorsal horn (Mense et al. 1985).

About 40% of the sensitive mechanoreceptors with axons conducting at less than 40 m/s had response properties of secondary endings from muscle spindles. The morphology of the stained collaterals (conduction velocity of afferent axons: 21 – 30 m/s) appeared to be similar to that described by Fyffe (1979) and Brown (1981) for secondary endings of faster conduction velocity (54 – 61 m/s). The collaterals of the slowly conducting secondaries likewise had terminations in laminae IV – V and VI – VII, with some additional projections to the more ventrally located motor nuclei. A certain difference between fast and slowly conducting secondaries may exist in the mediolateral extent of the terminations in laminae VI – VII. The slowly conducting units possessed boutons that were distributed over almost the whole width of lamina VI; whereas, the terminations of the fast-conducting units appeared to be more restricted to the middle portion of this lamina (Brown 1981). Somatotopical factory may be of importance for these differences, since most of the slowly conducting secondaries of the study by Mense et al. 1985 originated in the flexor muscles of the toes and not in the GS muscle, as in Fyffe's work.

The majority (about 60%) of the sensitive mechanoreceptors differed from muscle spindles in that they were rapidly adapting and/or did not respond to stretching of muscles. The most effective stimuli were local pressure or joint movements. The spinal areas of termination of these mechanosensitive receptors showed more interindividual variation than those of nociceptors. Some of the units possessed terminal arborizations with bouton-like structures concentrated in lamina IV (Fig. 8 A); others had rather diffuse projections to great portions of the dorsal horn, e.g., laminae II, IV, and V (Fig. 8 B). Lamina III was usually free of terminations. Probably, the traces of reaction product observed in lamina IV after application of HRP to a severed muscle nerve (cf. Sect. 2.3.1) were identical to the terminal arborizations of these mechanosensitive afferents.

It is not yet clear whether the various patterns of termination correlate with the location of the receptive ending in one of the deep tissues, e.g., muscle, tendon, or joint capsule, or whether the type of receptor supplied by the fiber (e.g., paciniform corpuscle, Ruffini ending, or free nerve ending) determines the pattern of termination. The results of a similar study on cutaneous afferent units of small diameter (Light and Perl 1979b) speak more in favor of the latter assumption, since all the units studied had receptive fields in the skin but showed

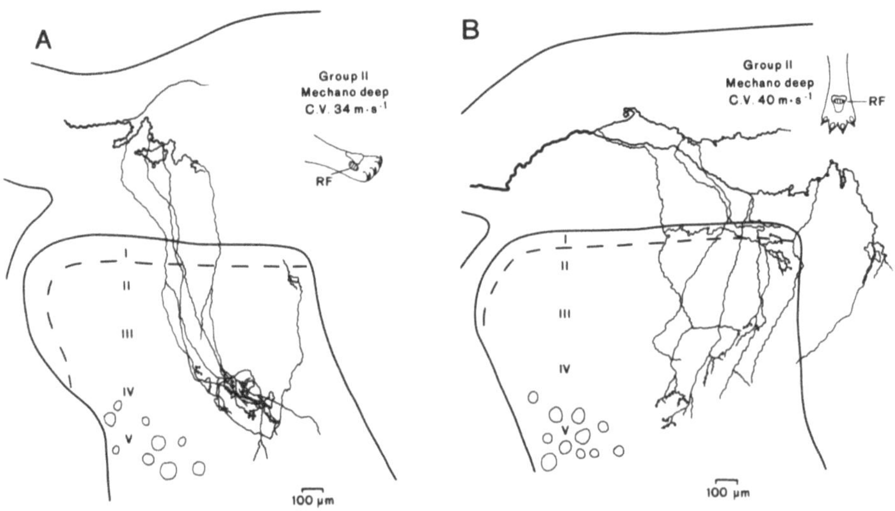

Fig. 8A, B. Terminal branching pattern of afferent fibers supplying deep sensitive mechanoreceptors. The unit shown in **A** possessed a circumscribed field of bouton-like swellings mainly in lamina IV; the unit in **B** had terminal aborizations in large areas of the dorsal horn. Note that lamina III is almost free of terminals. (From Mense et al. 1985)

differences in their spinal projection, depending on the receptor type supplied by the stained fiber.

When the termination patterns of deep afferent units from the hindlimb are compared with those of deep units from the tail (Mense et al. 1981; cf. Fig. 6), some differences, mainly concerning the nociceptive fibers, become apparent. One is that no contralateral projections were observed in the fibers from the hindlimb; whereas, they were frequent in high-threshold mechanoreceptive units from the deep tissues of the tail and also from the skin of the tail (Light and Perl 1979b). In general, contralaterally projecting primary afferent fibers appear to be more common in the sacral and caudal spinal cord (Matsushita and Tanami 1983) than in the lumbosacral enlargement, although they are also present at other segmental levels in the cat (Réthelyi et al. 1979) and rat (Smith 1983).

Another difference existed in the projections to the gray matter around and dorsal to the central canal, which were common in afferent units from the tail, but not in those from the hindlimb. Recently, the area around the central canal (lamina X) has attracted new interest; it contains neurons that appear to be involved in nociception in the cat (Honda and Perl 1981) and rat (Nahin et al. 1983).

The results of the experiments with intraaxonal staining of identified fine fibers from the deep tissues of the hindlimb show that these units project strongly to laminae I and IV/V of the dorsal horn of the cat; whereas, lamina II (particularly IIi) and III do not receive a substantial projection from deep afferent units. This means that nociceptive information from the deep tissues of the cat hindlimb, which has been shown to influence cells of origin of the spinocervival tract (SCT) (Lundberg and Oscarsson 1961, Hong et al. 1979) in laminae IV (and V) (Bryan et al. 1973, Craig 1976) and of the spinothalamic tract (STT) (Craig and Kniffki 1982) in lamina I (Trevino et al. 1972, Carstens and Trevino 1978), may be transmitted via monosynaptic connections. Since only one nociceptive fiber has been found to project to laminae VII – VIII, where most of the spinothalamic cells are located in the cat (Trevino et al. 1972, Carstens and Trevino 1978), it is likely that the bulk of nociceptive input to these cells via group III fibers is transmitted by interneurons.

Likewise, the low-threshold mechanosensitive input via small-diameter fibers from deep tissues could reach monosynaptically the cells of origin of the SCT in lamina IV. The muscle spindle secondary endings with group III afferent fibers might possess monosynaptic contacts with cells of the ventral spinocerebellar tract which are located in the dorsolateral portion of the ventral gray (Ha and Liu 1968).

The few projections to lamina II observed in some low-threshold mechanosensitive afferent units (and also in some nociceptive ones) may provide an input to the local circuitry of the substantia gelatinosa and/or long ascending tracts which have been shown to originate in cells of lamina II in the rat (Giesler et al. 1978) and monkey (Willis et al. 1978, 1979).

3 Considerations Concerning the Distinction Between Different Receptor Types

In the following sections, an attempt has been made to classify the fine deep afferents according to their response properties, and to assess the role these units might play in various central nervous functions.

When tested with physiological stimuli, individual group III and IV units show marked differences in their responsiveness (cf. Mense 1977b, Mense and Stahnke 1983, Schaible and Schmidt 1983a, b). These differences are so great – the mechanical thresholds, for example, span the whole range from light touch to overtly noxious pressure – that the assumption that the units form a single receptor population with a statistical variability in sensitivity appears to be unlikely. Therefore, the alternative explanation, namely, that in deep tissues receptors with fine afferent fibers exist which are functionally diverse, has been favored in this review. Admittedly, the evidence for the existence of functionally different receptor types is indirect since, up to now, no studies with selective stimulation of one of the types and recording of the ensuing central nervous effects has been performed. For the moment, the concept of a specificity of fine deep afferents appears to be more compatible with the experimental data than that of a polymodality (cf. Kumazawa and Mizumura 1977). It has to be acknowledged, however, that some of the findings – e.g., the broad responsiveness of many units when tested with mechanical and chemical algesic stimuli – could also be explained by the polymodality concept. It will depend on future results whether the assumption of a specificity is tenable for fine deep afferents. In this respect, microneurographic studies in awake human beings should be particularly suitable for providing additional data.

For classification purposes, it is essential to use only stimuli that are likely to occur in the natural environment of the receptor and, if possible, to use all of them. The latter requirement is impossible to meet (considering all possible stimulus combinations) and the first one is a problem, particularly in studies on deep receptors, since touching or pressing upon an exposed deep tissue is not a "natural" stimulus. In addition, for receptors in muscles, tendons, and joints, muscle contractions and joint movements are the most common stimuli, but they are not easy to employ in experiments with recordings from fine deep afferents. In studies on cutaneous receptors, the mechanical threshold has been used as a criterion for distinguishing between nociceptive and nonnociceptive endings. If a receptor possesses a high threshold in the tissue-threatening range, its nociceptive nature is probable (unless it is a thermoreceptor; see below). Receptors giving liminal responses to moderate (innocuous) intensities of stimulation are considered to be nociceptive if they reach their maximal discharge rate upon noxious stimulation. In contrast, nonnociceptive endings usually show a saturation or even a decline in discharge frequency when the noxious range is approached (cf. Perl 1984). For deep receptors, the absolute mechanical threshold is not a suitable criterion, since sensitive endings lying deep within a tissue simply require strong deformations of the tissue to be activated. Thus, the thresholds of articular mechanoreceptors, determined with von Frey hairs, were found to be considerably higher than those of cutaneous mechanoreceptors (Schaible and Schmidt 1983a). The occurrence of the maximal response upon noxious stimulation is not an unequivocal sign for a deep nociceptor either, since even tendon organs and muscle spindles require noxious local pressure to fire with their maximal frequency (unpublished observations). Therefore, for the purpose of this review, receptors were considered to be nonnociceptive if they reached a considerable proportion (more than about 20%) of their maximal discharge rate in response to innocuous stimulation (weak pressure, midrange joint movements, physiological temperature changes). Most of these receptors showed maximal discharge frequencies when stimulated with noxious local pressure. Only if a unit did not react to various forms of innocuous stimulation (or showed only a liminal response), but could be vigorously excited by noxious stimuli, was it considered to be a nociceptor.

A serious problem for the interpretation of the experimental data is the possible sensitization of deep nociceptors brought about by the unsterile surgical exposure of the tissue containing the receptors or by repetitive testing with stimuli of tissue-damaging intensity. Sensitized nociceptors could be mistaken for low-threshold mechanosensitive receptors. Although for a particular low-threshold unit under study, a sensitization cannot be ruled out, it appears unlikely that all the low-threshold receptors described in the following sections could be sensitized nociceptors. The reasons are as follows:

1. In a study, in which the spinal terminations of single identified afferents from the deep tissues of the cat hindlimb were visualized with HRP (cf. Figs. 7, 8), a great number of group III units were found which responded to innocuous deformation of the deep tissues through the intact skin.
2. Unless repeatedly stimulated with noxious intensities, the group III and IV receptors in muscle and tendon did not show an increase in the background discharge or a lowering of the mechanical threshold in the course of experiments of more than 10 h duration. Apparently, the exposure of the tissue does not lead inevitably to a sensitization of nociceptors. In contrast, cutting through the tissue was sometimes found to sensitize deep nociceptors which had their receptive fields close to the wound (unpublished observations).
3. The existence of background activity in many fine deep afferents does not appear to be an indication of a sensitization. There was no correlation between mechanical threshold and frequency of background discharge. In fact, group III units in muscle and joint which usually had lower mechanical thresholds than group IV units, showed less background activity (Mense 1978, Schaible and Schmidt 1983b).
4. Since some group III afferents have been shown to supply paciniform corpuscles, low-threshold mechanosensitive units with fine afferents must exist in deep tissues – in addition to secondary endings of muscle spindles, which likewise can be supplied by group III fibers (Barker 1962, Stacey 1969).

However, an unwanted sensitization of receptors surely poses an important problem in experiments in which deep tissues have to be surgically exposed. It can be expected that further studies of the impulse activity from artificially inflamed tissues – such as those performed by Schaible and Schmidt (1983b) and Iggo et al. (1984) – will help in deciding whether a particular unit is sensitized or not.

The application of strong chemical stimuli, such as high doses of bradykinin or hypertonic solutions does not appear to be suitable for classification purposes. Although bradykinin has been shown to activate almost exclusively the fine deep afferents and not the encapsulated stretch receptors (Mense 1977a), an excitability by bradykinin cannot be taken as an indication of the nociceptive nature of a particular receptor. It is well known that slowly adapting mechanoreceptors of the skin also respond to bradykinin (Fjällbrant and Iggo 1961, Beck and Handwerker 1974). Thus, the impression of a "polymodality" might arise, if sensitive receptors are tested with strong chemical stimuli.

4 Central Nervous Functions Probably Related to Activity in Fine Deep Afferents

4.1 Nociception

4.1.1 Response Behavior of Deep Receptors with Nociceptive Properties

Skeletal muscle. Since the early work of Lewis and coworkers (Lewis et al. 1931, Lewis 1932), the pain of intermittent claudication has attracted much interest. For technical reasons, the recording of the impulse activity in single muscle afferents during ischemic contractions is difficult. Therefore, Bessou and Laporte (1958) employed the method of antidromic collision in order to find out whether nonmyelinated fibers contribute to ischemic pain. As the amplitude of the compound action potential of these fibers in a muscle nerve was reduced directly after ischemic work, it was concluded that muscle receptors with nonmyelinated afferent fibers were active under these conditions and, thus, might be involved in the production of pain from an ischemically working muscle.

The first recordings of the impulse activity in single slowly conducting muscle afferents were published by Paintal (1960) in his study on group III fibers in the cat. The afferent units reacted quite differently to the various stimuli applied, some responding mainly to noxious stimulation (strong local pressure, injection of hypertonic saline) and some being sensitive to weak, probably physiological, stimuli (moderate local pressure, tetanic contractions). Owing to their elevated mechanical threshold – in comparison to muscular stretch receptors – and their sensitivity to noxious chemical stimulation, these endings were called "pressure-pain receptors" by Paintal. The observation that many group III receptors respond rather specifically to local pressure or strong stretch, was confirmed by Bessou and Laporte (1961).

Concerning the response behavior of single group IV receptors, Paintal (1960) mentioned (as unpublished observations) that these units can be activated by strong local pressure, but not by stretch. A study performed by Iggo (1961) supported and extended these findings by showing that nonmyelinated muscle afferents responded to a variety of mechanical, thermal, and chemical stimuli, particularly if the stimulus intensity was in the noxious range.

Subsequent systematic recordings from a great number of group III and IV receptors in the GS muscle of the cat demonstrated that the injection of endogenous algesic substances (bradykinin, serotonin, histamine, and potassium) into the muscle artery is a potent stimulus for group III and IV muscle receptors (Mense and Schmidt 1974, Franz and Mense 1975, Fock and Mense 1976, Hiss and Mense 1976, Kumazawa and Mizumura 1976, 1977); whereas, muscle spindles and tendon organs do not exhibit comparable responses (Mense 1977a). Dose-response curves for the effects of bradykinin on muscle group IV receptors are shown in Fig. 9. The activation of the slowly conducting afferent units by algesic agents was considered to support the assumption that at least some of the

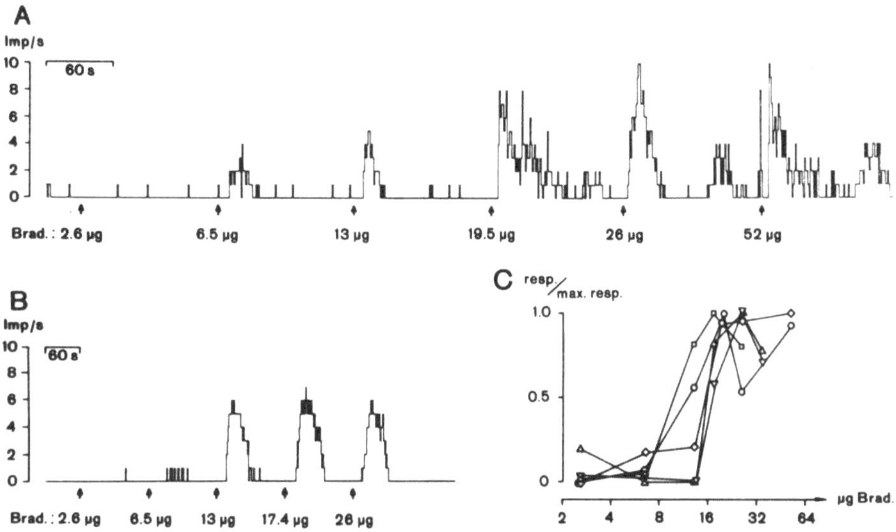

Fig. 9A – C. Responses of single group IV muscle afferents to increasing doses of bradykinin. **A** Appearance of separate afterdischarges at a higher dosage (19.5 – 52 µg). Unit from lateral GS, conduction velocity 0.76 m/s. Recording from sciatic nerve. *Arrows* indicate start of injection, which was completed in 10 – 15 s. Bin width of time histogram, 1 s. **B** Occurrence of saturation in the magnitude of response at a bradykinin dosage exceeding 13 µg. Unit from lateral GS, conduction velocity 0.64 m/s. Recording from sciatic nerve. Bin width of time histogram, 1 s. **C** Log dose-response plots of five group IV afferent units from GS muscle, all recorded from sciatic nerve. *Ordinate*, relative magnitude of response in terms of impulses evoked per injection. *Abscissa*, doses of bradykinin applied as single injections at intervals of 2 min (logarithmic scale). (From Mense 1977a)

units fulfill nociceptive functions. In addition, the bradykinin-induced activation could be significantly reduced by the peripherally acting analgetic acetylsalicylic acid (Mense 1982). However, from similar observations in the dog, Kumazawa and Mizumura (1977) arrived at a different conclusion, namely, that the group IV muscle receptors are polymodal and possibly participate in the mediation of exercise hyperpnea.

In a recent study (Mense et al. 1983, Mense and Meyer 1985), group III and IV endings in the GS muscle of the cat were subjected to a variety of graded stimuli in order to find out whether different receptor types can be distinguished among this population. A great proportion of the units required noxious (tissue damaging, presumably painful) stimuli to be clearly activated and, thus, behaved like nociceptors. Most receptors of this type responded to both noxious pressure and injections of bradykinin (Fig. 10). In these experiments, the algesic solution (86 µg bradykinin per ml saline) was injected directly into the receptive field in order to make sure that the ending was reached by a high concentration of the substance. Some of the high-threshold mechanosensitive receptors did not react to the infiltration of the receptive field with bradykinin. In these cases, it is not known whether the unit was really chemoinsensitive, since other algesic substances were not tested. Some of the units with a high mechanical threshold behaved similarly to the polymodal nociceptors of the skin in that they responded

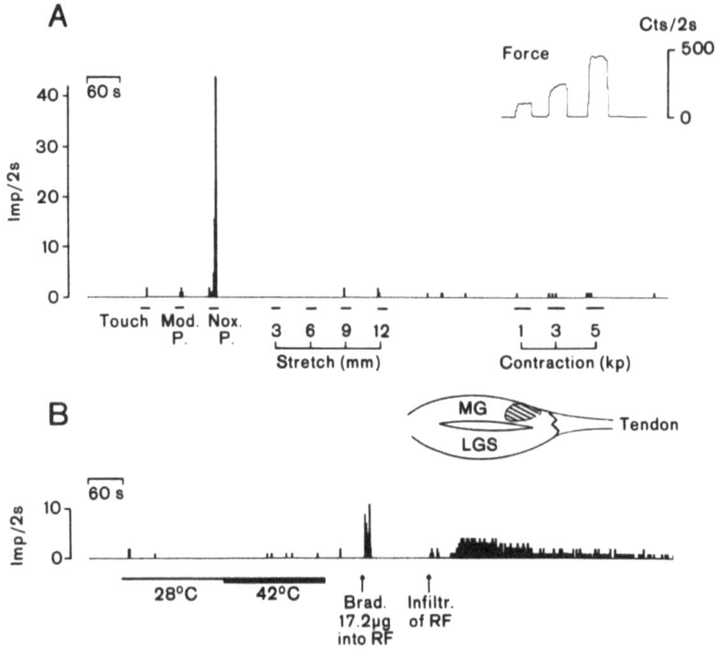

Fig. 10 A, B. Nociceptive unit from the cat skeletal muscle with group III afferent fiber, conduction velocity 12 m/s. The receptive field was situated in the distal third of the medial gastrocnemius (*MG*) muscle (hatched area in inset drawing). The peristimulus time histogram of the fiber's activity was constructed using a bin width of 2 s. The duration of the various stimuli used is indicated by the length of the *bars* underneath the histogram. **A** Local pressure applied by hand was effective only if it reached noxious levels (*Touch*, touching the receptive field with a painter's brush; *Mod.P.*, moderate pressure leading to an innocuous deformation of the tissue; *Nox.P.*, noxious pressure). Stretching the muscle by up to 12 mm and active contractions of a force of up 5 kp were ineffective. The force registration was obtained from a strain gauge; the scale of the ordinate is given in arbitrary units. Owing to the long bin width, the registration is continuous although the contractions were intermittent (a tetanic contraction of 0.5 s duration every second). **B** Cold and warm stimulation was performed with large thermodes in contact with the receptive field. A single injection of bradykinin (*Brad.*) 17.2 µg into the receptive field (*RF*) produced only a short-lasting increase in activity; the infiltration of the receptive field with bradykinin was followed by an activation of several min duration. (From Mense and Meyer 1985)

not only to algesic substances, but also to noxious heat (Iggo 1961, Kumazawa and Mizumura 1977, Mense 1978). Interestingly, group III receptors in the neck muscles of the cat were found to be rather insensitive to bradykinin; most of these had a high threshold upon mechanical stimulation (Abrahams et al. 1984). About 40% of the slowly conducting afferent units from the GS muscle in the cat were found to have nociceptive properties (Mense and Meyer 1985). A much smaller proportion of the group III and IV units (about 10%) were found to be activated during ischemic contractions (Mense and Stahnke 1983). Thus, not all the nociceptive endings appear to be sensitive to all painful conditions of muscular tissue. This might be an indication that different subtypes exist among the nociceptors.

An interesting feature of many of the nociceptive units was that their mechanical thresholds were lowered after application of bradykinin. Thus, a receptor that responded only to noxious pressure prior to injection of the substance was likely to be activated by innocuous pressure or contractions after infiltration of the receptive field with bradykinin (Fig. 11). Almost all nociceptors that were excited by bradykinin showed this behavior; in addition, some units not responding to the algesic substance were more sensitive to local pressure after infiltration of the receptive field with bradykinin. Apparently, bradykinin sensitized the nociceptors towards mechanical stimuli; the concentration required for sensitization

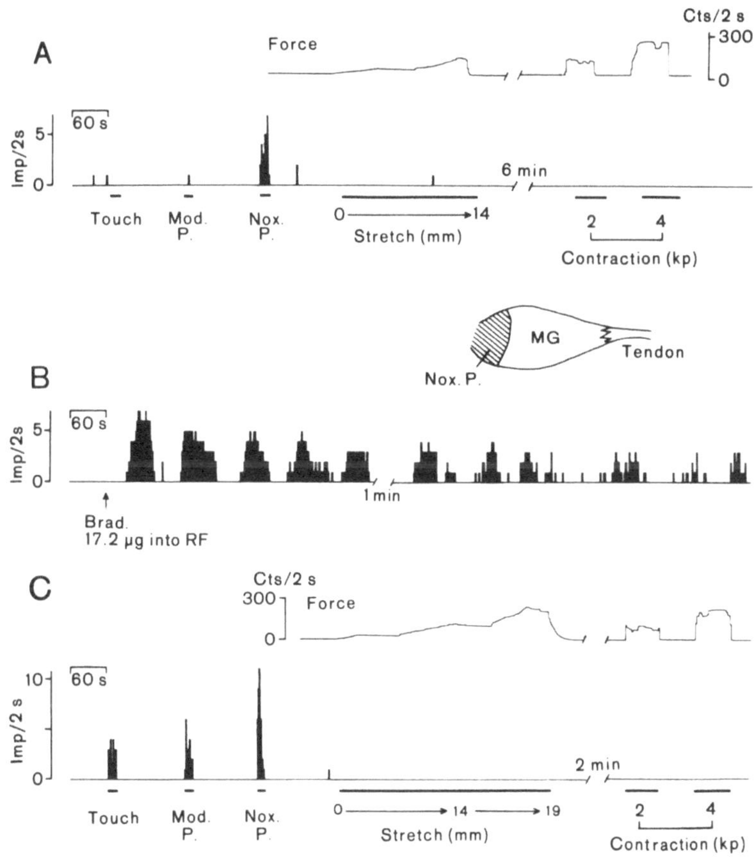

Fig. 11 A – C. Nociceptive unit from the cat MG muscle with group III afferent fibers, conduction velocity 6.2 m/s. Labeling of receptive field and stimuli as in Fig. 10. Note that after the response to bradykinin had died off, the mechanical threshold of the unit was lowered so that it was excited by touch and moderate pressure (**C**), i.e., by stimuli which were totally ineffective before application of bradykinin (cf. **A**). Despite the sensitization to pressure stimuli in **C**, muscle stretch and contractions had no excitatory effect, although this time the muscle was stretched 19 mm. (From Mense and Meyer, to be published)

of a receptive unit seemed to be lower than that for direct activation. Sensitizing phenomena have also been described for cutaneous nociceptors, particularly if heat stimuli are applied repeatedly (Bessou and Perl 1969, Perl et al. 1976, Fitzgerald and Lynn 1977, Campbell et al. 1979, La Motte et al. 1982). The local release of bradykinin might be involved in heat sensitization; injections of this substance have been shown to produce transient increases in the heat responses of nociceptors (Beck and Handwerker 1974). Another group of substances possibly producing sensitization in nociceptors are the prostaglandins of the E type (Juan et al. 1984). Leukotrienes are likewise known to mediate inflammatory processes (for a review, see Piper 1984), but their influence on nociceptive endings has not yet been tested systematically. The receptors possessing a low mechanical threshold (see below) usually showed no signs of sensitization. They exhibited, rather, a reduced sensitivity to mechanical stimuli after infiltration of the receptive field with bradykinin. The sensitization observed in single nociceptors might constitute the neurophysiological correlate for the hyperalgesia and allodynia of an inflamed or mechanically lesioned muscle.

Studies performed on group III and IV muscle afferents in other species yielded similar results, but the conclusions drawn by the authors differed somewhat. Kumazawa and Mizumura (1976, 1977) recorded the activity from slowly conducting AF from a hindlimb muscle in the dog. The stimuli used and the responses obtained from single receptors were comparable with the data from the cat, with the exception that no units were found that reacted to weak mechanical stimulation. However, since most receptors were found to respond to more than one type of stimulus (mechanical, chemical, thermal), they were considered not to be nociceptive, but polymodal. The high bradykinin sensitivity of group III and IV muscle receptors observed in the cat was also present in the dog. Many of the receptors responded not only to endogenous algesic agents, such as bradykinin, serotinin or potassium, but also to capsaicin, the pungent substance derived from the red pepper (Kaufman et al. 1982).

The slowly conducting muscle afferents in the monkey were found to behave similarly to those in the cat. Their chemical excitability was tested by Foreman et al. (1979b). The proportion of units being activated by endogenous pain-producing substances, the effective dosage, and the time course of the activations were almost identical to the data obtained with the same experimental set-up in the cat (cf. Mense 1977a). Thus, it appears that the general properties of group III and IV muscle receptors are comparable in the cat, dog, and monkey. It is probable that in the human being these receptors will behave in a similar manner.

The bulk of information about afferent activity in muscle nerves stems from studies on muscles involved in locomotion and maintenance of posture; the slowly conducting afferents from the diaphragm have been less extensively investigated. Whole nerve recordings in anaesthetized cats showed that definitely noxious stimulation, such as application of acetic acid to the abdominal surface of the diaphragm, activated group III and IV fibers, but not afferents with large action potentials (Gernandt 1946). Therefore, diaphragmatic pain was assumed to be due to afferent activity in fine fibers.

In the author's laboratory, an isolated rat phrenic nerve-diaphragm preparation in vitro was used to study systematically the discharges in single group III and IV receptors in response to defined chemical stimuli, such as increased levels of catecholamines (adrenaline and noradrenaline) and hypoxia (Mense et al. 1983). These stimuli were chosen since they have been discussed as possible factors influencing the excitability of receptors during periods of increased sympathetic activity (cf. Loewenstein 1956, Akoev 1981).

Upon mechanical stimulation, low- and high-threshold mechanosensitive endings could be distinguished among the slowly conducting afferent units. The former responded to gentle stroking of the diaphragm with a fine painter's brush. This stimulus produced only a small deformation of the tissue and was not considered to be noxious. Probably, movements of the stomach or gut against the diaphragm would exert a similar pressure in the intact animal. Other units required definitely noxious stimuli to be activated, i.e., the receptive field had to be squeezed with the blunt tip of a glass rod. The receptors responding to light stroking were considered to represent sensitive mechanoreceptors, those requiring noxious pressure, nociceptors. The von Frey thresholds of the latter were usually above 0.7 p, in some cases as high as 2 p. Routinely, a test for thermosensitivity was performed with injections of warm or cold fluid into the bath containing the diaphragm. Although some units displayed a certain thermosensitivity (cf. Fig. 12), they were not considered to be true thermoreceptors, since the magnitude of the response to strong thermal stimuli was small and a differential behavior during temperature changes was missing.

Adrenaline in concentrations of 0.1 – 1 mg/l (5.46×10^{-7} to $5.46 \times 10^{-6}\,M$/l) had a differential effect on mechanoreceptors and nociceptors: about half of the

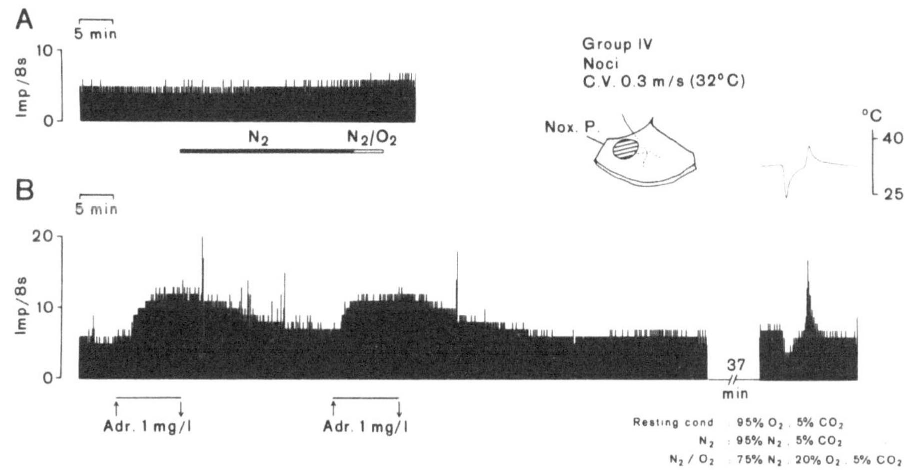

Fig. 12A,B. Response to adrenaline in a nociceptive group IV unit recorded from the rat phrenic nerve in vitro, conduction velocity 0.3 m/s at 32 °C. The unit required noxious mechanical stimulation to be activated; the location and size of the receptive field is shown in the *inset*. **A** Hypoxia produced by gassing the organ bath with 95% nitrogen and 5% carbon dioxide (*N_2*, *filled bar*) or 75% nitrogen, 20% oxygen, and 5% carbon dioxide (N_2/O_2, *open bar*) evoked only a slight increase in the background discharge. **B** During the periods indicated by the *arrows* and the *thin bar*, adrenaline (*Adr.*) 1 mg/l (corresponding to $5.46 \times 10^{-6}\,M$/l) was added to the perfusion solution. Strong temperature stimulation (fall to below 25 °C, rise to about 40 °C) led to relatively small changes in discharge frequency. (J. Kieschke and S. Mense unpublished)

nociceptors responded with an increase in discharge frequency (Fig. 12); whereas, sensitive mechanoreceptors did not show comparable reactions. The responses to adrenaline observed in nociceptive afferents were of small magnitude and had a slow time course. Similar effects of adrenaline were described for feline stretch receptors of the stomach, probably supplied by fine afferent fibers (Paintal 1954). Therefore, a sensitivity to adrenaline cannot be considered as being a special property of nociceptors (cf. also Paintal 1959, Eldred et al. 1960, Akoev 1981, Hunt et al. 1982, Passatore and Filippi 1982, 1983). On the other hand, the effects of adrenaline on nociceptive endings suggest that changes in the efferent sympathetic activity might modify the discharge frequency in nociceptive afferents and thus the magnitude of painful sensations.
Hypoxia (PO_2 of about 20 mmHg) had no clear effects on mechanosensitive and nociceptive units. In some receptors, the mechanical threshold determined with von Frey hairs was lowered during hypoxia. Possibly the loss of potassium, which has been found to occur in this preparation during hypoxia (Paul 1961), could account for this phenomenon.

Joint. In the single-fiber studies on receptors of the cat knee joint carried out by Burgess and Clark (1969) and Clark (1975), a few endings have been described as possessing a high threshold to local pressure and to bending of the joint. These receptors, which were classified as "weakly activated" or "nonactivated" by passive joint movements, were considered to be nociceptors; many of them had AF conducting in the group III range (group IV units were not recorded in these studies). In his report on the cat wrist joint, Tracey (1979) likewise described many receptors with afferent fibers conducting at less than 30 m/s which could not be excited by joint movements.
A systematic investigation of the response properties of single group III and IV afferent units from the cat knee joint performed by Schaible and Schmidt (1983b) yielded four categories of receptors: (a) units activated by nonnoxious movements; (b) units weakly activated by nonnoxious movement, but responding strongly to noxious movements; (c) units responding only to noxious movements; and (d) units not excitable by joint movements. The third type was considered to be a nociceptor. The nociceptors did not respond to nonnoxious movements, but fired constantly to noxious ones. Forceful rotations in the long axis of the tibia were found to be particularly effective stimuli (Fig. 13). Upon local mechanical stimulation, many group III and IV joint afferents had a high threshold in the tissue-threatening range (several grams with stimulating hairs of the von Frey type) and, thus, exhibited properties of nociceptors (Schaible and Schmidt 1983a). Generally, units possessing a high threshold to local pressure were also relatively insensitive to passive joint movements (Schaible and Schmidt 1983b). A certain correlation between conduction velocity and mechanical threshold was found insofar as group IV units more often had a high threshold than group III receptors. In this respect, a parallel to muscular group III and IV units exists (Mense 1977b).
Artificial inflammation of the knee joint by injection of kaolin and carrageenan led to a dramatic increase in the proportion of group III and IV receptors that could be activated by innocuous movements, namely, from 31% to 90% in group III, and 13% to 73% in group IV units (Coggeshall et al. 1983). An important

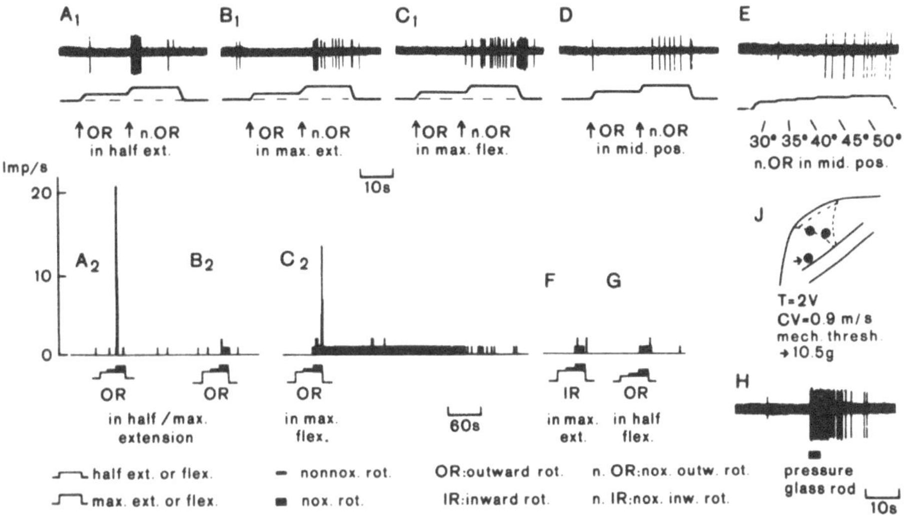

Fig. 13A–J. Responses of a group IV afferent unit of the cat medial articular nerve to passive movements of the knee joint. **A_1, B_1, C_1, D, E,** and **H** are specimen discharges; **A_2, B_2, C_2, F,** and **G** are peristimulus time histograms recorded while the indicated movements were performed (**A_2, B_2, C_2** correspond to **A_1, B_1, C_1**). Note the slower time base of histograms. The address advance rate was 1 s. This unit, activated only by noxious movements, had three receptive fields (**J**). The lowest von Frey threshold of 10.5 p was in the field marked by an *arrow*. The conduction velocity (*CV*) was 0.9 m/s, the electrical threshold (*T*) 2 V. Extreme rotations (**C_1, C_2**), as well as intense local stimulation (**H**), induced afterdischarges. (From Schaible and Schmidt 1983b)

finding was that, in an inflamed joint, receptors in the patellar region responded to innocuous movements. In the normal joint, units situated in this area required noxious movements to be activated or did not react at all (Schaible and Schmidt 1983b). This means that originally movement-insensitive units must have been sensitized. This assumption was supported by the finding that other signs of sensitization were also present, namely, an increase in the percentage of units possessing a background activity, an increase in the resting discharge frequency of individual units, and an enlargement in the size of the receptive fields upon local pressure stimulation (Coggeshall et al. 1983). A similar enhancement of the sensitivity of group III and IV articular receptors during adjuvant-induced arthritis has been reported in the rat (Guilbaud et al. 1984). The sensitization of nociceptors might account for the spontaneous pain and the pain during movement and weak mechanical stimulation of an inflamed or mechanically lesioned joint.

4.1.2 Effects of Activation of Deep Nociceptors on Spinal Neurons

Influence on Dorsal Horn Neurons. Up to now, only a few studies dealing specifically with the responses of dorsal horn neurons to discharges in fine deep afferents have been undertaken. In the cat, cells receiving input from muscular group III fibers were found mainly in lamina V (Pomeranz et al. 1968). The cells were not specific with regard to muscle input, since they also responded to discharges in cutaneous fibers. In the cited study, electrical stimulation of peripheral nerves was used to produce the input. Therefore, the receptor populations evoking the responses could not be specified. It could only be inferred from the high electrical threshold of the dorsal horn cells that they were not influenced by receptors supplied by thick AF, e.g., muscle spindles and tendon organs. Later studies, using both electrical and natural stimulation, showed that lamina I was an important region for nociception from deep tissues (Cervero et al. 1976, Cervero and Iggo 1980). Many of the lamina I cells were considered to be segmental interneurons (Cervero et al. 1979), but a considerable proportion – including units responding to algesic stimulation of skeletal muscle – was found to project to the thalamus, i.e., belonged to the STT (Craig and Kniffki 1982).

Some of the cells that are situated in or near lamina V and receive nociceptive input from deep tissues are probably cells of origin of the SCT (for a review of the SCT see Brown 1973). In tests with electrical stimulation of a muscle nerve, SCT cells have been shown to require relatively high intensities to be activated; this is an indication of the absence of an input via group I fibers to these cells (Hamann et al. 1978). An excitatory input from group III muscle AF onto SCT cells has been described by Lundberg and Oscarsson (1961) and Norrsell and Wolpow (1966). That cells of this pathway receive a powerful drive from small-diameter muscle afferents also has been shown with chemical algesic stimulation of muscle receptors (Hong et al. 1979). The neurons were found to be subject to a strong descending inhibition which could be reduced by cooling of the rostral spinal cord (Fig. 14).

Information on the functional properties of spinoreticular (SRT) cells in the cat is scarce. At least some of them appear to be influenced by afferent input from deep tissues (Fields et al. 1977, Maunz et al. 1978). The same is true for cells of the dorsal column postsynaptic system (Angaut-Petit 1975).

Data obtained from animals other than the cat have supported the above findings. In monkeys, volleys in muscle nerves were found to activate STT neurons, particularly at stimulus intensities exceeding the threshold of group II fibers (Foreman et al. 1979a). The field potential produced by muscular group III fibers was maximal in the neck of the dorsal horn. A complex convergence of nociceptive and nonnociceptive inputs from skin and muscle was noted in STT cells (Foreman et al. 1979b). Dorsal horn cells in the rat were found to form functional categories similar to those in the cat; one of these responded best to movement of joints or pressure on deep tissues (Menétrey et al. 1977). In this study, neurons responding specifically to noxious stimulation of deep tissues were not observed, but, in the rat, nociceptive cells were also mainly located in the marginal layer and in the neck of the dorsal horn. An interesting outcome of another study performed on rats was that the SRT contained a portion which carried information predominantly from high-threshold receptors in deep tissues. The cells of origin of this part of the SRT were situated in the dorsolateral funiculus rather than in the dorsal horn (Menétrey et al. 1980).

These physiological data fit well with histological findings obtained with intraaxonal iontophoresis of HRP. Nociceptive primary afferent units from the deep tissues of the cat tail or hindlimb invariably were found to terminate in laminae I and/or IV/V (see Sect. 2.3.2), i.e., in the same laminae in which nociceptive dorsal horn cells are located. This close spatial arrangement might be indicative

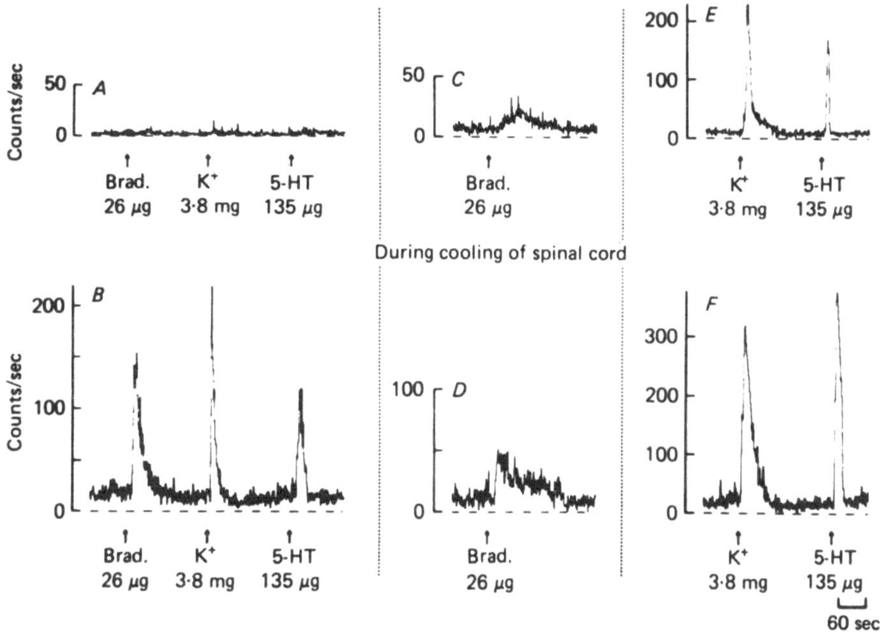

Fig. 14. Responses of three spinocervical tract neurons to algesic chemical stimulation of the GS muscle in decerebrate cats before cooling of spinal cord (**A**, **C**, **E**) and the responses of the same cells during cooling of spinal cord (**B**, **D**, **F**). The algesic agents (*Brad.*), potassium (K^+) and 5-hydroxytryptamine (*5-HT*) were injected into the arterial circulation of the GS muscle at the *arrows*. The histograms were constructed with a dwell time of 1 s; temperature of the cooling thermode, −1 °C. (From Hong et al. 1979)

of a monosynaptic connection between nociceptive primary afferents and projecting cells processing nociceptive information.

Elicitation of the Flexion Reflex. In this paragraph, only experiments using noxious stimulation for studying the reflex effects of fine deep afferents will be discussed. Data concerning more generally the influence of these units on motoneurons are presented in Sect. 4.5 (Locomotion).

Whereas, the occurrence of a flexion reflex upon activation of cutaneous nociceptors is a common experience, the elicitation of the reflex by deep nociceptors has been less well studied. Sherrington (1910) pointed out that a "nociceptive flexion-reflex" following noxious stimulation of deep tissues can be distinguished from a "locomotor flexion-reflex" as the result of weak, proprioceptive stimulation. O'Leary et al. (1935) were among the first to demonstrate that electrical stimulation of a muscle nerve strong enough to activate the small-diameter afferent fibers led to a reflex contraction of flexor and extensor muscle. The flexion reflex elicited by muscular group III afferents was shown to be mediated by a multineuron-arc by Lloyd (1943). The existence of a flexion reflex evoked by nociceptive muscle afferents was proven by Paintal (1961), who demonstrated that strong (presumably noxious) pressure stimulation of a skeletal muscle, or injection of noxious chemicals, leads to a facilitation of the flexion reflex via noci-

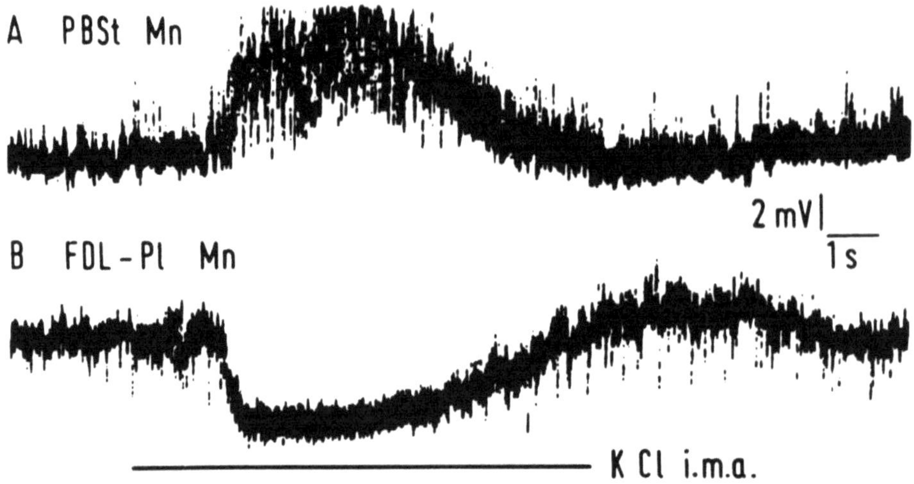

Fig. 15 A, B. Responses of a flexor and an extensor α-motoneuron (*Mn*) to intraarterial injection of KCl into the GS muscle in a spinal cat. Intracellular records of the membrane potential (depolarization upwards) of a posterior biceps-semitendinosus (*PBSt*) motoneuron (**A**) and a flexor digitorum and hallucis longus-plantaris (*FDL-Pl*) motoneuron (**B**). DC recording. The duration of the injection of 3.8 mg KCl (0.32 *M*) is marked by the *bar* below **B**. (From Kniffki et al. 1979)

ceptive group III (and probably group IV) muscle afferent units. Electrical stimulation of articular nerves of group III – IV strength has likewise been shown to elicit a flexion reflex (Gardner 1950).

Very little information is available concerning specifically the influence of deep group III and IV afferents on α-motoneurons. Kniffki et al. (1981) reported that in decerebrate and spinal cats most of the GS motoneurons were inhibited by noxious chemical stimulation of group III and IV GS afferents; whereas, flexor motoneurons were facilitated. In intracellular recordings, these effects were associated with the appearance of EPSPs and IPSPs, respectively (Fig. 15). The authors concluded that the reflex might be of importance not only for nocifensive motor reactions, but also for locomotion in general. As to fusimotor neurons, a strong input via group III and IV muscle afferents produced by intraarterial injection of algesic chemicals, has been shown to lead to an activation of both extensor and flexor γ-motoneurons in decerebrate cats (Hong et al. 1978). Muscular nociceptors can influence γ-motoneurons via long spinal reflex pathways. Thus, nociceptors of the neck muscles in the cat have been reported to affect the discharge of hindlimb (GS) γ-motoneurons (Ellaway and Murthy 1984). In awake, freely moving cats, the same stimuli elicited a clear flexion reflex pattern with excitation of ipsilateral flexor muscles, inhibition of ipsilateral extensors, and excitation of contralateral extensors (Schmidt et al. 1981).

4.1.3 Effects of Activation of Deep Nociceptors on Supraspinal Centers

Influence on Thalamic and Cortical Neurons. Little is known about the supraspinal processing of the information from fine deep afferents. In experiments with electrical stimulation of muscle nerves and recording of thalamic evoked potentials, Mallart (1968) found a sole input from high-threshold (group III) muscle afferents from the cat hindlimb to the ventrobasal complex (VB) and the medial thalamus; whereas, from the forelimb muscles, an additional input via group II fibers was present. The role of the VB in the processing of nociceptive information in the rat, cat, and monkey has been a matter of controversy for many years and the reader is referred to the literature dealing with this topic (Gaze and Gordon 1954, Perl and Whitlock 1961, Harris 1970, Berkley 1973, Mitchell and Hellon 1977, Guilbaud et al. 1980, Kenshalo et al. 1980, Perl 1984). Recently, nociceptive neurons driven by slowly conducting cutaneous and muscular afferent units have been found in the ventral and dorsolateral periphery of the ventroposterolateral nucleus (VPL) and in the transitional zone between VPL and ventrolateral nucleus of the cat (Figs. 16, 17) (Honda et al. 1983, Kniffki and Mizumura 1983), i.e., in one of the regions to which STT cells project in this species (Carstens and Trevino 1978, Craig and Burton 1979).

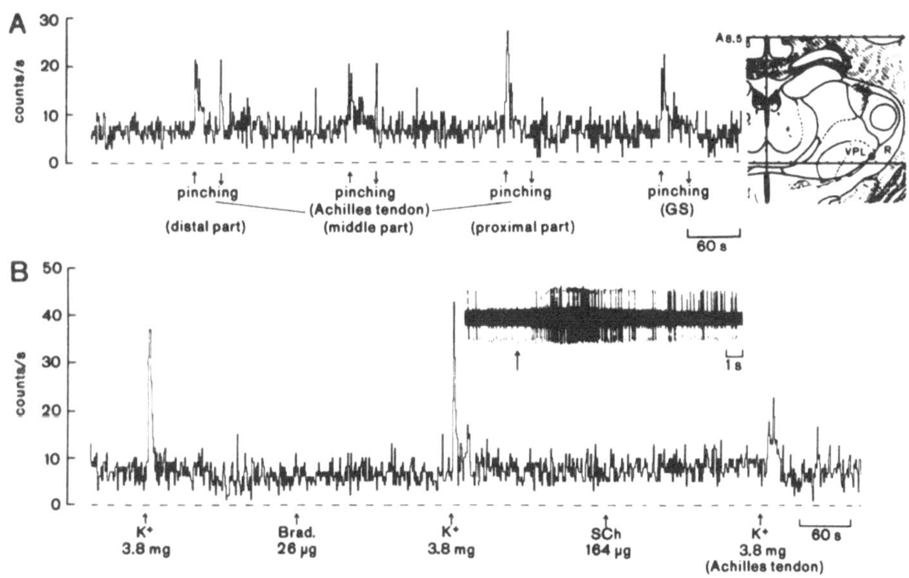

Fig. 16. Responses of a VPL neuron with musculotendinous input to noxious mechanical stimulation (**A**) and to algesic chemical stimulation (**B**). *Coordinates:* peristimulus time histogram of the neuronal activity (*ordinate*) with a time bin width of 1 s (*abscissa*). No cutaneous receptive field could be found for this unit, although it could be excited by electrical stimulation of the common peroneal and the sural nerves. The electrical threshold for GS nerve-evoked responses was 10 times the peripheral threshold (*T*). The thalamic neuron did not respond to bradykinin and succinylcholine or to innocuous pressing of muscle and tendon (not shown), but it was activated by strong pressure (pinching) and by injection of K^+ into the GS muscle and the Achilles tendon. *Arrows* mark the start and end of stimulation period. For the single injection (duration about 10 s) only the beginning is marked. Unless noted, chemicals were injected into the GS muscle. *Inset* in **B** shows a specimen record of extracellular action potentials during the response to the second potassium injection. The recording site was at the ventrolateral edge of the VPL at A 8.5 (*inset*, right side in **A**). (From Kniffki and Mizumura 1983)

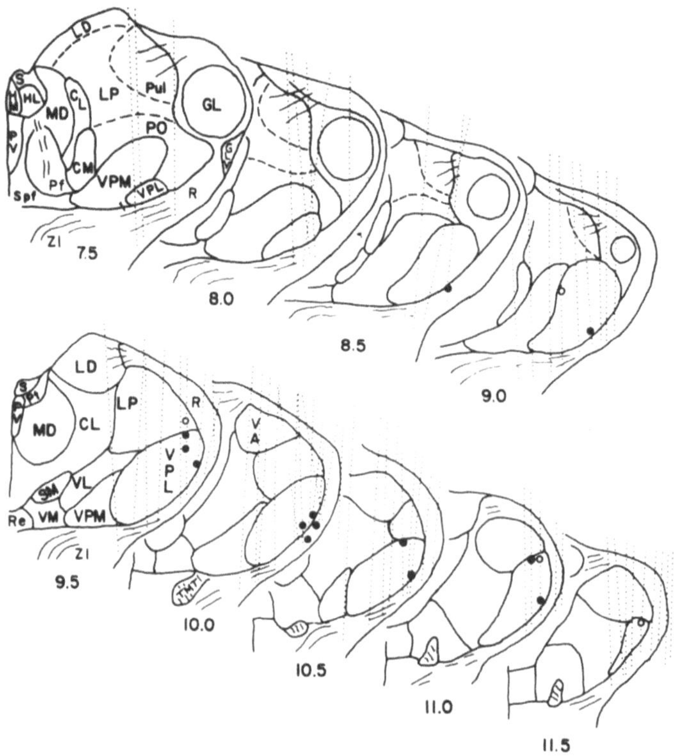

Fig. 17. Locations of recording sites of 17 HT units. *Filled circles* indicate recording sites of unitary discharges excitable from peripheral receptive fields: *open circles* indicate sites for those that could only be driven electrically from the contralateral sciatic nerve. The 65 penetrations from reconstructions of penetrations in 10 animals illustrate the extent to which the thalamus was explored. *Numbers* indicate the frontal plane anterior to the interaural line. *AV*, anterior ventral n.; *CL*, central lateral n.; *CM*, centrum medianum n.; *GL*, lateral geniculate n.; *GLv*, ventral part of lateral geniculate n.; *HM*, medial habenular n.; *HL*, lateral habenular n; *LD*, lateral dorsal n.; *LP*, lateral posterior n.; *MD*, medial dorsal n.; *Pf*, parafascicular n.; *Pt*, paratenial n.; *Pul*, pulvinar n.; *PV*, periventricular n.; *R*, reticular n.; *Re*, reuniens n.; *S*, stria terminalis; *Sm*, submedian n.; *Spf*, subparafascicular n.; *TMT*, mamillothalamic tract; *VA*, ventral anterior n; *VL*, ventral lateral n.; *VM*, ventral medial n.; *VPL*, ventral posterior lateral n.; *VPM*, ventral posterior medial n.; *ZI*, zona incerta. (From Honda et al. 1983)

In the VB of the rat, cells have been described as being specifically sensitive to noxious stimuli (Hellon and Mitchell 1975). These nociceptor-driven cells are intermingled with VB neurons processing tactile information. No morphological differences could be found between these functionally different neurons (Peschanski et al. 1984).

In arthritic rats, cells of the VB were reported to have unusual response characteristics with long afterdischarges and lowered mechanical thresholds which could be raised again by application of aspirin (Gautron and Guilbaud 1982, Iggo et al. 1984).

In human beings, an area ventral to the VB – the ventrocaudal parvocellular nucleus of Hassler – is considered to be specific for the mediation of pain, particularly for painful sensations elicited by activity in thin myelinated fibers (Hassler 1976). It is possible, however, that other nuclei of the thalamus (e.g., the posterior and medial nuclei) also process information from fine deep afferents, at least from nociceptive ones. Thus, the posterior nuclei in the cat have been found to contain cells with nociceptive properties (Guilbaud et al. 1977) and some of the units in the medial thalamus were reported as responding to noxious stimulation of deep tissues in this species (Dong et al. 1978). With anatomical methods, a projection of STT cells to the medial thalamus has been demonstrated in the cat (Boivie 1971, Craig and Burton 1981) and primate (Mehler 1969).
At the cortical level, neuronal responses to volleys in muscle nerves were found in the first somatosensory area contralaterally and the second bilaterally (Mountcastle et al. 1950). The authors pointed out that activity in group I fibers from deep tissues does not reach the cerebral cortex (this conclusion was corrected later, see McIntyre 1974); whereas, activity in group II and group III afferents does. This was taken as an indication that these cortical areas might be of importance for sensations of deep pain. The existence of an input from group III (high-threshold) muscle afferents to the primary and secondary somatosensory cortex in the cat was confirmed by Gardner and Haddad (1953), Oscarsson and Rosén (1963), and Norrsell and Wolpow (1966).
In a systematic search for cortical neurons driven by muscular group III and IV afferent units, Iwamura et al. (1981) applied noxious stimuli (strong pressure and injection of algesic substances) to the GS muscle in cats and recorded the impulse activity of single cortical cells. They found only a small fraction (less than 10%) of the units in area 3a, with input from the deep tissues of the contralateral hindlimb, to be responsive to these stimuli. Thus, up to now, animal experiments have not offered a clear answer to the old clinical question as to where in the cerebrum pain is consciously felt and whether a "cerebral pain center" exists. Penfield and Boldrey (1937) studied the effects of electrical stimulation of the pre- and postcentral gyri in awake patients; in only 11 out of 800 tests did the patients report the sensation of pain. Therefore, the authors arrived at the conclusion that "the sense of pain has little if any representation in the cortex" (cf. also Marshall 1951). However, recent investigations have shown that, in areas 3b and 1 of the monkey, neurons exist which may process nociceptive information from the skin (Kenshalo and Isensee 1983). Likewise, the rat somatosensory cortex has been demonstrated as receiving a strong nociceptive input: of 292 cells with peripheral receptive fields, 91 were found to respond to noxious stimulation; among these, 13 units were driven from deep tissues (Lamour et al. 1983). Since patients with thalamic lesions may have cutaneous analgesia, but still feel deep pain elicited by i.m. injection of hypertonic saline (Marshall 1951), it is possible that deep pain is processed differently from cutaneous pain. Recent experimental evidence obtained with neuromagnetic recording methods suggests that the anterior end of the secondary somatosensory cortex may be the integrative site for dental pain (Hari et al. 1983).

Pseudaffective Responses. In lightly anesthetized animals, the injection of bradykinin or other algesic agents into a somatic or visceral artery has been shown to

elicit a pseudaffective response (or reflex, cf. Woodworth and Sherrington 1904) which consists of vocalization, increase in blood pressure and heart rate, and hyperpnea (Moore and Moore 1933, Moore and Singleton 1933, Guzman et al. 1962, Taira et al. 1968). The response can also be evoked by electrical stimulation of the abdominal surface of the diaphragm (Pollock and Davis 1935).
Vocalization is considered to be most closely related to nociception; whereas, the concomitant cardiopulmonary (CP) reflexes can also be elicited by nonnociceptive afferents (cf. Sect. 4.2.3). Since vocalization depends upon the integrity of the thalamus and, thus, involves a highly specialized nervous center, it has been assumed to be the most specific sign of pain in animals (Guzman et al. 1962). The afferent limb of the reflex arc for chemically induced pseudaffective responses was supposed to be formed by afferent units that are closely associated with small blood vessels, particularly those in deep tissues (Moore and Moore 1933, Moore and Singleton 1933, Moore et al. 1934). However, at that time the fiber types mediating the responses were unknown.
In light of more recent results from experiments with chemical stimulation of muscle afferents and single-fiber recordings (Mense and Schmidt 1974, Kumazawa and Mizumura 1977, Mense 1977a, Foreman et al. 1979b), it is obvious that activation of group III and IV muscle receptors must have been the main cause of the pseudaffective response, since the thicker afferent units are not excited or they show variable and inconsistent reactions to algesic chemical stimulation. This assumption is supported by experiments using electrical stimulation of peripheral nerves and recordings of somatosympathetic or CP reflexes. A wealth of data shows that activation of small-diameter (electrically high-threshold) afferent fibers from muscle is required to produce the reflexes. In these experiments, it is not possible to differentiate between CP reflexes elicited by nociceptive afferents and those due to activation of nonnociceptive fibers of equal diameter and, hence, electrical threshold. The actions of an increased activity in fine deep afferents on CP centers in general will be discussed in Sect. 4.2.3, together with their possible significance for the capacity of a living organism to perform physical work.

4.1.4 Deep Pain

Muscular Pain. Using the newly developed method of electrical stimulation of nerve fascicles in awake human beings, Torebjörk et al. (1984a, b) were able to demonstrate that the occurrence of deep pain is dependent on the activation of small-diameter afferent fibers. The pain sensations were projected distally into the innervation area of the stimulated fascicle and were often associated with referred pain in muscles or other deep tissues proximal to the site of stimulation. Evidence was obtained showing that the referred pain is elicited by afferent activity in muscular group III (and possibly also group IV) fibers. Similar to cutaneous pain, deep pain has been reported as possessing two components corresponding to activation of group III and group IV fibers, respectively, but from deep tissues both types of afferents were said to elicit the same subjective sensation of dull, aching pain (Landau and Bishop 1953). On the other hand, observations have been made suggesting that deep pain exhibits different qualities, namely, localized pain from fascia and tendons and diffuse, referred pain from muscle (Kellgren 1938b).

A clinical pain syndrome with a more complex pathogenesis is that of cold pain arising from deep tissues after exposure of an extremity to normally nonpainful cold (about 20 °C) (Kellgren et al. 1948). Apparently, the syndrome requires both a hyperalgesic state and a reduction in blood flow; in some cases, blocking of sympathetic efferent fibers was found to relieve the pain.

Intramuscular Injections of Algesic Substances. Both local muscle pain and referred pain could be elicited with i.m. injections of hypertonic (5% – 6%) saline; the evoked sensations of referred pain were similar to those associated with diseases of visceral organs (Kellgren 1938b, Lewis and Kellgren 1939 – 1942, Lewis 1942). In clinical cases of painful muscle disorders, areas of referred pain central to the affected muscle can likewise be found (Kellgren 1938a).

A systematic study of the effectiveness of various endogenous substances to produce pain upon i.m. injection was performed by Lindahl (1970). He tested hypo- and hypertonic solutions, high concentrations of potassium ions, and solutions of high and low hydrogen ion concentrations. The conclusion drawn from these experiments was that hydrogen ions were the only factor eliciting muscle pain in concentrations which are likely to occur in living organism. Maison (1939) presented evidence that lactate and ammonium ions in physiological concentrations are ineffective in producing muscular pain.

Ischemic Contractions. Ischemia of a resting extremity, even of long duration (about 20 min), is known not to produce pain (Lewis et al. 1931, Lewis 1942) and not to evoke cardiovascular (CV) reflexes (Staunton et al. 1964). However, if the muscle is forced to contract rhythmically under ischemic conditions, pain develops within about 1 min (Lewis et al. 1931, Lewis 1932, Harpuder and Stein 1943). The mechanisms underlying this type of ischemic pain, which is the leading symptom in intermittent claudication, are still a matter of controversy. Accumulation of acidic metabolites (Moore et al. 1934) or potassium ions (Harpuder and Stein 1943) was considered to be a causal factor, but the weak effectiveness of the substances to excite group III and IV muscle receptors and to evoke pain (see Sect. 4.2.2, Maison 1939) speak against this assumption.

An interesting observation is that patients suffering from severe anemia experience muscle pain during rhythmic contractions without occlusion of the blood supply. They also show symptoms of intermittent claudication and/or angina pectoris, although no pathological alterations of the blood vessels can be detected. Therefore, not the accumulation, but the lack of oxidation of metabolic products, was considered to cause ischemic pain (Pickering and Wayne 1933 – 1934). Lactate appears not to be of importance for the occurrence of ischemic pain since patients with McArdle's disease, who cannot produce lactate owing to an enzyme deficiency, tend to have particularly strong anginal pain (Rodbard 1975).

Another substance possibly involved in the production of ischemic pain is bradykinin. It is supposed to be released in the blood during ischemia (Sicuteri et al. 1964) and, thus, might contribute to the pain of intermittent claudication. This assumption is supported by the finding that application of a proteinase inhibitor, which prevents the release of bradykinin from its precursor molecule kallidin, prolongs the distance patients with intermittent claudication are able to walk without pain (Digiesi et al. 1975).

Of course, combinations of factors have also to be taken into consideration. As early as 1931, Lewis and coworkers proposed a physicochemical mechanism as the cause of ischemic muscle pain and labeled the hypothetical pain-producing stimulus "factor P". A possible mechanism would be that chemical agents released from muscle or blood sensitize muscular nociceptors which then respond to the chemical and/or mechanical stimuli present in an ischemically contracting muscle. A sensitization of nociceptors during ischemic work is suggested by the observation that under these conditions the muscle becomes tender upon mechanical stimulation (Lewis et al. 1931). The time course of the responses of group IV nociceptors during ischemic contractions speaks in favor of this assumption, since it appeared to be dependent on both the duration of the exercise period (determining the degree of chemical alterations) and the force of contraction (Mense and Stahnke 1983).

In cases of traumatic lesion or inflammation of skeletal muscle and tendons, similar sensitizing mechanisms might exist leading to a lowered threshold of nociceptors and, thus, provoking pain upon stimulation at intensities which are normally innocuous (pressure, movement). It has been shown that prostaglandins of the E type are especially effective in sensitizing nociceptors (Juan 1978) and in enhancing the algesic action of bradykinin (Lembeck et al. 1976).

Cramp and Soreness. A cramp has been defined as an "involuntary painful contraction of voluntary muscles" (Joekes 1982). EMG activity has been reported to be increased during cramp (Layzer 1981, Lanari et al. 1973); as possible reasons for the increased electrical activity, excessive activation of α- and γ-motoneurons by descending impulses, decreased inhibition of spinal neurons, and impulse generation in the motoaxon itself, have been proposed (Layzer and Rowland 1971). An interesting hypothesis has been put forward by Lundberg (1979), who stated that a positive feedback between motoneurons and flexion reflex afferents activated by movements might lead to cramps.

Cramps are most easily provoked if the muscle is contracted unstretched, i.e., with a minimal distance between its insertion points, so that only slight tension builds up within the tendons. This suggests that lack of activation of tendon organs, which are known to inhibit homonymous motoneurons (Granit 1950, Eccles et al. 1957), may promote the development of cramps (Lanari et al. 1973, Weiner and Weiner 1980).

Muscle soreness or tenderness is a common experience after hard and unaccustomed exercise, but its origin is still unclear. It is characterized by pain at rest or during movement and palpation, and typically occurs with a delay of more than 10 h after the end of the exercise. Owing to this long delay, it is very unlikely that an increased lactate concentration is a causal factor for the pain since, after exhausting work, the muscle lactate decreases exponentially with a half-time of 9.5 min (Sahlin et al. 1976). This means that within about 1 h the lactate concentration will be close to normal. However, the accumulation of metabolic substances and/or the development of high osmotic tension might be the cause of the pain and soreness which occurs during tetanic contractions of long duration.

In contrast to the long-latency soreness, this type of pain subsides within 3 – 4 h (Hough 1902, Staton 1951).

Several theories have been proposed as an explanation for muscle soreness. The "torn-fiber-theory" of Hough (1902) states that, during exercise, muscle fibers and fibers of the connective tissue are ruptured by mechanical stress owing to overloading and/or poor coordination between neighboring fibers during contraction. This could lead to an aseptic inflammation (Staton 1951) in the affected muscle. The importance of mechanical overloading versus chemical factors was emphasized by Asmussen (1956), who found that negative work leads to soreness, whereas positive work of the same intensity does not. This finding was explained by assuming that, during negative work, a smaller number of motor units is active, so that the mechanical stress for the single unit is higher. Results from animal experiments have fully supported this view by showing that exhausting exercise is followed by necrosis of muscle fibers (Schumann 1972). The muscular injury was found to be greater after eccentric exercise than after concentric contractions (Armstrong et al. 1983). DeVries (1966) developed the "spasm theory" which is based on his findings that increased electrical activity is present in a tender muscle. As with cramps, static stretching was found to be of therapeutic value, reducing the EMG activity and relieving the pain. Recently, the current hypotheses have been summarized in German by Wietoska and Böning (1979). One of the more likely mechanisms appears to be that, during unaccustomed exercise, mechanical damage is inflicted on muscular or connective tissue which leads to repair processes in the sense of an aseptic inflammation with local swelling, sensitization of nociceptors and, possibly, reflex contraction of the affected region. This local contraction could impair the blood flow and maintain the painful state.

Local muscle hardenings associated with deep tenderness, from which fasciculations and referred pain can be elicited by local pressure, were labeled "trigger points" by Simons and Travell (Simons 1975, 1976, Simons and Travell 1981). Again, the pain could be relieved by static stretching of the affected muscle. The stretching was supposed to have a local action within the muscle by removing the overlap between actin and myosin, which reduces the energy consumption and, thus, leads to a normalization of the local metabolism. The muscle fibers of the trigger point are electrically silent; therefore, a rupture of the sarcoplasmic reticulum with release of the stored calcium was assumed to be the cause of the local contraction. The possible mechanisms leading to the formation of trigger points and their treatment have been published recently in extenso (Travell and Simons 1983).

For the pain of both cramps and tenderness, no evidence is available as to which receptors mediate the pain. A cramp appears to differ from a maximal voluntary contraction in that only portions of the cramping muscle contract (Layzer and Rowland 1971). This may result in shearing forces within the muscle which could excite mechanonociceptors. The tenderness or soreness of a muscle after strenuous exercise might be due to a chemical sensitization of nociceptors, caused by local release of endogenous substances such as metabolites, prostaglandins, or bradykinin during the repair process. Thus, the evidence suggesting that group III and IV receptors of muscle and tendon may be involved is only indirect.

Tenderness of bone may occur in cases of vitamin D deficiency with ensuing osteomalacic myopathy. The pain is mainly present during movement; it is thought to be due to traction of the muscle on the bone at the site of myotendinous insertions (Morgan-Hughes 1979).
As to painful sensations from the diaphragm, information is scarce. The problems start with the identification of the diaphragm as a skeletal muscle or a visceral organ (Woollard et al. 1932, Pollock and Davis 1935). A possible explanation might be that it consists of two muscles, the peripheral (costal) portion being derived from thoracic muscles, the central (crural) one from the esophagus (De Troyer et al. 1981).
In clinical experience, the diaphragm appears to behave more like a viscus, since stimulation of the phrenic nerve in awake human beings has been reported to elicit predominantly referred pain in the shoulder (Woollard et al. 1932, Lewis 1942). Pollock and Davis (1935) proposed a complicated chain of neurones, including the phrenic nerve, sympathetic efferent fibers and fibers of "the ordinary cerebrospinal system", as the pathway for diaphragmatic pain. On the other hand, diaphragmatic nociception persisted in the experiments of Hinsey and Phillips (1940) after cutting of all connections to the vagosympathetic system. Therefore, the phrenic nerve is the most probable pathway for afferent impulses mediating diaphragmatic pain. As to the afferent fibers supplying nociceptors in the diaphragm, it is apparent from the results of Gernandt (1946) and Mense et al. (1983) that group III and IV fibers are the only elements in the phrenic nerve which are capable of transmitting information about noxious stimulation of the diaphragm to the central nervous system (CNS).

Articular Pain. Pain with origin in joints being situated directly underneath the skin, such as the knee, has been reported to be well localized, while pain from more deeply placed joints, such as the hip, is diffuse and often referred to remote areas (Lewis 1942). Thus, the situation appears to be similar to muscular pain where stimulation of more superficial parts – the fascia – gives rise to localized pain and pain from the muscle itself is diffuse and referred (Kellgren 1938b, Lewis 1942). Differences in innervation density might be of importance for this phenomenon (Feindel et al. 1948). The pain-sensitive structures of a joint have been explored when opened under local anesthesia or through cannulae. The articular surfaces were found to be insensitive to noxious mechanical stimulation and to cautery, but the synovialis close to ligaments proved to be very sensitive to scratching and to hypertonic saline (Lewis 1942).
From experiments with recordings of single articular afferent units in cats (Coggeshall et al. 1983, Schaible and Schmidt 1983a, b) and rats (Iggo et al. 1984), it can be inferred that the pain of mechanical trauma and of inflammation is most probably mediated by nociceptors having group III and IV afferent fibers. These receptors can be directly activated by a traumatic lesion or can be sensitized in the course of an inflammation, so that normally innocuous pressure stimuli or movements become painful.

Reflex Sympathetic Dystrophies. Some forms of chronic deep pain (e.g., Sudeck's syndrome) show signs of an involvement of sympathetic efferent activity, in that the painful state is associated with disturbances in sweat production and local blood supply and can be abolished or ameliorated by a sympathectomy or blocking of the sympathetic trunk. It was assumed that the underlying mechanism is a reflex activation of sympathetic postganglionic fibers by afferent (nociceptive) activity originating in a peripheral trauma, mostly a nerve lesion. The efferent activity was thought to cause dystrophic changes in the affected area, thus maintaining the painful state. Therefore, this clinical syndrome was called reflex sympathetic dystrophy or sympathetic algodystrophy (cf. Bonica 1979, Nathan 1980). The participation of catecholamines has been suggested by Wallin et al. (1976), who showed that topical application of noradrenaline to the skin in patients suffering from causalgia reproduced the causalgic pain.

In animal experiments, in which the afferent activity in single fibers from a neuroma was recorded, injection of small quantities of adrenaline, or electrical stimulation of the sympathetic trunk, led to an increase in afferent activity (Devor and Jänig 1981, Scadding 1981, Devor 1983). All these findings point to the possibility that afferent activity is influenced or induced by efferent sympathetic impulses. This is the case not only in neuromas but also in normal tissue, as has been shown by the results of the experiments with recordings from single diaphragmatic afferents and from muscle spindles (see Sect. 4.1.1). In chronic dystrophic tissues, these influences may be enhanced owing to the concomitant disturbances in blood supply with ensuing alterations of the chemical environment, possibly leading to a sensitization of nociceptors. Another hypothesis is that areas of demyelination or growing nerve sprouts form sites of ephaptic impulse transmission, or become sensitive to catecholamines, so that transmitter release from sympathetic efferent fibers is followed by discharges in damaged afferent fibers (Seltzer and Devor 1979, Wiesenfeld-Hallin and Hallin 1984).

4.1.5 Conclusions

The presented data leave little doubt that a considerable proportion of the fine deep afferents have nociceptive properties. Only a few of the nociceptive endings are supplied by group II fibers (cf. Mense et al. 1981), and there is no evidence supporting the assumption that muscle spindles or tendon organs are involved in the mediation of deep pain, as was suggested by Schmidt (1916) and Matthews (1933). Therefore, all types of deep pain with origin in the peripheral tissues will be associated with an increased activity in small-diameter primary afferents. These units are the only ones that have the proper functional characteristics and central nervous connections to evoke pain and induce nocifensive reactions.

Whether different types of nociceptors exist in the deep tissues is still uncertain. However, there are some differences in the response behavior between group III and group IV nociceptors. For instance, group III units in skeletal muscle respond more vigorously to noxious mechanical stimuli and to bradykinin; whereas, strong activations during ischemic contractions were only observed among group IV receptors. Thus, it is possible that the pain of intermittent claudication is mainly due to an afferent inflow via group IV fibers.

4.2 Ergoreception

The term ergoreception will be used for all peripheral neuronal mechanisms capable of informing the CNS about the amount of work performed by a muscle (cf. Kao 1963) and leading to circulatory and respiratory adjustments during physical exercise.

4.2.1 Receptors Responding to Muscle Contractions

Several authors have described muscular group III and IV receptors that were activated by muscular contractions (Paintal 1960, Kniffki et al. 1978, Ellaway et al. 1982). In the context of this article (cf. Mense and Meyer 1985), units were considered to be contraction-sensitive only if contractions of moderate force were the most effective of all the innocuous stimuli tested (other receptors showed weak responses to contractions, but were strongly excited by light, local pressure; these units were considered as constituting a different type, see below). Many of the contraction-sensitive units were found to be also sensitive to stretch, although with a threshold by far exceeding that of muscle spindles and tendon

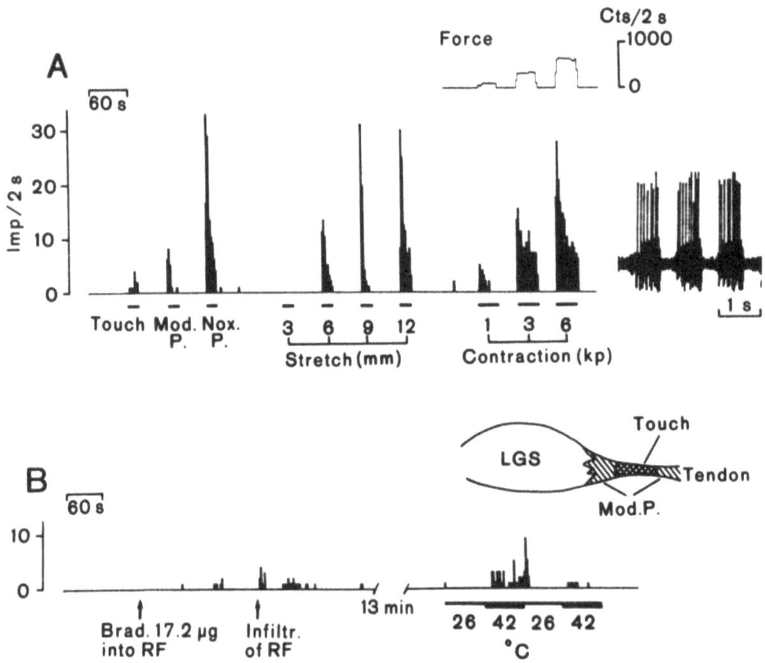

Fig. 18 A, B. Contraction-sensitive unit from the cat skeletal muscle with group III afferent fiber, conduction velocity 26 m/s. Location of the receptive field on the calcaneal tendon. Labeling of stimulations as in Fig. 10. The original registration shows the fiber's activity during rhythmic contractions with a force of 6 kp. The response to warming of the tendon in **B** was not reproducible and, therefore, not considered to indicate thermoreceptive properties of the unit. (From Mense and Meyer 1985)

organs (Fig. 18) (Kaufman et al. 1983, Mense and Stahnke 1983). These units differed from the low-threshold pressure-sensitive ones by their higher threshold to local pressure. They required noxious pressure to produce responses of a magnitude similar to those elicited by contractions. Thus, it appears that these units are sensitive predominantly to forces acting in the direction of the long axis of the muscle. Most of them also responded to bradykinin. It is unlikely, though, that they are nociceptive, since their magnitude of response was unaffected or decreased if the contractions were made ischemic (Mense and Stahnke 1983), i.e., they are not able to differentiate between normal and painful contractions.

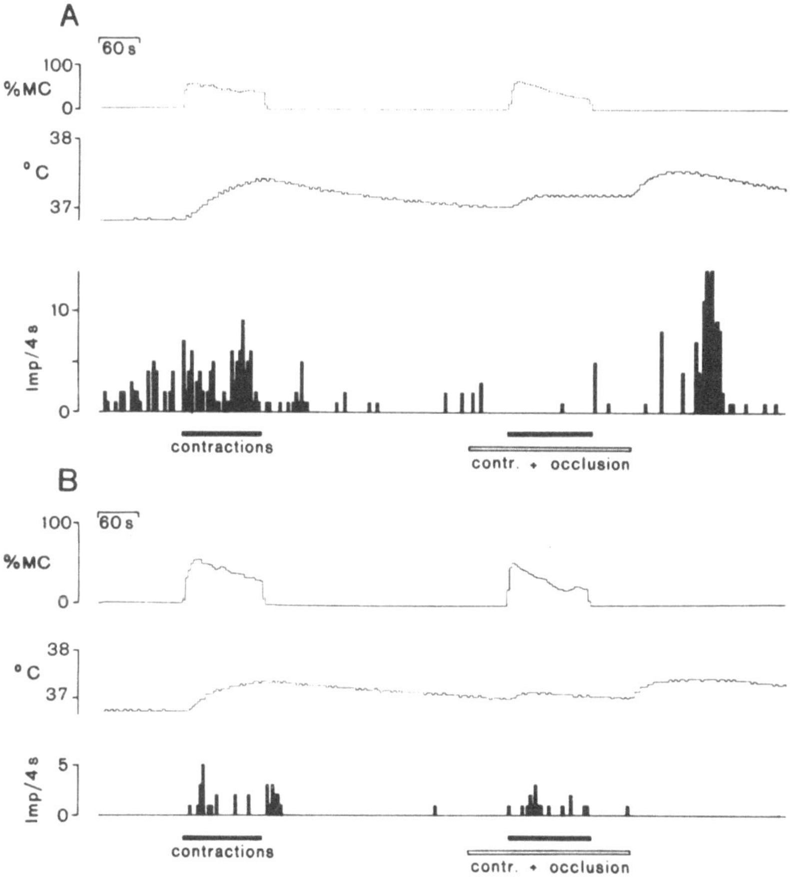

Fig. 19 A, B. Contraction-sensitive receptors with presumably nonmechanical mechanisms of activation. The *upper trace* shows the force developed by the muscle in percent of maximal contraction (*MC*); the *middle trace* is the intramuscular temperature; and the *lower trace* the histogram of the fiber activity. The *filled bars* underneath the histograms indicate the periods of rhythmic contractions; occlusion of the popliteal artery (resulting in ischemic contractions) is marked by *open bars*. **A** Group III unit possibly reacting to the thermal changes associated with the contraction, *c.v.* 6.6 m/s. **B** Group IV receptor possibly activated by chemical factors, *c.v.* 1.1 m/s. Neither ending responded to stretching the muscle. Bin width of histograms, 4 s. (From Mense and Stahnke 1983)

About 20% of the group III and IV units from the cat GS muscle which were tested with rhythmic contractions belonged to this type (Mense and Meyer 1985). Most of the contraction-sensitive group III receptors showed a sudden onset of their response, a slow or missing adaptation during repetitive contractions, no afterdischarges, and a graded response to increasing forces of contraction. Therefore, it is likely that their mechanism of activation is mechanical (Mense and Stahnke 1983), although an additional sensitivity to metabolites cannot be ruled out (Kaufman et al. 1983). In contrast, most of the group IV units that were activated by muscular contractions had a delayed response followed by afterdischarges. This time course of the response is suggestive of a chemical (or combined mechanochemical) mechanism of activation (Fig. 19B). A few units were

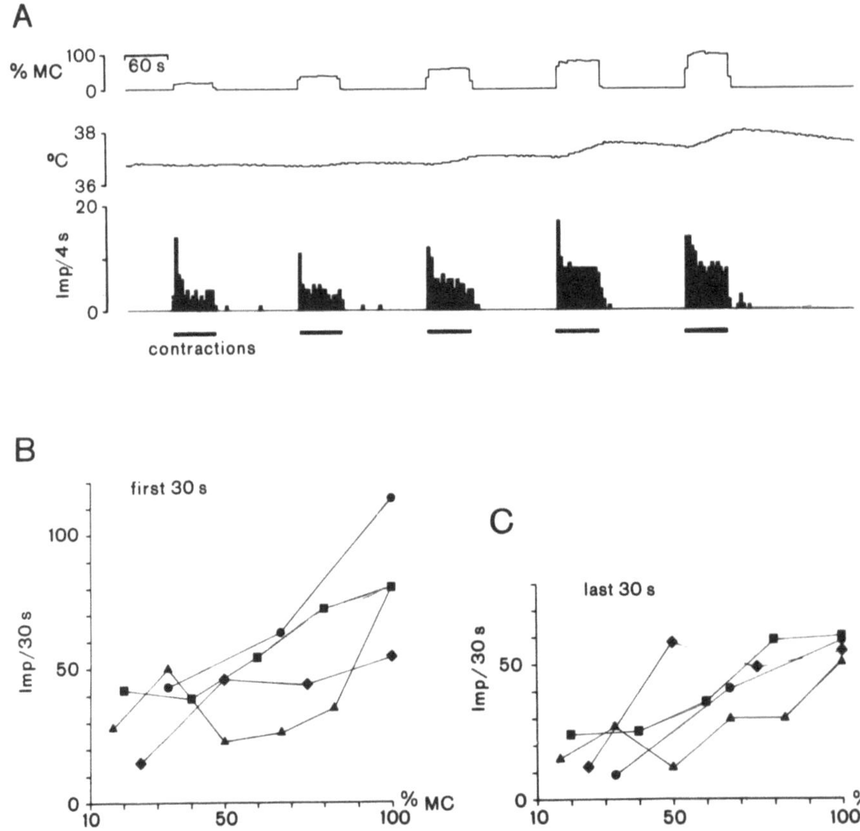

Fig. 20A–C. Responses of contraction-sensitive group III receptors to rhythmic exercise of graded force. **A** The force of contraction was adjusted to approximately 20%, 40%, 60%, 80%, 100% of maximal contraction (*upper trace*, from left to right). *Middle trace* shows intramuscular temperature; *lower trace* histogram of fiber activity. Bin width of histogram 4 s. **B, C** Stimulus-response relations of four group III receptors tested with exercise periods of 60 s duration as shown in **A**. The sum of impulses elicited during the first (**B**) or last (**C**) 30 s of the working period is plotted on the ordinate against relative muscle force (% maximal contraction) on the *abscissa*. (From Mense and Stahnke 1983)

found which behaved as if thermal changes might contribute to their excitation (Fig. 19A). The group III and IV muscle receptors of the dog, termed "polymodal" by Kumazawa and Mizumura (1976, 1977), responded to increases in temperature and osmotic pressure as well as to endogenous agents (potassium, histamine, acetylcholine, bradykinin), some of which are released in a contracting muscle. The polymodal receptors were thought to participate in exercise hyperpnea, although an activation of these receptors during contractions has not yet been demonstrated.

The majority of the contraction-sensitive receptors of the cat GS muscle were located near the proximal insertion of the muscle or in the calcaneal tendon and, thus, showed a spatial distribution similar to the other types of group III and IV muscle receptors (Mense and Meyer 1985).

During static (sustained) contractions, the proportion of group III and IV receptors being activated was found to be much higher (63% and 68%, respectively) (Kaufman et al. 1983) than during rhythmic (intermittent) contractions (about 20%, see above). The reason for this difference may be that static contractions are likely to impair the blood supply to the muscle, thus leading to an accumulation of metabolites which may contribute to the excitation of the receptors during exercise. The contraction-sensitive receptors are good candidates for the "ergoreceptors" postulated by Kao (1963) and other authors. Owing to their graded response to contractions of up to 100% maximal force, the probably mechanically driven units should be particularly suited to signal the amount of work performed by the muscle (Fig. 20).

To what extent joint afferents are activated during muscular contractions has not yet been tested. As some of these units are sensitive to mechanical deformation of the joint capsule, produced by local pressure or joint movement (see below), it is likely that their discharge rate increases during physical exercise (cf. Coote 1975). Electrical stimulation of joint afferents has been shown to elicit CP reflexes of small amplitude (Barron and Coote 1973, Sato et al. 1983). Thus, the existence of ergoreceptive endings in joint nerves cannot be excluded.

4.2.2 Possible Receptive Mechanisms

The mechanism(s) underlying the activation of muscular group III and IV receptors during muscular contractions is (are) still unknown. Probably, at least two different types of contraction-sensitive receptors are present in skeletal muscle, namely, one being activated by the mechanical forces of a contraction and another one responding to factors other than mechanical (Mense and Stahnke 1983). While the evidence for a mechanically excited "ergoreceptor" appears to be clear-cut – mechanical sensitivity to stretch and local pressure, instantaneous response, and no afterdischarge during rhythmic contractions – the possible factors activating the other type of contraction-sensitive receptors are numerous. Of course, chemical alterations in the contracting muscle were considered as being the principal stimulus, but up to now all attempts to identify the chemical substance have failed. Hník et al. (1969) studied the responses of a limited number of group III muscle afferents to intraarterial infusions of potassium. They found that these afferents were activated by potassium concentrations

which are likely to occur during muscular exercise (cf. Hník et al. 1976). Therefore, the units were considered to be candidates for the hypothetical receptors mediating CP adjustments during exercise.
In the few investigations in which the chemical excitability of contraction-sensitive muscle receptors was tested, only a limited number of metabolic products was studied. The stimulating effect of potassium ions on these receptors was found to be rather unspecific, since other muscle receptors with slowly conducting afferent fibers were activated in a similar proportion and with a similar response magnitude (Mense 1978). The same is true for bradykinin and capsaicin. The latter substance stimulated predominantly group IV receptors, but again both contraction-sensitive and contraction-insensitive units were excited (Kaufman et al. 1983). Activation of group IV muscle receptors by intraarterial injections of lactate and phosphate was tried by Kniffki et al. (1978). The metabolites were found to be effective only in high concentrations which are unlikely to occur under physiological conditions. The low pH of the phosphate injections, and the high osmotic pressure of other stimulating solutions likewise did not constitute an effective stimulus. Thus, among the metabolic products tested so far, none can be considered a specific stimulant for contraction-sensitive muscle receptors. The temperature increase in a working muscle might also be a factor for the activation of ergoreceptors. In fact, thermosensitive muscle receptors have been found to exist, particularly among the group IV units (Hertel et al. 1976, Mense and Meyer 1985). However, of 34 contraction-sensitive group III and IV muscle receptors in a recent study (Mense and Stahnke 1983) only four behaved as if they were influenced by muscle temperature. Therefore, the heat produced by a contracting muscle appears to be of minor importance for the excitation of contraction-sensitive receptors. It should be kept in mind, however, that in a working muscle, all the factors discussed above change simultaneously. Thus, the failure to identify an adequate chemical stimulus for ergoreceptors might be due to the fact that, under physiological conditions, a certain combination of factors is required for activating these receptors. Another possibility could be that the real stimulant for chemically activated ergoreceptors has not yet been found. Among the substances released during exercise and, therefore, possibly of significance for the activation of muscle receptors (besides being involved in the functional vasodilatation) are histamine (Anrep and Barsoum 1935), catecholamines (Hutter and Loewenstein 1955), kinins (Webster et al. 1967), adenosine (Bockman et al. 1976, Tominaga et al. 1980), adenine nucleotides and inosine monophosphate (Sahlin et al. 1978), prostaglandins (Young and Sparks 1980), acetate (Steffen et al. 1982), and ammonia (Banister et al. 1983).

4.2.3 Elicitation of Cardiopulmonary Reflexes

Electrical Stimulation of Peripheral Nerves. O'Leary et al. (1935) reported that electrical stimulation of a muscle nerve in the cat at two times threshold strength led to a slight fall in blood pressure; whereas, stimulation at 4 – 5 times threshold strength was followed by an increase in blood pressure, probably due to the activation of fine muscle afferents (2- to 6-μm diameter). Electrical stimulation of the central ends of the cut phrenic nerves yielded similar results, low intensity stimulation leading to depressor and high intensity stimulation to pressor responses (Kohrman et al. 1947). Changes in respiration in cats and rabbits upon phrenic nerve stimulation occurred only at high stimulation strengths (Grandjean 1943). An input via muscular group IV fibers proved to be particularly effective in evoking excitatory CV reflexes (Laporte et al. 1960, Johansson 1962, Sato et al. 1981); whereas, stimulation of presumably all group I fibers in a dorsal root was ineffective (Laporte et al. 1962). Additional evidence for an involvement of small-diameter afferents in the reflexes was obtained in experiments employing cold blockage of the large-diameter fibers (Tibes 1977). The effects of group III fiber activation on blood pressure were found to be dependent on the frequency of stimulation, a low frequency leading to a fall and a high frequency to a rise in blood pressure (Fernandez et al. 1953, Laporte et al. 1960). However, depressor responses upon high-frequency stimulation of muscular group III afferents were also observed (Johansson 1962). The importance of the frequency of stimulation for the occurrence of CV reflexes had already been pointed out by Colle and Gybels (1957), who added the observation that a fall in blood pressure upon low-frequency stimulation of small-diameter afferent fibers was present in anesthetized, but absent in awake animals.

While all studies with electrical stimulation of peripheral nerves agree that group I fibers do not have an excitatory effect on CV centers, the situation may be different for respiratory reflexes. An increase in ventilation upon activation of group I muscle afferents was observed by Bessou et al. (1959) and Carcassi et al. (1984), although in these investigations an admixture of effects elicited by group II fibers cannot be excluded. Experiments with recordings of the impulse activity of respiratory neurones in the brain stem yielded similar evidence for the existence of a facilitatory influence of group I muscle afferents on inspiratory cells (Koizumi et al. 1961).

Interesting results with regard to a possible dissociation of the afferent fibers mediating a rise in blood pressure and those eliciting an increase in heart rate were obtained by Longhurst et al. (1980). They were able to demonstrate that the afferent information leading to the pressor response was conducted via both dorsal and ventral roots, whereas afferent fibers eliciting an increase in heart rate were only present in the dorsal roots.

A functional part of the CP adjustments are somatosympathetic reflexes which were reported as possessing spinal (Beacham and Perl 1964) and supraspinal components (Sell et al. 1958, Koizumi et al. 1968). They can be elicited by electrically induced volleys in cutaneous, articular, and muscular nerves (Sato and Schmidt 1966, 1973, Sato et al. 1983). A consistent finding was that activation of group I muscle afferents did not evoke such reflexes (Sato et al. 1969, Schmidt and Schönfuss 1970, Sato and Schmidt 1973) and often also a group II fiber volley in a muscle nerve proved to be ineffective (Sato et al. 1981). Group III and IV muscle afferents were found to be powerful in eliciting sympathetic reflexes, but while group IV fibers consistently had an excitatory influence on sympathetic efferent neurons (Coote and Pérez-González 1970, Sato and Schmidt 1973), the effects of group III fiber activation from skeletal muscle were variable. Coote and Pérez-González (1970) reported that stimulation of electrically low threshold group III muscle afferents inhibited sympathetic neurons, while an input via electrically high-threshold group III fibers excited the sympathetic cells. The authors concluded from their results that the group III and IV muscle receptors were responsible for the mediation of the pressor reflex during exercise. Sato et al. (1981) produced tachycardiac or bradycardiac responses with stimulation of group III muscle afferents. The direction of the effect appeared to depend on the experimental situation, but the factors involved could not be determined. The variability of the effects was attributed to the fact that group III fibers supply different kinds of muscle receptors which are activated in different proportions in different experiments. This assumption is in strong agreement with the results of single-fiber studies presented in this review (cf. Figs. 10, 18, 22). In general, the results of the studies on somatosympathetic reflexes support the hypothesis that among the muscle afferents the group III and IV fibers are the ones mediating CV adjustments during exercise.

Injection of Chemical Stimulants. One of the first substances injected or infused intraarterially to elicit CP reflexes was potassium (Wildenthal et al. 1968, Liu et al. 1969). In comparison with the extracellular potassium concentration measured in a contracting muscle (8 – 9 m*M*/l, cf. Hník et al. 1976), the stimulating solutions used contained a much higher concentration, namely, about 50 – 3500 m*M*/l. This raises the question whether the hypertonicity of the solution and/or activation of nociceptors contributed to the observed reflexes (increase in blood pressure, heart rate, and ventilation). Both Wildenthal et al. (1968) and Liu et al. (1969) stated that nociceptors were most probably not involved, although the higher doses applied have been shown to elicit pseudaffective responses (Guzman et al. 1962) and to activate presumably nociceptive muscle receptors (Kniffki et al. 1978).

Another powerful stimulus for evoking CP reflexes is capsaicin. Capsaicin-sensitive muscle receptors were considered to form the afferent limb of the reflex arc mediating circulatory adjustments during exercise (Webb-Peploe et al. 1972). In a recent study, in which the stimulant was injected into a vascularly isolated hindlimb of the dog, capsaicin application was shown to evoke an increase in blood pressure, heart rate, cardiac output, and respiratory minute volume (Crayton et al. 1981) and, thus, mimicked many aspects of the reflexes during exercise. It should be kept in mind, however, that capsaicin was reported to activate predominantly group IV muscle receptors (Kaufman et al. 1983). Therefore, the method of capsaicin injection does not seem to be suitable for studying reflex effects caused by activation of group III receptors.

The stimulating effect of bradykinin on respiratory reflexes in the dog was compared with other substances by Mizumura and Kumazawa (1976), who found 4.5% NaCl to be superior to equal injection volumes of bradykinin (1 μg/ml) and KCl (60 m*M*/l). In these experiments, no excitatory actions of group I muscle afferents on respiratory functions could be detected. Following application of stronger chemical (or electrical) stimuli, a sequence of respiratory facilitation and suppression was observed, the latter lasting for several minutes (Kumazawa et al. 1983). In rabbits, i.a. injection of bradykinin was reported to elicit hypotension and bradycardia in low doses, but hypertension and tachycardia in high doses (Tallarida et al. 1979). In contrast, with hypertonic solutions (NaCl, glucose) only the latter reflex pattern could be evoked. The authors put forward the hypothesis that in skeletal muscle two functionally distinct types of receptors are present which are differently sensitive to chemical substances and elicit different patterns of CV reflexes. In their study on the effects of chemical algesic stimulation on heart rate and blood pressure, Sato et al. (1982) arrived at a similar conclusion, namely, that the various responses observed (pressor, depressor, acceleration, deceleration) were due to the activation of different receptor populations by the different substances.

Muscle Warming and Mechanical Stimulation. Warming a whole extremity to about 42 °C has been reported as reflexly increasing ventilation, but if the temperature rise was restricted to the muscles themselves the rate of ventilation did not change significantly (Morgan et al. 1955). Therefore, muscle temperature was not considered to be an effective respiratory stimulus (Kao 1963).
Manually squeezing a hindlimb muscle proved to be a potent stimulus for eliciting hyperpnea in dogs. It was inferred from indirect evidence that group III and group IV muscle receptors were stimulated in these experiments (Kao et al. 1963, Parida et al. 1969, Kalia et al. 1972).
Most attempts to elicit CV reflexes by longitudinal vibration of the GS muscles in decerebrate cats failed (Hodgson and Matthews 1968, McCloskey et al. 1972). This result was taken as an indication that an input via Ia fibers – which are vigorously excited by this stimulus – does not activate respiratory or circulatory centers. However, reflex increases in ventilation during muscle vibration have also been reported (Leitner and Dejours 1971). In the latter study, the GS muscles of both hindlimbs were vibrated in order to produce a stronger input to the CNS. The reflex muscle contractions evoked by the vibration are a problem for the interpretation of the data, since the contraction-sensitive group III and IV receptors could contribute to the observed effect.
Passive movements of an extremity are likewise effective in eliciting hyperventilation, probably due to activation of receptors in skeletal muscle or tendon (Grandpierre et al. 1951). The effect was found always to be smaller than that of active movements of an identical amplitude and frequency. Joint afferents appear to contribute to the CP reflexes during passive movements, since electrical stimulation of a joint nerve evoked similar reflexes (Barron and Coote 1973, Sato et al. 1983). In comparison with the strong effect of muscle afferents, the influence of joint receptors on CP centers appears to be small and is often not present in anesthetized animals (Gardner and Jacobs 1948). The CV reflexes during passive movements were supposed to originate in mechanosensitive group III and IV joint receptors, according to Coote (1975).

Muscle Contractions. After it had been established that CP reflexes could be elicited by electrically induced muscular work in anesthetized animals (Kao and Ray 1954a, b), the first data concerning exercise hyperpnea indicated that the reflex could be brought about either by centrally acting metabolic substances produced by a contracting muscle or by neuronal activity originating in joint receptors. These receptors were supposed to be the only ones mediating the reflex, since it was abolished after denervation or local anesthesia of the knee joint. Species differences were reported as existing in that the chemical mechanism was present in cats, but absent in dogs (Comroe and Schmidt 1943). An elegant cross-circulation technique was used by Kao and coworkers in order to separate neurally mediated reflexes from those induced by centrally acting substances released from the contracting muscle into the circulation. It was found that both mechanisms influence cardiac output and ventilation in dogs during exercise (Kao and Ray 1954b, Kao et al. 1963). The peripheral muscle receptors inducing exercise hyperpnea were not sensitive to the well-known respiratory stimuli (increase in PCO_2, decrease in PO_2 and pH) and were different from muscular stretch receptors. They were labeled "ergoreceptors" by Iaria et al. (1959). Exercise hyperpnea was assumed to be primarily due to activation of these receptors and, to a lesser degree, caused by metabolic factors (mainly CO_2) acting on chemosensitive structures in the head (Kao 1963). The chemical control of breathing during exercise has been emphasized by Cross et al. (1982), who showed that transection of the spinal cord did not abolish the steady-state ventilatory response.

As to the afferent fiber types evoking the CP reflexes, Coote et al. (1969) concluded from indirect evidence that group III and IV receptors might be involved. Direct experimental support for this assumption was obtained by McCloskey and Mitchell (1972) who showed that a conduction block of the small-diameter afferent fibers abolished the reflex, whereas blocking of the large-diameter afferent fibers did not (Fig. 21). These results were confirmed by Kalia et al. (1981), who added histological data which were interpreted as indicating that ergoreceptive

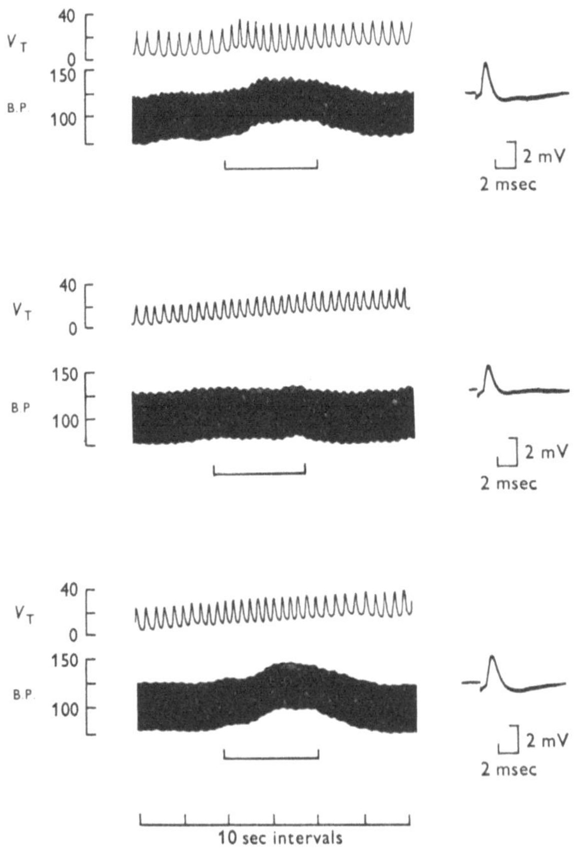

Fig. 21. Records of tidal volume (V_T), arterial blood pressure (B.P.), and dorsal root compound action potential from three periods of isometric hindlimb exercise in a chloralose-anesthetized cat. From top downwards are shown a control period of exercise; a period of exercise which commenced $3\frac{1}{2}$ min after application of a few drops of 0.125% lignocaine solution to the dorsal roots. Note that this period of exercise produced no pressor or ventilatory response, although the A-wave of the compound action potential was little, if at all, reduced; and in the *bottom* set of records, a further control period of exercise begun some minutes after the lignocaine had been washed away with warm saline. Ventilation and pressure are retouched U.V. records. (From McCloskey and Mitchell 1972)

afferent fibers from the GS muscle project to the nucleus tractus solitarii, which in turn is connected to CV regions of the brain stem. However, as the anterograde transport of i.m. injected HRP was used to trace the connections of the primary afferents, nothing can be said about the type of afferents stained in these experiments.

Interestingly, the pressor reflex during static contractions was found to be mainly due to an increase in stroke volume and left ventricular contractility with small changes in peripheral vascular resistance. It, thus, showed a totally different pattern in comparison with the baroreceptor-induced pressor reflex which results predominantly from an increased systemic resistance (Fisher and Nutter 1974). In fact, the baroreceptor reflex was shown to be depressed (Coote and Dodds 1976) or even turned off during exercise (McRitchie et al. 1976).

Concerning circulatory and respiratory adjustments to exercise in man, Paterson (1928) pointed out that the changes appeared to be reflex in nature. It soon became apparent that the reflex reponses varied, depending on the type of exercise performed. Sustained (static) contractions were found to lead to increases in heart rate, cardiac output, and blood pressure, while rhythmic (dynamic) exercise was accompanied by an increase in heart rate and cardiac output, but little or no change in systemic blood pressure (Lind and McNicol 1967). This finding was supported by the observation that sustained handgrip movements influenced mainly the blood pressure and rhythmic contractions mainly the heart rate. Moreover, if a vascular occlusion was maintained at the end of an exercise period, the heart rate dropped rapidly, whereas the blood pressure remained at an elevated level (Cunningham et al. 1972). Thus, different mechanisms for the control of the various parameters of the CV reflexes might exist. This assumption is supported by the finding that muscle ischemia preceded by exercise can raise the blood pressure without affecting the ventilation (Rowell et al. 1976). The ventilation itself is differentially influenced by different types of contractions, dynamic exercise leading to an increase in respiration frequency and static exercise to an increase in the depth of breathing (Tallarida et al. 1983). The situation is further complicated by the fact that differences in the reflex pattern may exist depending on the muscle type involved (fast-twitch or slow-twitch, cf. Mitchell et al. 1983) and that descending influences from higher centers modulate respiration and circulation in awake human beings (Goodwin et al. 1972a, Schibye et al. 1981). In any case, the amplitude of CP reflexes was found to be graded and increased with the amount of work performed (Lind et al. 1964, Staunton et al. 1964, Lind and McNicol 1967, McCloskey and Streatfeild 1975).

As to the peripheral factors activating the "ergoreceptors", some discrepancies exist in the literature. Alam and Smirk (1937, 1938) favored the accumulation of metabolites in a working muscle as the cause for the circulatory reflexes. The local H^+-concentration or PCO_2 were named by Stegemann (1963) as likely stimuli. Both mechanical and chemical factors have been discussed by Asmussen and Nielsen (1964). In his study on the control of respiration during exercise, Dejours (1964) described a fast and slow component of the regulation and pointed out that the fast component must not necessarily be elicited by mechanoreceptors, but could be due to activation of chemosensitive muscle receptors, since changes of the chemical milieu can occur very rapidly. On the other hand,

the available experimental evidence supporting the hypothesis of muscular chemoreceptors was considered not to be convincing.
Other authors concluded that muscular chemoreceptors may well exist, since muscle ischemia was found to influence strongly the time course of blood pressure changes during exercise (Rowell et al. 1976). As a logical consequence, persons who do not do much physical exercise should exhibit greater changes in the chemical milieu of the muscle and, therefore, have CP reflexes of greater amplitude. However, no unequivocal experimental evidence proving or disproving this assumption appears to be available (Rowell 1974).
With regard to the afferent fibers mediating CP reflexes in man, some authors assumed that fibers from muscle spindles (Asmussen et al. 1965) or tendon organs (Paulev 1973) might be involved, but the bulk of the data points to group III and IV muscle receptors as the afferent limb of the reflex arc (for a more comprehensive review see Mitchell et al. 1983, Mitchell and Schmidt 1983). Reviews dealing specifically with the regulation of blood pressure (Ludbrook 1983) and ventilation (Whipp 1983) during exercise have appeared recently.

4.2.4 Conclusions

Although some evidence exists that group I muscle afferents might be involved in the regulation of respiration, the bulk of the data clearly indicates that fine deep afferents have to be activated in order to evoke CP reflexes. Single unit studies show that a considerable proportion of the muscle receptors with small-diameter afferent fibers is activated by contractions and, thus, could fulfill the function of ergoreceptors. At least two different types of contraction-sensitive receptors appear to exist, one responding without delay (probably mechanically excited), the other one having a delayed response (probably chemically excited). The former might contribute to the fast component of CP adjustments at the beginning of physical work, the latter could maintain the reflex adjustments during prolonged exercise.

4.3 Mechanoreception

4.3.1 Low-Threshold Mechanosensitive Receptors

Skeletal Muscle. Some group III and IV receptors of the cat GS muscle have been found to possess such a low mechanical threshold that they react readily to the tissue being touched with a painter's brush (Fig. 22; cf. also Mense and Meyer 1985). In the skeletal muscle of the dog, such units have not been observed (Kumazawa and Mizumura 1977). As can be seen from Fig. 22, these receptors were quite insensitive to stretch and contractions. (The liminal responses during the onset of stretch and rhythmic contractions were due to the movement and shaking of the receptive field rather than to the stretch or contraction itself; a similar effect could be produced by tapping the muscle outside the receptive field.) In contrast to the contraction-sensitive receptors (cf. Fig. 18), these endings appeared to respond predominantly to forces acting perpendicular to the surface of the muscle or tendon; therefore, they were labeled low-threshold pres-

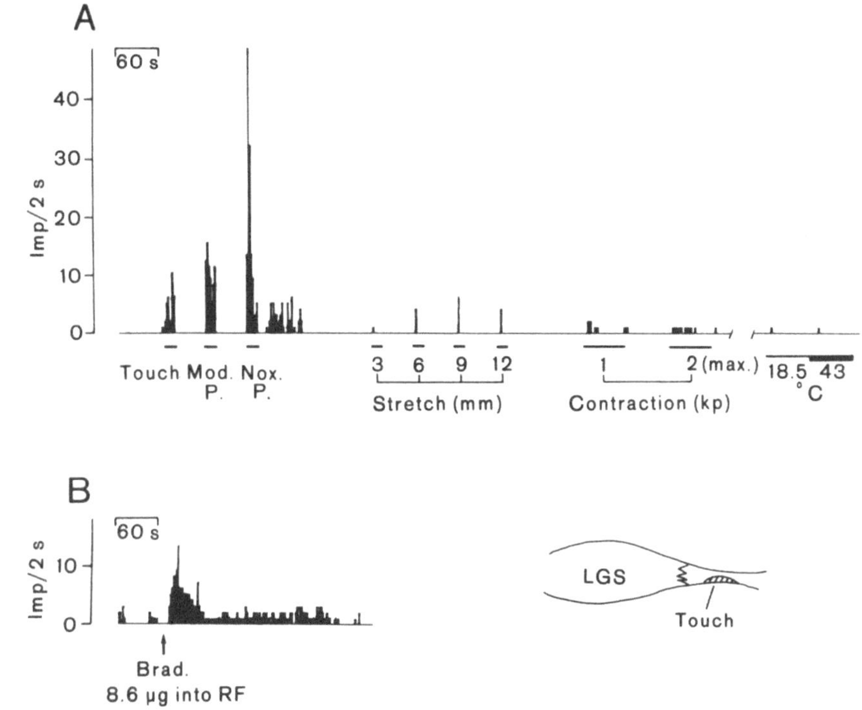

Fig. 22 A, B. Low-threshold pressure-sensitive unit from the cat tendon with group III AF, conduction velocity 9 m/s. Location of the receptive field was on the lateral edge of the calcaneal tendon. Labeling of stimulation procedures as in Fig. 10. Note that even strong stretch did not elicit a clear response, while slightly touching the receptive field did evoke a response. (From Mense and Meyer 1985)

sure-sensitive receptors (Franz and Mense 1975, Mense and Meyer 1985). The receptive fields of the units were often situated superficially in the connective tissue surrounding the muscle or tendon; this might explain their relative insensitivity to stretch and contractions, since forces which build up within the muscle or tendon would only weakly distort the surrounding connective tissue. Many of the units subsumed under this category required an innocuous deformation (moderate pressure) of the muscle or tendon to be activated. In these cases, it was not clear whether the units were less sensitive to mechanical stimulation or whether the endings were situated deeper within the muscle. For a few endings, evidence supporting the latter assumption was obtained, as they did not respond to compression of superficial layers of the muscle with a fine forceps.

Many of the low-threshold pressure-sensitive receptors could also be activated by bradykinin (cf. Fig. 22). In spite of their bradykinin sensitivity, these units were not considered to be nociceptors, since they were clearly activated by weak mechanical stimuli (touch and moderate pressure). In contrast to nociceptors, some of which have been described as showing liminal responses upon nonnoxious stimulation (cf. Perl 1984), the low-threshold pressure-sensitive receptors reached a considerable percentage of their maximal discharge rate (20% – 40%) during nonnoxious deformation of the receptive field.

The activation induced by bradykinin was considered to be the result of an inadequate stimulation, i.e., due to the action of a strong chemical stimulus on a sensitive mechanoreceptive ending. Similar reactions have also been observed in slowly adapting low-threshold mechanoreceptors of the skin (Fjällbrant and Iggo 1961, Beck and Handwerker 1974).

The proportion of low-threshold pressure-sensitive endings among group III and IV receptors in the GS muscle of the cat was found to be about 30% (Mense and Meyer 1985).

In the diaphragm of the rat, low-threshold pressure-sensitive receptors with group III and IV AF were likewise found (Kieschke and Mense, unpublished results). They gave clear responses to light stroking with a fine painter's brush and in some cases had von Frey threshold below 0.1 p. In contrast to muscle spindles, they could not be activated by slightly stretching the diaphragm.

Joint. The response properties of mechanically low-threshold joint receptors with thick myelinated afferent fibers have been described by several authors (Boyd and Roberts 1953, Skoglund 1956, Burgess and Clark 1969, Clark 1975, Millar 1975, Grigg 1976). Whereas, the earlier works emphasized a role of these endings in signaling joint position and movement, it later became clear that most of the units were active only in the extremes of the joint positions and/or responded to both extension and flexion. Thus, they appeared not to be suitable for signaling unequivocally the joint angle over the whole working range of a joint (Burgess and Clark 1969, Clark 1975, Millar 1975, Grigg 1976). However, there may be differences between the properties of articular afferents in different joints. Tracey (1979) reported that wrist joint receptors in the cat responded to (extreme) flexion, but not to extension, as do most of the knee joint receptors.

In the first studies that included recordings from group III joint afferents, the units were described as being either weakly activated or not activated by weak local pressure or movements in the physiological range (Burgess and Clark 1969, Tracey 1979) and were, therefore, considered to be nociceptors (Clark 1975; cf. Sect. 4.1.1). Among the 41 group III afferents from the cat knee joint sampled by Clark (1975), only eight appeared to have a low mechanical threshold since they responded to passive nonnoxious movements.

In their systematic study of group III and IV afferents from the feline knee joint, Schaible and Schmidt (1983a) found many units that responded to weak local pressure stimulation with von Frey hairs. The mechanical threshold of these receptors, which were particularly numerous among group III units, was in the same range as that of group II articular afferents. Therefore, this type of unit was assumed not to fulfill a nociceptive function. Again, upon stimulation with passive joint movements, low-threshold units were found (Schaible and Schmidt 1983b) which showed clear responses to nonnoxious movements. Typically, these receptors were activated by more than one movement, i.e., they were sensitive to flexion and extension and also to twisting of the tibia. Most of the units did not show a distinct phasic response during changes of joint angle (Fig. 23). Owing to these response characteristics, it appears unlikely that receptors of this type could be involved in position and movement sense. Movement-sensitive receptors of

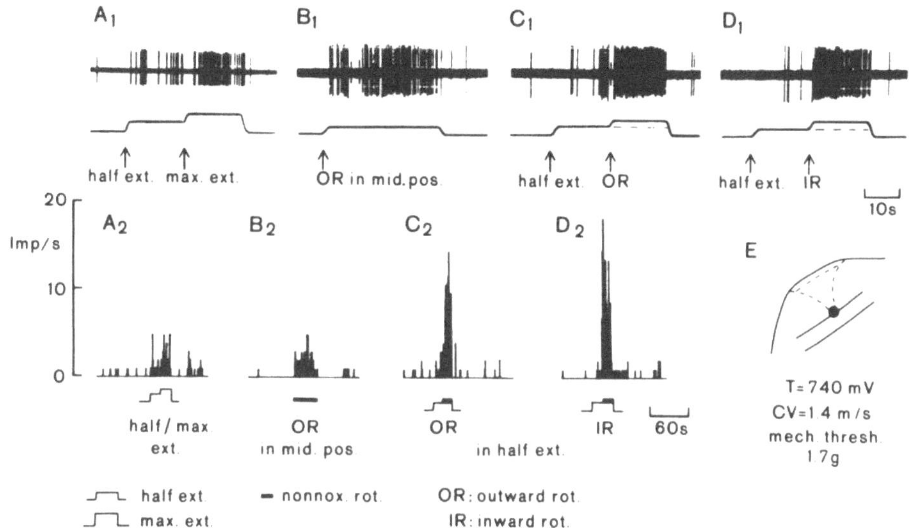

Fig. 23A–E. Responses of a group IV afferent unit of the cat medial articular nerve to passive movements of the knee joint recorded and displayed as in Fig. 13. All rotations were in the nonnoxius range. The receptive field of the unit is shown in **E.** The von Frey threshold was 1.7 p. (From Schaible and Schmidt 1983b)

this kind were rare among group IV units (13%), but quite common among group III units (31%). The general function assigned by the authors to the group III and IV joint receptors is that of a warning system providing information about the limits of the normal working range of a joint. Thus, the function of these endings appears to be similar to that of joint receptors supplied by group II fibers, with the exception that nociceptive units have not been described as existing among the latter.

The existence of a separate system of mechanically low-threshold afferent units in the wall of veins has been described by Thompson et al. (1982). These units have a conduction velocity in the group II range; as a possible function, the elicitation of reflex muscle contractions upon distension of the vein has been suggested.

4.3.2 Central Nervous Effects of Raised Activity in Deep Low-Threshold Mechanoreceptors

No studies dealing specifically with the input from deep sensitive mechanoreceptors with fine afferent fibers to central nervous centers have been performed as yet. In the few investigations that mention (marginally) the processing of information from deep tissues, the effects of deep pressure and joint movements have been mostly described together, so that no distinction can be made between cells responding to mechanoreceptive or proprioceptive input (cf. Sect. 4.4).

At the spinal level, 11% of monkey STT cells were found to be fired by joint movement or by manipulation of muscles. Most of them were located in laminae

V and VII/VIII (Willis et al. 1974). Possibly, the proportion of cells reported as responding to deep input would be greater if manipulation of joints and muscles had been used as a searching stimulus more often. In the monkey, as in the cat, the information from deep low-threshold mechanoreceptors that respond to movement of joints or compression of muscles appears to be processed in laminae ventral to the neck of the dorsal horn (Wagman and Price 1969, Applebaum et al. 1979). The location in the ventrolateral spinal cord of STT axons with input from deep tissues differs from that of other STT axons in that deep STT fibers form a separate band along the ventral surface of the cord (Applebaum et al. 1975).

At the thalamic level, cells responding to deep pressure stimulation or joint movement have been reported to be situated in the dorsal and rostral portions of the VB of the cat and monkey, but no distinction could be made between input via thick and fine primary afferent fibers (Mountcastle et al. 1963, Poggio and Mountcastle 1963, Maendly et al. 1981). Recently, a separate shell of neurons, situated anterodorsal to the core of the VB and responsive only to stimulation of deep tissues, has been described in the monkey (Jones and Friedman 1982). In man, neuronal responses to muscle contraction and joint movements were recorded in an area corresponding to the nucleus ventralis intermedius (Hardy et al. 1980).

As has been pointed out in Sect. 4.1.3, an input from group III muscle afferents to the cerebral cortex of the cat has been shown to exist in experiments using electrical stimulation of muscle nerves. Light mechanical stimulation of a skinned extremity also proved effective in producing cortical neuronal responses. In a study with single cell recordings, Mountcastle (1957) found 152 out of 685 cells in the cat somatosensory cortex to be drivable from deep tissues. Of these, the majority responded to joint movement, the rest were sensitive to pressure stimulation of deep tissues. No input from muscular stretch receptors to cortical cells could be detected. This latter observation was confirmed by Levitt and Levitt (1968), who discussed a possible input via group III fibers to the cortical cells driven by deep innocuous pressure and joint movements. In the monkey's postcentral gyrus, similar cells were observed exhibiting a high sensitivity to joint position and movement or to slight pressure directed to deep tissues (Mountcastle and Powell 1959).

However, the existence of a short-latency projection from group I muscle afferents to the cerebral cortex of the cat, monkey, and man (mainly to a region corresponding to area 3a) has been demonstrated by other authors (Amassian and Berlin 1958, Oscarsson and Rosén 1963, Landgren and Silfvenius 1969, Phillips et al. 1971, Gandevia et al. 1984). Neurons in this area responded to movements of joints, pulling on tendons or kneading of muscles (Maendly et al. 1981, Jones and Friedman 1982) and, thus, showed properties which might be of importance for mediating proprioceptive sensations (cf. Sect. 4.4). Another region processing the information from low-threshold deep receptors was found to be cortical area 2 in the monkey (Mountcastle and Powell 1959, Dreyer et al. 1974). In this area, cortical neurons were encountered that responded to the slightest mechanical distortion of deep tissues without being sensitive to muscle stretch (Mountcastle and Powell 1959). These findings suggest that in deep tissues low-threshold mechanosensitive receptors exist which are different from muscle spindles and tendon organs and are capable of activating cells of the postcentral gyrus.

As to the subjective sensations possibly elicited by activation of deep mechanoreceptors with group III and IV afferent fibers, it can only be speculated that they might mediate deep pressure sensations. In the older literature, the existence of such nonpainful sensations is a matter of controversy. For instance, Schmidt (1916) stated that innocuous pressure applied to an exposed muscle is not perceived at all, while Goldscheider (1920) described such sensations as occurring in patients upon slight tapping of the periosteum. Experiments on human beings, in which the nerves to the skin of the forearm were blocked or cut in order to study the subjective sensations upon local deep pressure, yielded results that support the assumption that a deep pressure sense exists (Head et al. 1905, Kellgren and McGowan 1948).

Interestingly, the function of mediating deep pressure sensations has also been ascribed to the articular receptors with group II afferent fibers that respond only in the extremes of a joint position (Clark 1975, Burgess et al. 1982) or are sensitive to innocuous local pressure (Grigg et al. 1982).

4.4 Proprioception

The term proprioception will be used for the movement sense (kinesthesia in the strict sense) and position sense (cf. Goodwin 1976). In addition, the sensation of force or tension will be dealt with under this heading. As early as 1914, von Frey suggested that the force sense depends upon signals coming from muscle and tendon receptors, and that this sense is used to assess the resistance against which a movement is performed.

As to the receptors mediating movement and position sense, the first studies were performed on thick myelinated afferents from joints. The results were thus interpreted: slowly adapting joint receptors (e.g., Ruffini and Golgi endings) could provide information about the relative position and slow movements of the bones forming a joint, while rapidly adapting units (e.g., Vater-Pacini corpuscles) could signal changes of the joint angle (Boyd and Roberts 1953, Skoglund 1956). In Skoglund's article it had already been made clear that many of the joint receptors responded best to extreme flexion or extension, with little sensitivity to intermediate positions (cf. also Burgess and Clark 1969). Similar observations were made in a single-fiber study on afferent units from the feline elbow joint (Millar 1975). (Recently, however, the existence of afferent units signaling midrange positions has been emphasized for the cat elbow joint (Baxendale and Ferrell 1983).) Since many of the joint receptors responded only to extreme joint positions, and/or were sensitive to both flexion and extension, many authors arrived at the conclusion that the slowly adapting receptors may have little to do with position sense. This view was supported by Clark (1975) and by Grigg (1976), who concluded that "the sense of position in the knee must depend upon receptors other than joint afferents". An observation strongly supporting this assumption is that after total replacement of a joint with a prosthesis, patients usually retain the sense of position in the affected joint (Cross and McCloskey 1973, Grigg et al. 1973, Barrack et al. 1983).

In his review, Goodwin (1976) stated that receptors in various tissues surrounding and forming a joint (in joint, muscle and skin) might cooperate in order to provide information about the position and movement of that joint. Since most of the joint receptors appeared not to be suitable for signaling steady joint position and joint movements (cf. Sect. 4.3.1), other groups favored muscle receptors (muscle spindles, in particular) as the endings mediating position and movement sense (Clark and Burgess 1975, McCloskey 1978, 1983, Burgess et al. 1982).

The tendon organs were supposed to be mainly responsible for the sensation of force (Roland and Ladegaard-Pedersen 1977, Matthews 1982). For this last function, corollary discharges from motor centers have been assumed to be of importance, particularly in a fatiguing muscle (Jones and Hunter 1983). The contention of other authors (Gelfan and Carter 1967) that "there is no muscle sense in man" has been refuted by the experiments of Goodwin et al. (1972b), who were able to produce illusions of movement by vibrating muscles longitudinally, thus inducing a strong input via Ia afferents.

Besides articular and muscular receptors, cutaneous receptors could be capable of providing information about the position and movement of a joint. An observation supporting this assumption is that patients with local anesthesia of the skin overlying a joint and with intact innervation of the muscle moving that joint, often perform movements without noticing them (Moberg 1983). Possibly, the three sources of afferent information for kinesthesia (joints, muscles, and skin) have to be regarded as different aspects of one system, each being specialized to provide data on a particular component of the movement (McCloskey 1978). It has to be kept in mind that the kinesthetic information might use different channels in different joints. There is experimental evidence indicating that the position of the neck in the cat is signaled mainly by muscle receptors (Abrahams 1977), while position and movement of the temporomandibular joint in rabbits (Lund and Matthews 1981) and of the costovertebral joint in rabbits (Godwin-Austen 1969) may be signaled by joint receptors.

Slowly conducting afferent units from muscles, tendons, and joints are quite insensitive to passive movements in the midrange position of a joint. Therefore, it appears unlikely that they are involved in movement and position sense. The only function which might be mediated by small-diameter muscle afferents is the sensation of force. It has been mentioned above that many contraction-sensitive units were also activated by muscle stretch. The response to stretch was graded as was the response to contractions. Therefore, these receptors would be capable of informing the CNS about the tension building up within the muscle. In this respect, the units behaved quite similarly to tendon organs, although operating with a higher mechanical threshold. However, whether the information from the contraction-sensitive receptors is really used for mediating the sensation of force, and whether the receptors are different from those which induce CP adjustments during exercise, is unknown at present.

4.5 Locomotion

Electrical stimulation of various nerves at an intensity exceeding the threshold of the fastest-conducting afferent fibers has been shown to facilitate flexor α-motoneurons and inhibit extensor α-motoneurons (Lloyd 1943). Irrespective of their response behavior upon natural stimulation, all afferent units producing this reflex were labeled "flexion reflex afferents" (FRA) (Eccles and Lundberg 1959b, Holmqvist and Lundberg 1961). The FRA concept describes the functional organization of reflex pathways from muscle, joint, and skin afferents to α-motoneurons and contains no statements as to whether the observed flexion reflex pattern is identical to that occurring during a nocifensive reaction. Among the fibers found to produce the reflex were not only (electrically) high-threshold muscle, skin, and joint afferents, but also muscular group II fibers, the great majority of which do not subserve a nociceptive function (cf. Sect. 2.3.2). On the other hand, application of strong natural stimuli, such as pressing a muscle or pinching the skin, elicited a similar reflex pattern in the spinal cat. Since mainly electrical stimulation was used in these studies, it is not clear which subpopulation (e.g., the nociceptive or low-threshold mechanosensitive) of the group III and II fibers had to be activated in order to produce the flexion reflex pattern. Probably, nociceptive and nonnociceptive afferents are included in the FRA. Both appear to have access to the spinal motoneuron pool, the former possibly producing the nocifensive flexion reflex and the latter being involved in the locomotor control of flexor muscles. As has been shown by Sherrington (1910), both flexion reflexes can be elicited from deep tissues, the locomotor flexion reflex being the result of the stimulation of low-threshold, proprioceptive fibers. This implies that there are nonnociceptive FRA which are activated during normal movements (Lundberg 1979).

The reflex effects of the FRA were found to be subject to a powerful descending control from the brain stem (Eccles and Lundberg 1959a). Thus, in spinal cats, the flexion reflex pattern was clear-cut while in the decerebrate state the effects of the FRA were depressed and, after a low pontine lesion, an inhibitory path from the FRA to flexor motoneurons was released in addition to the inhibition of extensor motoneurons (Holmqvist and Lundberg 1961). Later, it was shown that in the decerebrate and spinal cat, group II muscle afferents or secondary endings of muscle spindles, respectively, can also have an excitatory influence on extensor motoneurons (Matthews 1969, Fromm et al. 1977, Lundberg et al. 1977, Tan 1983). In electrophysiological experiments, evidence was obtained showing that this effect was partly produced via monosynaptic connections (Kirkwood and Sears 1974, Stauffer et al. 1976, Munson et al. 1982). This finding fits well with the morphological data derived from single-stained group II fibers which possessed collaterals projecting into the ventral horn (see Sect. 2.3.2).

As to the effects of fine deep afferents on γ-motoneurons, it was originally assumed that the action of group II and group III afferent fibers conformed to the FRA concept (Grillner et al. 1969). Later, it became clear that increased activity in group II muscle fibers not only inhibits, but also excites, extensor γ-motoneurons (Appelberg et al. 1977, Noth and Thilmann 1980). A recent study showed that especially the dynamic γ-motoneurons possess a high responsiveness to activity in group II fibers from muscle, the overall incidence of excitation outweighing that of inhibition (Appelberg et al. 1983a). In contrast, group III muscle afferents were found to act preferentially on static γ-motoneurons. Again, excitation of extensor fusimotoneurons by the so called FRA was frequently observed. These results were taken as an indication that the FRA concept is not valid for γ-motoneurons (Appelberg et al. 1983b). A great part of the excitatory input from muscular group III fibers to an extensor muscle appears to stem from nonnociceptive group III afferents which can be activated by muscle contractions (Ellaway et al. 1982).

In decerebrate, spinalized cats, in which rhythmic activity of motoneuron pools was induced by nialamide or L-DOPA (fictive spinal locomotion), input via group III and IV GS-muscle afferents elicited by noxious chemical stimulation had complex effects. They consisted partly of an accentuation of the rhythmic activity and partly of a tonic hyperactivity of the motoneurons which could be observed in nerves to both flexor and extensor muscles (Kniffki et al. 1980).

4.6 Thermoreception

4.6.1 Thermosensitive Receptors

Many of the fine deep afferents can be influenced by temperature stimuli (Hertel et al. 1976, Kumazawa and Mizumura 1977, Mense 1978). A response to temperature does not necessarily mean that a unit under study is a thermoreceptor, as certain cutaneous mechanoreceptors have been shown to possess a weak thermosensitivity (Hensel and Zotterman 1951, Iggo 1969). Additional requirements are that the unit has a high threshold upon mechanical stimulation and shows well-graded and reproducible responses to physiological changes in temperature (Hensel 1973, Iggo 1969). In a recent study (Mense and Meyer 1985), only a few of the fine afferents from the GS muscle of the cat appeared to supply true thermoreceptors; these exhibited a response behavior very similar to that of specific thermoreceptors in the skin (Hensel 1973, 1974). An example of a warm-sensitive receptor is given in Fig. 24. With the exception of noxious local pressure, all the mechanical stimuli and active contractions were ineffective. The response to noxious pressure in Fig. 24A was not considered to indicate a nociceptive nature of the unit, but to represent an activation of a thermoreceptor by a strong mechanical (i.e., inadequate) stimulus. Such responses are likely to occur in warm and cold receptors if the stimulus strength exceeds "reasonable limits of intensity" (Hensel 1973, 1974). The high thermal sensitivity of the unit is demonstrated by the fact that an increase in temperature from 32 °C (initial temperature of the paraffin pool containing the muscle) to 35 °C was sufficient to elicit a clear response (Fig. 24B). The unit reacted in a graded fashion to increasing temperature stimuli and showed a marked differential-proportional response characteristic.

Cold-sensitive units were also found. They usually showed a background activity which could be raised by cooling and abolished by warming the receptive field (Fig. 25). All the thermosensitive receptors were supplied by group IV fibers; they comprised about 10% of the muscle receptors with slowly conducting afferent fibers. This proportion is much smaller than that of group III and IV muscle receptors which showed a certain sensitivity to strong thermal stimuli, as noted in a previous report (Hertel et al. 1976). In the latter study, the mechanical sensitivity of the units could not be tested, since the muscle was totally covered by large thermodes. Therefore, no distinction could be made between thermal re-

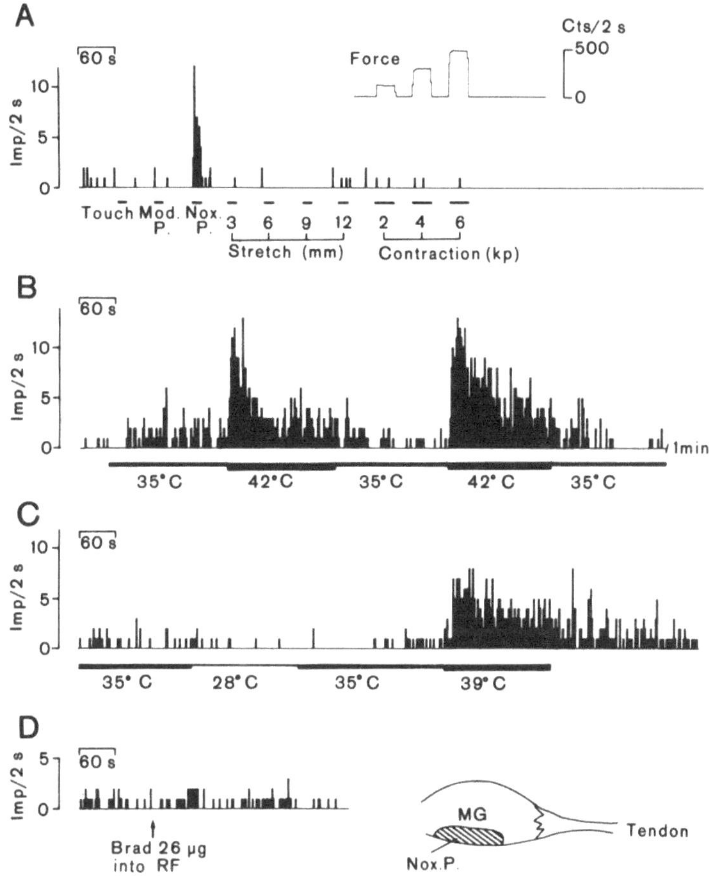

Fig. 24A – D. Thermosensitive unit from the cat skeletal muscle with group IV afferent fibers, conduction velocity 1.1 m/s. The unit could be mechanically activated by squeezing the lower edge of the medial gastrocnemius muscle. Labeling of stimulations as in Fig. 10. Starting temperature of the pool containing the muscle was 32 °C. Note that in **B** and **C** warming to 35 °C activated the unit, and cooling to 28 °C abolished the background activity almost completely (**C**). (From Mense and Meyer 1985)

sponses from mechanoreceptors and true thermoreceptors. Apparently, many of the thermally induced responses in slowly conducting muscle afferents are derived from mechanoreceptors which behave like "spurious thermoreceptors" (Iggo 1969).

The thermosensitive receptors appeared to be the only group that did not respond to bradykinin. Only one cold-sensitive unit could be activated by the algesic substance; this effect was not reproducible.

Evidence for the existence of thermosensitive receptors has also been obtained in studies of other deep tissues, e.g., the femoral vein of the cat (Thompson and Barnes 1969).

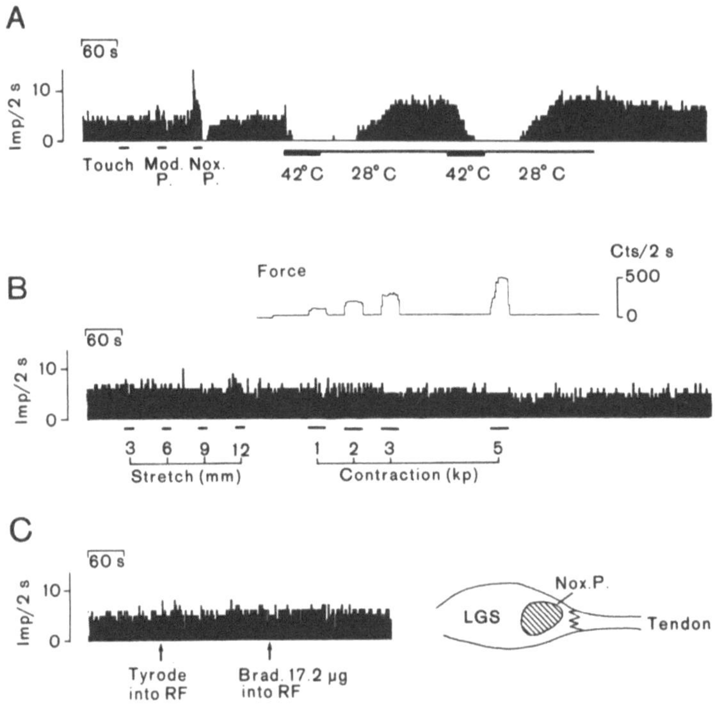

Fig. 25A–C. Cold-sensitive unit with group IV afferent fibers, conduction velocity 0.75 m/s. Labeling of receptive field and stimuli as in Fig. 10. Starting temperature of the pool containing the muscle was 31 °C. In **B** the histogram during contractions of 3 and 5 kp was smoothed by hand in order to remove movement artifacts. (From Mense and Meyer 1985)

4.6.2 Central Nervous Effects of Raised Activity in Deep Thermoreceptors

There are reports in the older clinical literature that thermal stimulation of a muscle in awake patients does not elicit subjective sensations (Schmidt 1916). Similarly, the rise in muscle temperature during exercise has been shown not to influence respiratory centers (Morgan et al. 1955, Kao 1963). The only possible function of deep thermoreceptors for which experimental evidence is available is that of an input channel for thermoregulation. Jessen et al. (1983) have shown that cooling of the bone marrow cavity in goats leads to a reduction in evaporative heat loss, i.e., to a thermoregulatory effect. Thus, it is possible that in addition to central nervous and cutaneous thermoreceptors, a third input to thermoregulatory centers exists which has its origin in deep thermoreceptors.

4.7 Other Functions

Lately, it has been discussed that fine deep afferents might be involved in acupuncture. Unfortunately, the available experimental data do not allow this hypothesis to be proved or disproved, so that the following considerations are admittedly speculative. However, several observations have been made that point to the fine deep afferents as being the afferent limb for some acupuncture effects. For instance, some of the common acupuncture methods (needling, acupressure) apparently require activation of deep afferents since they aim at deep tissues: at the stomach point 36, the needle is introduced to a depth of approximately 4 cm (from *An Outline of Chinese Acupuncture* 1975). The acupuncture effects produced by needling of the point Ho-ku on the hand have been shown to be abolished after infiltration of the deep tissues with a local anesthetic, while anesthesia of the skin overlying the acupuncture point was without effect (Chiang et al. 1973). An input via large-diameter deep afferents is apparently not required; such an input could easily be evoked by stretching muscles and moving joints, but these procedures are not part of acupuncture treatment. On the other hand, no direct proof for the involvement of fine deep afferents in acupuncture is available.

The mechanism(s) by which the acupuncture effects are brought about is (are) unknown. In a recent review, Chang (1980) stated that impulses via fine deep afferents could gain access to the brain stem nuclei which produce descending inhibition of spinal nociceptive neurons. Impulse activity in fine deep afferents has been described as eliciting cardiopulmonary reflexes via neurons in the lateral reticular formation and raphe nuclei (Mitchell et al. 1983), i.e., via brain stem regions that are close to the sites of origin of descending pathways which have an inhibitory effect on nociceptive dorsal horn neurons (Basbaum et al. 1978). There are data indicating a release of endorphins during acupuncture-like transcutaneous nerve stimulation in humans (Sjölund et al. 1977), but unequivocal experimental evidence showing that acupuncture really functions by activation of the descending pain-inhibiting system is still missing.

In this context, observations by Thorén and coworkers concerning changes in pain threshold during muscular exercise are of interest. They found that rats running in a wheel exhibited an increased threshold for vocalization upon electrical stimulation of the tail (Shyu et al. 1984). The authors also presented evidence that the increase in pain threshold during and after exercise was due to the release of endorphin (Shyu et al. 1982). The accompanying decrease in blood pressure could also be produced by acupuncture-like electrical stimulation of the sciatic nerve; it was probably mediated by an opiate receptor different from that responsible for the hypalgesia (Yao et al. 1982). From the intensity of electrical stimulation required to produce the effects, it is apparent that somatic group III and IV fibers had to be activated. Further studies are needed, however, to answer the question whether receptors in deep tissues form the afferent limb for the reflex depressor response and for the hypalgesia.

5 Summarizing Conclusions

The bulk of the available data suggests that among the small-diameter afferent units from deep tissues several distinct types exist which probably fulfill different functions. The types are:

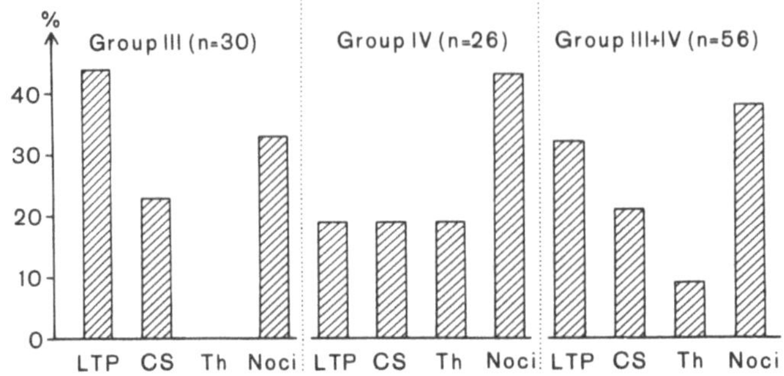

Fig. 26. Distribution of receptor types among group III and IV afferent units. *LTP*, low-threshold pressure-sensitive units; *CS*, contraction-sensitive units; *Th.*, thermo-sensitive units; *Noci.*, nociceptive units. (From Mense and Meyer 1985)

1. Nociceptors, which probably mediate the different forms of deep pain and the accompanying reflexes.
2. Low-threshold mechanoreceptors, which possibly mediate deep pressure sensations.
3. Contraction-sensitive receptors, which probably mediate cardiopulmonary adjustments during exercise and, thus, possibly function as the hypothetical "ergoreceptors" (Kao 1963), "metabolic receptors" (Stegemann and Kenner 1971), or "exercise receptors" (Pérez-Gonzáles and Coote 1972).
4. Thermoreceptors, which are probably involved in thermoregulation.

The frequencies of the different receptor types within group III and IV units as found in a recent study (Mense and Meyer 1985) are shown in Fig. 26. Up to now, contraction-sensitive and thermosensitive receptors have been found only in skeletal muscle and tendon; the other types appear to also be present in other deep tissues.

The experimental evidence supporting the conclusion that fine deep afferents do not form a functionally homogenous population stems from three independent approaches: (a) From single-fiber recordings of muscular, articular, and diaphragmatic afferent units showing differences in response behavior. (b) From investigations with a combined physiological-histological approach showing that the response behavior of a fine deep afferent is associated with a characteristic termination pattern of its spinal projection. (c) From electron microscopic studies showing morphological differences between free nerve endings.

The existence of different receptor types among group III and IV afferent units would offer a simple explanation for the variability of cardiovascular effects upon electrical stimulation at group III strength observed by many authors. It could likewise explain the everyday experience that localized deep pain can occur without cardiopulmonary reflexes and that, during exercise, these reflexes are usually present without pain.

With the use of conventional neurophysiological techniques in animal experiments, it will remain difficult to obtain unequivocal evidence for or against the polymodality or specificity of deep receptors. Two conceivable ways of further pursuing this issue could be to study single fibers with immunohistochemical methods (functionally different units may have different transmitters), or to perform microneurographic studies in conscious human beings who could report their subjective sensations (if present) upon stimulation of a single fine deep afferent.

Acknowledgements. I am greatly indebted to my colleagues, P. Grigg, H.-G. Schaible, R. F. Schmidt, and W. D. Willis for helpful comments on the manuscript. I am grateful to B. Howaldt for her help in the preparation of the references, and to H. Dawson for typing the manuscript.

6 References

Abrahams VC (1977) The physiology of neck muscles; their role in head movement and maintenance of posture. Can J Physiol Pharmacol 55:332 – 338

Abrahams VC, Lynn B, Richmond FJR (1984) Organization and sensory properties of small myelinated fibers in the dorsal cervical rami of the cat. J Physiol (Lond) 347:177 – 187

Adrian ED, Zotterman Y (1926) The impulses produced by sensory nerve-endings. 2. The response of a single end-organ. J Physiol (Lond) 61:151 – 171

Akoev GN (1981) Catecholamines, acetylcholine and excitability of mechanoreceptors. Prog Neurobiol 15:269 – 294

Alam M, Smirk FH (1937) Observations in man upon a blood pressure raising reflex arising from the voluntary muscles. J Physiol (Lond) 89:372 – 383

Alam M, Smirk FH (1938) Observations in man on a pulse-accelerating reflex from the voluntary muscles of the legs. J Physiol (Lond) 92:167 – 177

Amassian VE, Berlin L (1958) Early cortical projection of group I afferents in the forelimb muscle nerves of cat. J Physiol 143:61P

An outline of chinese acupuncture (1975) The academy of traditional Chinese medicine, Foreign Language Press, Peking

Andres KH, Düring M von, Schmidt RF (1985) Sensory innervation of the achilles tendon by group III and IV afferent fibers. Anat Embryol, to be published

Angaut-Petit D (1975) The dorsal column system: II. Functional properties and bulbar relay of the postsynaptic fibers of the cat's fasciculus gracilis. Exp Brain Res 22:471 – 493

Anrep GV, Barsoum GS (1935) Appearance of histamine in the venous blood during muscular contraction. J Physiol (Lond) 85:409 – 420

Appelberg B, Johansson H, Kalistratov G (1977) The influence of group II muscle afferents and low threshold skin afferents on dynamic fusimotor neurones to the triceps surae of the cat. Brain Res 132:153 – 158

Appelberg B, Hulliger M, Johansson H, Sojka P (1983a) Actions on γ-motoneurones elicited by electrical stimulation of group II muscle afferent fibers in the hind limb of the cat. J Physiol (Lond) 335:255 – 273

Appelberg B, Hulliger M, Johansson H, Sojka P (1983b) Actions on γ-motoneurones elicited by electrical stimulation of group III muscle afferent fibers in the hind limb of the cat. J Physiol (Lond) 335:275 – 292

Applebaum AE, Beall JE, Foreman RD, Willis WD (1975) Organization and receptive fields of primate spinothalamic tract neurons. J Neurophysiol 38:572 – 586

Applebaum AE, Leonard RB, Kenshalo DR, Martin RF, Willis WD (1979) Nuclei in which functionally identified spinothalamic tract neurons terminate. J Comp Neurol 188:575 – 586

Armstrong RB, Ogilvie RW, Schwane JA (1983) Eccentric exercise-induced injury to rat skeletal muscle. J Appl Physiol 54:80 – 93
Asmussen E (1956) Observation on experimental muscular soreness. Acta Rheumatol Scand 2:109 – 116
Asmussen E, Nielsen M (1964) Experiments on nervous factors controlling respiration and circulation during exercise employing blocking of the blood flow. Acta Physiol Scand 60:103 – 111
Asmussen E, Johansen SH, Jørgensen M, Nielsen M (1965) On the nervous factors controlling respiration and circulation during exercise. Acta Physiol Scand 63:343 – 350
Banister EW, Allen ME, Mekjavic IB, Singh AK, Legge B, Mutch BJC (1983) The time course of ammonia and lactate accumulation in blood during bicycle exercise. Eur J Appl Physiol 51:195 – 202
Barker D (1962) The structure and distribution of muscle receptors. In: Barker D (ed) Symposium on muscle receptors. Hong Kong University Press, Hong Kong, pp 227 – 240
Barker D (1974) The morphology of muscle receptors. In: Hunt CC (ed) Muscle receptors. Springer, Berlin Heidelberg New York (Handbook of sensory physiology, vol 3/2)
Barker D, Ip MC, Adal MN (1962) A correlation between the receptor population of the cat's soleus muscle and the afferent fibre-diameter spectrum of the nerve supplying it. In: Barker D (ed) Symposium on muscle receptors. Hong Kong University Press, Hong Kong, pp 257 – 261
Barrack RL, Skinner HB, Cook SD, Haddad RJ (1983) Effect of articular disease and total knee arthroplasty on knee joint-position sense. J Neurophysiol 50:684 – 687
Barron W, Coote JH (1973) The contribution of articular receptors to cardiovascular reflexes elicited by passive limb movement. J Physiol (Lond) 235:423 – 436
Basbaum AI, Clanton CH, Fields HL (1978) Three bulbospinal pathways from the rostral medulla of the cat: an autoradiographic study of pain modulating systems. J Comp Neurol 178:209 – 224
Baxendale RH, Ferrell WR (1983) Discharge characteristics of the elbow joint nerve of the cat. Brain Res 261:195 – 203
Beacham WS, Perl ER (1964) Characteristics of a spinal sympathetic reflex. J Physiol (Lond) 173:431 – 448
Beck PW, Handwerker HO (1974) Bradykinin and serotonin effects on various types of cutaneous nerve fibres. Pflügers Arch 347:209 – 222
Berkley KJ (1973) Response properties of cells in ventrobasal and posterior group nuclei of the cat. J Neurophysiol 36:940 – 952
Bessou P, Laporte Y (1958) Activation des fibres afférentes amyéliniques d'origine musculaire. C R Soc Biol (Paris) 152:1587 – 1590
Bessou P, Laporte Y (1961) Étude des récepteurs musculaires innervés par les fibres afférentes du groupe III (fibres myélinisées fines) chez le chat. Arch Ital Biol 99:293 – 321
Bessou P, Perl ER (1969) Response of cutaneous sensory units with unmyelinated fibers to noxious stimuli. J Neurophysiol 32:1025 – 1043
Bessou P, Dejours P, Laporte Y (1959) Effets ventilatoires réflexes de la stimulation des fibres afférentes de grand diamètre d'origine musculaire chez le chat. C R Soc Biol (Paris) 153:447 – 481
Bishop GH, Heinbecker P (1935) The afferent functions of non-myelinated or C fibers. Am J Phsiol 114:179 – 193
Bockman EL, Berne RM, Rubio R (1976) Adenosine and active hyperemia in dog skeletal muscle. Am J Physiol 230:1531 – 1537
Boivie J (1971) The termination of the spinothalamic tract in the cat. An experimental study with silver impregnation methods. Exp Brain Res 12:331 – 353
Bonica JJ (1979) Causalgia and other reflex sympathetic dystrophies. In: Bonica JJ, Liebeskind JC, Albe-Fessard DG (eds) Advances in pain research and therapy, vol 3. Raven, New York, pp 141 – 166

Boyd IA, Davey MR (1968) Composition of peripheral nerves. Livingstone, Edinburgh
Boyd IA, Roberts TDM (1953) Proprioceptive discharges from stretch-receptors in the knee-joint of the cat. J Physiol (Lond) 122:38–58
Brock LG, Eccles JC, Rall W (1951) Experimental investigations on the afferent fibres in muscle nerves. Proc R Soc Lond [Biol] 138:453–475
Brown AG (1973) Ascending and long spinal pathways: dorsal columns, spinocervical tract and spinothalamic tract. In: Iggo A (ed) Handbook of sensory physiology, vol II. Somatosensory system. Springer, Berlin Heidelberg New York
Brown AG (1981) Organization in the spinal cord. The anatomy and physiology of identified neurones. Springer, Berlin Heidelberg New York
Brushart TM, Mesulam M-M (1980) Transganglionic demonstration of central sensory projections from skin and muscle with HRP-lectin conjugates. Neurosci Lett 17:1–6
Bryan RN, Trevino DL, Coulter JD, Willis WD (1973) Location and somatotopic organization of the cells of origin of the spino-cervical tract. Exp Brain Res 17:177–189
Burgess PR, Clark FJ (1969) Characteristics of knee joint receptors in the cat. J Physiol (Lond) 203:317–335
Burgess PR, Jen Yu WEI, Clark FJ, Simon J (1982) Signaling of kinesthetic information by peripheral sensory receptors. Ann Rev Neurosci 5:171–187
Campbell JN, Meyer RA, LaMotte RH (1979) Sensitization of myelinated nociceptive afferents that innervate monkey hand. J Neurophysiol 42:1669–1679
Carcassi AM, Concu A, Decandia M, Onnis M, Orani GP, Piras MB (1984) Effects of long-lasting stimulation of extensor muscle nerves on pulmonary ventilation in cats. Pflügers Arch 400:409–412
Carstens E, Trevino DL (1978) Laminar origins of spinothalamic projections in the cat as determined by the retrograde transport of horseradish peroxidase. J Comp Neurol 182:151–166
Cassirer R (1899) Über Veränderungen der Spinalganglienzellen und ihrer centralen Fortsätze nach Durchschneidung der zugehörigen peripheren Nerven. Dtsch Z Nervenheilkd 14:150–166
Cauna N (1966) Fine structure of the receptor organs and its probable functional significance. In: de Reuck AVS, Knight J (eds) Touch, heat and pain. Churchill, London, pp 117–127
Cervero F, Connell LA (1984) Fine afferent fibers from viscera do not terminate in the substantia gelatinosa of the thoracic spinal cord. Brain Res 294:370–374
Cervero F, Iggo A (1980) The substantia gelatinosa of the spinal cord. A critical review. Brain 103:717–772
Cervero F, Iggo A, Ogawa H (1976) Nociceptor-driven dorsal horn neurones in the lumbar spinal cord of the cat. Pain 2:5–24
Cervero F, Iggo A, Molony V (1979) Ascending projections of nociceptor-driven lamina I neurones in the cat. Exp Brain Res 35:135–149
Chang H-T (1980) Neurophysiological interpretation of acupuncture analgesia. Endeavour 4:92–96
Chang SC, Wei JY, Mao CP (1983) Deep innervation of sural nerve. Brain Res 279:262–265
Chiang C-Y, Chang C-T, Chu H-L, Yang L-F (1973) Peripheral afferent pathway for acupuncture analgesia. Sci Sin 16:210–217
Clark FJ (1975) Information signaled by sensory fibers in medial articular nerve. J Neurophysiol 38:1464–1472
Clark FJ, Burgess PR (1975) Slowly adapting receptors in cat knee joint: can they signal joint angle? J Neurophysiol 38:1448–1463
Coggeshall RE, Ito H (1977) Sensory fibres in ventral roots L_7 and S_1 in the cat. J Physiol (Lond) 267:215–235

Coggeshall RE, Maynard CW, Langford LA (1980) Unmyelinated sensory and preganglionic fibers in rat L_6 and S_1 ventral spinal roots. J Comp Neurol 193:41–47
Coggeshall RE, Hong KAP, Langford LA, Schaible H-G, Schmidt RF (1983) Discharge characteristics of fine medial articular afferents at rest and during passive movements of inflamed knee joints. Brain Res 272:185–188
Colle J, Gybels J (1957) Étude des réactions tensionelles, respiratoires et corticales produites par l'excitation électrique des fibres afférentes d'un nerf somatique. Arch Int Physiol Biochim 65:547–567
Comroe JH, Schmidt CF (1943) Reflexes from the limbs as a factor in the hyperpnea of muscular exercise. Am J Physiol 138:536–547
Coote JH (1975) Physiological significance of somatic afferent pathways from skeletal muscle and joints with reflex effects on the heart and circulation. Brain Res 87:139–144
Coote JH, Dodds WN (1976) The baroreceptor reflex and the cardiovascular changes associated with sustained muscular contraction in the cat. Pflügers Arch 363:167–173
Coote JH, Pérez-González JF (1970) The response of some sympathetic neurones to volleys in various afferent nerves. J Physiol (Lond) 208:261–278
Coote JH, Hilton SM, Pérez-González JF (1969) Muscle afferents responsible for the pressor response to exercise. J Physiol (Lond) 201:34–35P
Craig AD (1976) Spinocervical tract cells in cat and dog, labeled by the retrograde transport of horseradish peroxidase. Neurosci Lett 3:173–177
Craig AD, Burton H (1979) Spinothalamic terminations in the ventroposterolateral nucleus of the cat. Soc Neurosci Abstr 5:705
Craig AD, Burton H (1981) Spinal and medullary lamina I projection to nucleus submedius in medial thalamus: a possible pain center. J Neurophysiol 45:443–466
Craig AD, Kniffki K-D (1982) Spinothalamic lumbosacral lamina I cells responding to algesic muscle stimulation in the cat. Pflügers Arch 394:R57
Craig AD, Mense S (1983) The distribution of afferent fibers from the gastrocnemius-soleus muscle in the dorsal horn of the cat, as revealed by the transport of horseradish peroxidase. Neurosci Lett 41:233–238
Crayton SC, Mitchell JH, Payne FC (1981) Reflex cardiovascular response during injection of capsaicin into skeletal muscle. Am J Physiol 240:H315–H319
Cross MJ, McCloskey DI (1973) Position sense following surgical removal of joints in man. Brain Res 55:443–445
Cross BA, Davey A, Guz A, Katona PG, MacLean M, Murphy K, Semple SJG, Stidwell R (1982) The role of spinal cord transmission in the ventilatory response to electrically induced exercise in the anesthetized dog. J Physiol (Lond) 329:37–55
Culberson JL, Brown PB (1984) Projections of hindlimb dorsal roots to lumbosacral spinal cord of cat. J Neurophysiol 51:516–528
Cunningham DJC, Petersen ES, Peto R, Pickering TG, Sleight P (1972) Comparison of the effect of different types of exercise on the baroreflex regulation of heart rate. Acta Physiol Scand 86:444–455
De Groat WC, Nadelhaft I, Morgan C, Schauble T (1978) Horseradish peroxidase tracing of visceral efferent and primary afferent pathways in cat's sacral spinal cord using benzidine processing. Neurosci Lett 10:103–108
Dejours P (1964) Control of respiration in muscular exercise. In: Handbook of physiology, Section 3: Respiration, vol I. American Physiological Society, Washington, DC
De Troyer A, Sampson M, Sigrist S, Macklem PT (1981) The diaphragm: two muscles. Science 213:237–238
Devor M (1983) Nerve pathophysiology and mechanisms of pain in causalgia. J Auton Nerv Syst 7:371–384
Devor M, Jänig W (1981) Activation of myelinated afferents ending in a neuroma by stimulation of the sympathetic supply in the rat. Neurosci Lett 24:43–47

DeVries HA (1966) Quantitative electromyographic investigation of the spasm theory of muscle pain. Am J Phys Med 45:119–134

Digiesi V, Bartoli V, Dorigo B (1975) Effect of a proteinase inhibitor on intermittent claudication or on pain at rest in patients with peripheral arterial disease. Pain 1:385–389

Dogiel AS (1902) Die Nervenendigungen im Bauchfell, in den Sehnen, den Muskelspindeln und dem Centrum tendineum des Diaphragmas beim Menschen und bei Säugetieren. Arch Mikroskop Anat 59:1–31

Dong WK, Ryu H, Wagman IH (1978) Nociceptive responses of neurons in medial thalamus and their relationship to spinothalamic pathways. J Neurophysiol 41:1592–1613

Dreyer DA, Schneider RJ, Metz CB, Whitsel BL (1974) Differential contributions of spinal pathways to body representation in postcentral gyrus of Macaca mulatta. J Neurophysiol 37:119–145

Eccles, JC, Eccles RM, Lundberg A (1957) Synaptic actions on motoneurones caused by impulses in Golgi tendon organ afferents. J Physiol (Lond) 138:227–252

Eccles RM, Lundberg A (1959a) Supraspinal control of interneurones mediating spinal reflexes. J Physiol (Lond) 147:565–584

Eccles RM, Lundberg A (1959b) Synaptic actions in motoneurones by afferents which may evoke the flexion reflex. Arch Ital Biol 97:199–221

Eldred E, Schnitzlein HN, Buchwald J (1960) Response of muscle spindles to stimulation of the sympathetic trunk. Exp Neurol 2:13–25

Ellaway PH, Murthy KSK (1984) Reflex effects from high threshold neck muscle afferents on hindlimb extensor gamma motoneurones in the cat. Exp Brain Res 54:212–216

Ellaway PH, Murphy PR, Tripathi A (1982) Closely coupled excitation of γ-motoneurones by group III muscle afferents with low mechanical threshold in the cat. J Physiol (Lond) 331:481–498

Erlanger J, Gasser HS (1924) The compound nature of the action current of nerve as disclosed by the cathode ray oscillograph. Am J Physiol 70:624–666

Erlanger J, Gasser HS (1930) The action potential in fibers of slow conduction in spinal roots and somatic nerves. Am J Physiol 92:43–82

Feindel WH, Weddel G, Sinclair DC (1948) Pain sensibility in deep somatic structures. J Neurol Neurosurg Psychiatry 11:113–117

Felix W (1922) Anatomische, experimentelle und klinische Untersuchungen über den Phrenicus und über die Zwerchfellinnervation. Dtsch Z Chir 171:283–397

Fernandez de Molina A, Achard O, Wyss OAM (1953) Respiratory and vasomotor responses to stimulation of afferent fibers in somatic nerves. Helv Physiol Pharmacol Acta 11:1–19

Fields HL, Clanton CH, Anderson SD (1977) Somatosensory properties of spinoreticular neurons in the cat. Brain Res 120:49–66

Fisher ML, Nutter DO (1974) Cardiovascular reflex adjustments to static muscular contractions in the canine hindlimb. Am J Physiol 226:648–655

Fitzgerald M, Lynn B (1977) The sensitization of high threshold mechanoreceptors with myelinated axons by repeated heating. J Physiol (Lond) 265:549–563

Fjällbrant N, Iggo A (1961) The effect of histamine, 5-hydroxytryptamine and acetylcholine on cutaneous afferent fibers. J Physiol (Lond) 156:578–590

Fock S, Mense S (1976) Excitatory effects of 5-hydroxytryptamine, histamine and potassium ions on muscular group IV afferent units: a comparison with bradykinin. Brain Res 105:459–469

Foreman RD, Kenshalo DR, Schmidt RF, Willis WD (1979a) Field potentials and excitation of primate spinothalamic neurones in response to volleys in muscle afferents. J Physiol (Lond) 286:197–213

Foreman RD, Schmidt RF, Willis WD (1979b) Effects of mechanical and chemical stimulation of fine muscle afferents upon primate spinothalamic tract cells. J Physiol (Lond) 286:215 – 231
Franz M, Mense S (1975) Muscle receptors with group IV afferent fibers responding to application of bradykinin. Brain Res 92:369 – 383
Freeman MAR, Wyke B (1967) The innervation of the knee joint. An anatomical and histological study in the cat. J Anat (Lond) 101:505 – 532
Frey M von (1914) Studien über den Kraftsinn. Z Biol 63:129 – 154
Fromm C, Haase J, Wolf E (1977) Depression of the recurrent inhibition of extensor motoneurones by the action of group II afferents. Brain Res 120:459 – 468
Fyffe REW (1979) The morphology of group II muscle afferent fiber collaterals. J Physiol (Lond) 296:39P – 40P
Gandevia SC, Burke D, McKeon B (1984) The projection of muscle afferents from the hand to cerebral cortex in man. Brain 107:1 – 13
Gardner E (1944) The distribution and termination of nerves in the knee joint of the cat. J Comp Neurol 80:11 – 32
Gardner E (1950) Reflex muscular responses to stimulation of articular nerves in the cat. Am J Physiol 161:133 – 141
Gardner E, Haddad B (1953) Pathways to the cerebral cortex for afferent fibres from the hindlimb of the cat. Am J Physiol 172:475 – 482
Gardner E, Jacobs J (1948) Joint reflexes and regulation of respiration during exercise. Am J Physiol 153:567 – 579
Gardner E, Lenn NJ (1977) Fibers in monkey posterior articular nerves. Anat Rec 187:99 – 106
Gasser HS (1950) Unmedullated fibers originating in dorsal root ganglia. J Gen Physiol 33:651 – 690
Gasser HS, Grundfest H (1939) Axon diameters in relation to the spike dimensions and the conduction velocity in mammalian A fibers. Am J Physiol 127:393 – 414
Gautron M, Guilbaud G (1982) Somatic responses of ventrobasal thalamic neurones in polyarthritic rats. Brain Res 237:459 – 471
Gaze RM, Gordon G (1954) The representation of cutaneous sense in the thalamus of the cat and monkey. Q J Exp Physiol 39:279 – 304
Gelfan S, Carter S (1967) Muscle sense in man. Exp Neurol 18:469 – 473
Gernandt B (1946) Pain conduction in the phrenic nerve. Acta Physiol Scand 12:255 – 260
Giesler GJ, Cannon JT, Urca G, Liebeskind JC (1978) Long ascending projections from substantia gelatinosa Rolandi and the subjacent dorsal horn in the rat. Science 202:984 – 986
Gobel S, Binck JM (1977) Degenerative changes in primary trigeminal axons and in neurons in nucleus caudalis following tooth pulp extirpation in the cat. Brain Res 132:347 – 354
Gobel S, Falls WM, Humphrey E (1981) Morphology and synaptic connections of ultrafine primary axons in lamina I of the spinal dorsal horn: candidates for the terminal axonal arbors of primary neurons with unmyelinated (C) axons. J Neurosci 10:1163 – 1179
Godwin-Austen RB (1969) The mechanoreceptors of the costo-vertebral joints. J Physiol (Lond) 202:737 – 753
Goldscheider A (1920) Das Schmerzproblem. Springer, Berlin Heidelberg New York
Goodwin GM (1976) The sense of limb position and movement. Exerc Sport Sci Rev 4:87 – 124
Goodwin GM, McCloskey DI, Mitchell JH (1972a) Cardiovascular and respiratory responses to changes in central command during isometric exercise at constant muscle tension. J Physiol (Lond) 226:173 – 190

Goodwin GM, McCloskey DI, Matthews PBC (1972b) The contribution of muscle afferents to kinaesthesia shown by vibration induced illusions of movement and by the effects of paralysing joint afferents. Brain 95:705 – 748
Grandjean E (1943) Atmungsveränderungen bei Reizung der afferenten Phrenicusfasern. Helv Physiol Acta 1:205 – 220
Grandpierre R, Franck C, Violette F (1951) Hyperventilation provoquée par les mouvements passifs. C R Soc Biol (Paris) 145:1100 – 1104
Granit R (1950) Reflex self-regulation of muscle contraction and autogenetic inhibition. J Neurophysiol 13:351 – 372
Grant G, Ygge J (1981) Somatotopic organization of the thoracic spinal nerve in the dorsal horn demonstrated with transganglionic degeneration. J Comp Neurol 202:357 – 364
Grant G, Arvidsson J, Robertson B, Ygge J (1979) Transganglionic transport of horseradish peroxidase in primary sensory neurons. Neurosci Lett 12:23 – 28
Grigg P (1976) Response of joint afferent neurons in cat medial articular nerve to active and passive movements of the knee. Brain Res 118:482 – 485
Grigg P, Finerman GA, Riley LH (1973) Joint-position sense after total hip replacement. J Bone Surg 55A:1016 – 1025
Grigg P, Hoffman AH, Fogarty KE (1982) Properties of Golgi-Mazzoni afferents in cat knee joint capsule, as revealed by mechanical studies of isolated joint capsule. J Neurophysiol 47:31 – 40
Grillner S, Hongo T, Lund S (1969) Descending monosynaptic and reflex control of γ-motoneurones. Acta Physiol Scand 75:592 – 613
Guilbaud G, Caille D, Bessou JM, Benelli G (1977) Single units activities in ventral posterior and posterior group thalamic nuclei during nociceptive and non-nociceptive stimulations in the cat. Arch Ital Biol 115:38 – 56
Guilbaud G, Peschanski M, Gautron M, Binder D (1980) Neurones responding to noxious stimulation in VB complex and caudal adjacent regions in the thalamus of the rat. Pain 8:303 – 318
Guilbaud G, Iggo A, Tegnér R (1984) Sensory receptors in the joints of rats with adjuvant-induced arthritis. J Physiol (Lond) 346:58P
Guzman F, Braun C, Lim RKS (1962) Visceral pain and the pseudaffective response to intra-arterial injection of bradykinin and other algesic agents. Arch Int Pharmacodyn Ther 136:353 – 384
Ha H, Liu C-N (1968) Cell origin of the ventral spinocerebellar tract. J Comp Neurol 133:185 – 206
Hamann WC, Hong SK, Kniffki K-D, Schmidt RF (1978) Projections of primary afferent fibres from muscle to neurones of the spinocervical tract of the cat. J Physiol (Lond) 283:369 – 378
Hardy TL, Bertrand G, Thompson CJ (1980) Position and organization of thalamic cellular activity during diencephalic recording. II. Joint- and muscle-evoked activity. Appl Neurophysiology 43:28 – 36
Hari R, Kaukoranta E, Reinikainen K, Huopaniemie T, Mauno J (1983) Neuromagnetic localization of cortical activity evoked by painful dental stimulation in man. Neurosci Lett 42:77 – 82
Harpuder K, Stein I (1943) Studies on the nature of pain arising from an ischemic limb. Am Heart J 25:429 – 448
Harris FA (1970) Population analysis of somatosensory thalamus in the cat. Nature 225:559 – 562
Hassler R (1976) Wechselwirkungen zwischen dem System der schnellen Schmerzempfindung und dem des langsamen, nachhaltigen Schmerzgefühls. Langenbecks Arch Chir 342:47 – 61
Head H, Rivers WHR, Sherren J (1905) The afferent nervous system from a new aspect. Brain 28:99 – 115

Hellon RF, Mitchell D (1975) Characteristics of neurones in the ventrobasal thalamus of the rat which respond to noxious stimulation of the tail. J Physiol (Lond) 250:29P – 30P
Hensel H (1973) Cutaneous thermoreceptors. In: Iggo A (ed) Somatosensory system. Springer, Berlin Heidelberg New York (Handbook of sensory physiology, vol 2)
Hensel H (1974) Thermoreceptors. Ann Rev Physiol 36:233 – 249
Hensel H, Zotterman Y (1951) The response of mechanoreceptors to thermal stimulation. J Physiol (Lond) 115:16 – 24
Hensel H, Andres KH, Düring M von (1974) Structure and function of cold receptors. Pflügers Arch 352:1 – 10
Hertel H-C, Howaldt B, Mense S (1976) Responses of group IV and group III muscle afferents to thermal stimuli. Brain Res 113:201 – 205
Hinsey JC (1927) Some observations on the innervation of skeletal muscle of the cat. J Comp Neurol 44:87 – 195
Hinsey JC (1928) Observations on the innervation of the blood vessels in skeletal muscle. J Comp Neurol 47:23 – 60
Hinsey JC, Phillips RA (1940) Observations upon diaphragmatic sensation. J Neurophysiol 3:175 – 181
Hiss E, Mense S (1976) Evidence for the existence of different receptor sites for algesic agents at the endings of muscular group IV afferent units. Pflügers Arch 362:141 – 146
Hník, P, Hudlická O, Kučera J, Payne R (1969) Activation of muscle afferents by nonproprioceptive stimuli. Am J Physiol 217:1451 – 1457
Hník P, Holas M, Krekule I, Kříž N, Mejsnar J, Smieško V, Ujec E, Vyskočil F (1976) Work-induced potassium changes in skeletal muscle and effluent venous blood assessed by liquid ion-exchanger microelectrodes. Pflügers Arch 362:85 – 94
Hodgson HJF, Matthews PBC (1968) The ineffectiveness of excitation of the primary endings of the muscle spindle by vibration as a respiratory stimulant in the decerebrate cat. J Physiol (Lond) 194:555 – 563
Holmqvist B, Lundberg A (1961) Differential supraspinal control of synaptic actions evoked by volleys in the flexion reflex afferents in alpha motoneurones. Acta Physiol Scand 54:1 – 51
Holmqvist B, Lundberg A, Oscarsson O (1960) Supraspinal inhibitory control of transmission to three ascending spinal pathways influenced by the flexion reflex afferents. Arch Ital Biol 98:60 – 80
Honda CN, Perl ER (1981) Properties of neurons in lamina X and the midline dorsal horn of the sacrococcygeal spinal cord of the cat (Abstr). Soc Neurosci 7:610
Honda CN, Mense S, Perl ER (1983) Neurons in ventrobasal region of cat thalamus selectively responsive to noxious mechanical stimulation. J Neurophysiol 49:662 – 673
Hong SK, Kniffki K-D, Schmidt RF (1978) Reflex discharges of extensor and flexor gamma motoneurones by chemically induced muscle pain (Abstr). Pain 1:58
Hong SK, Kniffki K-D, Mense S, Schmidt RF, Wendisch M (1979) Descending influences on the responses of spinocervical tract neurones to chemical stimulation of fine muscle afferents. J Physiol (Lond) 290:129 – 140
Hough T (1902) Ergographic studies in muscular soreness. Am J Physiol 7:76 – 92
Hunt CC, Jami L, Laporte Y (1982) Effects of stimulating the lumbar sympathetic trunk on cat hindlimb muscle spindles. Arch Ital Biol 120:371 – 384
Hutter OF, Loewenstein WR (1955) Nature of neuromuscular facilitation by sympathetic stimulation in the frog. J Physiol (Lond) 130:559 – 571
Iaria CT, Jalar UH, Kao FF (1959) The peripheral neural mechanism of exercise hyperpnoea. J Physiol (Lond) 148:49 – 50P
Iggo A (1961) Non-myelinated afferent fibres from mammalian skeletal muscle. J Physiol (Lond) 155:52 – 53P
Iggo A (1969) Cutaneous thermoreceptors in primates and sub-primates. J Physiol (Lond) 200:403 – 430

Iggo A, Guilbaud G, Tégner R (1984) Sensory mechanisms in arthritic rat joints. In: Kruger L, Liebeskind JC (eds) Advances in pain research and therapy, vol 6. Raven, New York, pp 83–93
Iles JF (1976) Central terminations of muscle afferents on motoneurones in the cat spinal cord. J Physiol (Lond) 262:91–117
Iwamura Y, Kniffki K-D, Mizumura K, Wilberg K (1981) Responses of feline SI neurones to noxious stimulation of muscle and tendon. Pain [Suppl] 1:S213
Jessen C, Feistkorn G, Nagel A (1983) Temperature sensitivity of skeletal muscle in the conscious goat. J Appl Physiol 54:880–886
Joekes AM (1982) Cramp: a review. J R Soc Med 75:546–549
Johansson B (1962) Circulatory responses to stimulation of somatic afferents. Acta Physiol Scand [Suppl 198] 57:1–91
Jones EG, Friedman DP (1982) Projection pattern of functional components of thalamic ventrobasal complex on monkey somatosensory cortex. J Neurophysiol 48:521–544
Jones LA, Hunter IW (1983) Effect of fatigue on force sensation. Exp Neurol 81:640–650
Juan H (1978) Prostaglandins as modulators of pain. Gen Pharmacol 9:403–409
Juan H, Sametz W, Petronijevič S, Lembeck F (1984) Prostaglandin release and nociceptor stimulation by peptides. Naunyn-Schmiedeberg's Arch Pharmacol 326:64–68
Kalia M, Senapati JM, Parida B, Panda A (1972) Reflex increase in ventilation by muscle receptors with nonmedullated fibers (C fibers). J Appl Physiol 32:189–193
Kalia M, Mei SS, Kao FF (1981) Central projections from ergoreceptors (C fibers) in muscle involved in cardiopulmonary responses to static exercise. Circ Res [Suppl I] 48:48–62
Kao FF (1963) An experimental study of the pathway involved in exercise hyperpnoea employing cross-circulation techniques. In: Cunningham DJC, Lloyd BB (eds) The regulation of human respiration. Blackwell, Oxford, pp 461–502
Kao FF, Ray LH (1954a) Respiratory and circulatory responses of anesthetized dogs to induced muscular work. Am J Physiol 179:249–254
Kao FF, Ray LH (1954b) Regulation of cardiac output in anesthetized dogs during induced muscular work. Am J Physiol 179:255–260
Kao FF, Michel CC, Mei SS, Li WK (1963) Somatic afferent influence on respiration. Ann NY Acad Sci 109:696–711
Kaufman MP, Iwamoto GA, Longhurst JC, Mitchell JH (1982) Effects of capsaicin and bradykinin on afferent fibers with endings in skeletal muscle. Circ Res 50:133–139
Kaufman MP, Longhurst JC, Rybicki KJ, Wallach JH, Mitchell JH (1983) Effects of static muscular contraction on impulse activity of groups III and IV afferents in cats. J Appl Physiol 55:105–112
Kellgren JH (1938a) A preliminary account of referred pains arising from muscle. Br Med J 1:325–327
Kellgren JH (1938b) Observations of referred pain arising from muscle. Clin Sci 3:175–190
Kellgren JH, McGowan AJ (1948) On the behaviour of deep and cutaneous sensibility during nerve blocks. Clin Sci 7:1–11
Kellgren JH, McGowan AJ, Hughes ESR (1948) On deep hyperalgesia and cold pain. Clin Sci 7:14–27
Kenshalo DR, Isensee O (1983) Responses of primate SI cortical neurons to noxious stimuli. J Neurophysiol 50:1479–1496
Kenshalo DR, Giesler GJ, Leonard RB, Willis WD (1980) Responses of neurons in primate ventral posterior lateral nucleus to noxious stimuli. J Neurophysiol 43:1594–1614
Kirkwood PA, Sears TA (1974) Monosynaptic excitation of motoneurons from secondary endings of muscle spindles. Nature 252:243–244
Kniffki K-D, Mizumura K (1983) Responses of neurons in VPL and VPL-VL region of the cat to algesic stimulation of muscle and tendon. J Neurophysiol 49:649–661

Kniffki K-D, Mense S, Schmidt RF (1978) Responses of group IV afferent units from skeletal muscle to stretch, contraction and chemical stimulation. Exp Brain Res 31:511 – 522

Kniffki K-D, Schomburg ED, Steffens H (1979) Synaptic responses of lumbar α-motoneurons to chemical algesic stimulation of skeletal muscle in spinal cats. Brain Res 160:549 – 552

Kniffki K-D, Schomburg ED, Steffens H (1980) Action of muscular group III and IV afferents on spinal locomotor activity in cat. Brain Res 186:445 – 447

Kniffki K-D, Schomburg ED, Steffens H (1981) Synaptic effects from chemically activated fine muscle afferents upon α-motoneurones in decerebrate and spinal cats. Brain Res 206:361 – 370

Koerber HR, Brown PB (1980) Projections of two hindlimb cutaneous nerves to cat dorsal horn. J Neurophysiol 44:259 – 269

Koerber HR, Brown PB (1982) Somatotopic organization of hindlimb cutaneous nerve projections to cat dorsal horn. J Neurophysiol 48:481 – 489

Kohrman RM, Nolasco JB, Wiggers CJ (1947) Types of afferent fibers in the phrenic nerve. Am J Physiol 151:547 – 553

Koizumi K, Ushiyama J, McC Brooks C (1961) Muscle afferents and activity of respiratory neurons. Am J Physiol 200:679 – 684

Koizumi K, Sato A, Kaufman A, McC Brooks C (1968) Studies of sympathetic neuron discharges modified by central and peripheral excitation. Brain Res 11:212 – 224

Kruger L, Perl ER, Sedivec MJ (1981) Fine structure of myelinated mechanical nociceptor endings in cat hairy skin. J Comp Neurol 198:137 – 154

Kumazawa T, Mizumura K (1976) The polymodal C-fiber receptor in the muscle of the dog. Brain Res 101:589 – 593

Kumazawa T, Mizumura K (1977) Thin-fiber receptors responding to mechanical, chemical, and thermal stimulation in the skeletal muscle of the dog. J Physiol (Lond) 273:179 – 194

Kumazawa T, Tadaki E, Mizumura K, Kim K (1983) Post-stimulus facilitatory and inhibitory effects on respiration induced by chemical and electrical stimulation of thin-fiber muscular afferents in dogs. Neurosci Lett 35:283 – 287

Kuo DC, Nadelhaft I, Hisamitsu T, De Groat WC (1983) Segmental distribution and central projections of renal afferent fibers in the cat studied by transganglionic transport of horseradish peroxidase. J Comp Neurol 216:162 – 174

LaMotte RH, Thalhammer JG, Torebjörk HE, Robinson CJ (1982) Peripheral neural mechanisms of cutaneous hyperalgesia following mild injury by heat. J Neurosci 2:765 – 781

Lamour Y, Willer JC, Guilbaud G (1983) Rat somatosensory (SmI) cortex: I. Characteristics of neuronal responses to noxious stimulation and comparison with responses to non-noxious stimulation. Exp Brain Res 49:35 – 45

Lanari A, Muchnik S, Rey N, Semeniuk G (1973) Muscular cramp mechanism. Medicina 33:235 – 240

Landau BR, Akert K, Roberts TS (1962) Studies on the innervation of the diaphragm. J Comp Neurol 119:1 – 10

Landau W, Bishop GH (1953) Pain from dermal, periosteal, and fascial endings and from inflammation. Arch Neurol Psychiatry 69:490 – 504

Landgren S, Silfvenius H (1969) Projection to cerebral cortex of group I muscle afferents from the cat's hindlimb. J Physiol (Lond) 200:353 – 372

Langford LA (1983) Unmyelinated axon ratios in cat motor, cutaneous and articular nerves. Neurosci Lett 40:19 – 22

Langford LA, Schmidt RF (1983a) Afferent and efferent axons in the medial and posterior articular nerves of the cat. Anat Rec 206:71 – 78

Langford LA, Schmidt RF (1983b) An electron microscopic analysis of the left phrenic nerve in the rat. Anat Rec 205:207 – 213
Laporte Y, Bessou P, Bouisset S (1960) Action réflexe des différents types de fibres afférentes d'origine musculaire sur la pression sanguine. Arch Ital Biol 98:206 – 221
Laporte Y, Leitner L-M, Pagès B (1962) Absence d'effets réflexes circulatoires des fibres afférentes du groupe I. C R Soc Biol (Paris) 156:2130 – 2133
Layzer RB (1981) Leg muscle cramps. JAMA 245:2298
Layzer RB, Rowland LP (1971) Cramps. N Engl J Med 285:31 – 40
Lehmann-Willenbrock E, Mense S, Prabakhar NR (1984) Correlation between function and spinal projection in slowly conducting afferent units from deep tissues. Pflügers Arch [Suppl] 400:R16
Leitner L-M, Dejours P (1971) Reflex increase in ventilation induced by vibrations applied to the triceps surae muscles in the cat. Respir Physiol 12:199 – 204
Lembeck F, Popper H, Juan H (1976) Release of prostaglandins by bradykinin as an intrinsic mechanism of its algesic effect. Naunyn-Schmiedeberg's Arch Pharmacol 294:69 – 73
Levitt J, Levitt M (1968) Sensory hind-limb representation in SmI cortex of the cat. A unit analysis. Exp Neurol 22:259 – 275
Lewis T (1932) Pain in muscular ischemia. Arch Intern Med 49:713 – 727
Lewis T (1942) Pain. Macmillian, London. (Facsimilie edition, 1981)
Lewis T, Kellgren JH (1939 – 42) Observations relating to referred pain, visceromotor reflexes and other associated phenomena. Clin Sci 4:47 – 71
Lewis T, Pickering GW, Rothschild P (1931) Observations upon muscular pain in intermittent claudication. Heart 15:359 – 383
Light AR, Metz CB (1978) The morphology of the spinal cord efferent and afferent neurons contributing to the ventral roots of the cat. J Comp Neurol 179:501 – 515
Light AR, Perl ER (1979a) Reexamination of the dorsal root projection to the spinal dorsal horn including observations on the differential termination of coarse and fine fibers. J Comp Neurol 186:117 – 132
Light AR, Perl ER (1979b) Spinal termination of functionally identified primary afferent neurons with slowly conducting myelinated fibers. J Comp Neurol 186:133 – 150
Lind AR, McNicol GW (1967) Muscular factors which determine the cardiovascular responses to sustained and rhythmic exercise. Can Med Assoc J 96:706 – 713
Lind AR, Taylor SH, Humphreys PW, Kennelly BM, Donald KW (1964) The circulatory effects of sustained voluntary muscle contraction. Clin Sci 27:229 – 244
Lindahl O (1970) Experimental muscle pain produced by chemical stimulus. Acta Orthop Scand 40:741 – 750
Liu CT, Huggins RA, Hoff HE (1969) Mechanisms of intra-arterial K^+-induced cardiovascular and respiratory responses. Am J Physiol 217:969 – 973
Lloyd DPC (1943) Neuron patterns controlling transmission of ipsilateral hindlimb reflexes in cat. J Neurophysiol 6:293 – 315
Loewenstein WR (1956) Modulation of cutaneous mechanoreceptors by sympathetic stimulation. J Physiol (Lond) 132:40 – 60
Longhurst JC, Mitchell JH, Moore MB (1980) The spinal cord ventral root: an afferent pathway of the hind-limb pressor reflex in cats. J Physiol (Lond) 301:467 – 476
Ludbrook J (1983) Reflex control of blood pressure during exercise. Ann Rev Physiol 45:155 – 168
Lund JP, Matthews B (1981) Responses of temporomandibular joint afferents recorded in the gasserian ganglion of the rabbit to passive movements of the mandible. In: Kawamura Y, Dubner R (eds) Oral-facial sensory and motor functions. Quintessence Books, Tokyo, pp 153 – 160
Lundberg A (1979) Multisensory control of spinal reflex pathways. Prog Brain Res 50:11 – 28

Lundberg A, Oscarsson O (1961) Three ascending pathways in the dorsal part of the lateral funiculus. Acta Physiol Scand 51:1 – 16
Lundberg A, Malmgren K, Schomburg ED (1977) Comments on reflex actions evoked by electrical stimulation of group II muscle afferents. Brain Res 122:551 – 555
Maendly R, Rüegg DG, Wiesendanger M, Wiesendanger R, Lagowska J, Hess B (1981) Thalamic relay for group I muscle afferents of forelimb nerves in the monkey. J Neurophysiol 46:901 – 917
Maison GL (1939) Studies on the genesis of ischemic pain: the influence of the potassium, lactate and ammonium ions. Am J Physiol 127:315 – 321
Mallart A (1968) Thalamic projection of muscle nerve afferents in the cat. J Physiol (Lond) 194:337 – 353
Marfurt CF, Turner DF (1983) Sensory nerve endings in the rat oro-facial region labeled by the anterograde and transganglionic transport of horseradish peroxidase: a new method for tracing peripheral nerve fibers. Brain Res 260:1 – 12
Marshall J (1951) Sensory disturbances in cortical wounds with special reference to pain. J Neurol Neurosurg Psychiatry 14:187 – 204
Masumoto K (1934a) Histologische Studien über die peripheren Nerven des Zwerchfells. I. Mitteilung. Mitt Med Akad Kioto 10:1015 – 1018
Masumoto K (1934b) Histologische Studien über die peripheren Nerven des Zwerchfells. II. Mitteilung. Mitt Med Akad Kioto 11:1214 – 1217
Matsushita M, Tanami T (1983) Contralateral termination of primary afferent axons in the sacral and caudal segments of the cat, as studied by anterograde transport of horseradish peroxidase. J Comp Neurol 220:206 – 218
Matthews BHC (1931) The response of a single end organ. J Physiol (Lond) 71:64 – 110
Matthews BHC (1933) Nerve endings in mammalian muscle. J Physiol (Lond) 78:1 – 53
Matthews PBC (1969) Evidence that secondary as well as the primary endings of muscle spindles may be responsible for the tonic stretch reflex of the decerebrate cat. J Physiol (Lond) 204:365 – 393
Matthews PBC (1972) Mammalian muscle receptors and their central actions. Arnold, London
Matthews PBC (1982) Where does Sherrington's "muscular sense" originate? Muscles, joints, corollary discharges? Ann Rev Neurosci 5:189 – 218
Maunz RA, Pitts NG, Peterson BW (1978) Cat spinoreticular neurons: locations, responses and changes in responses during repetitive stimulation. Brain Res 148:365 – 379
McCloskey DI (1978) Kinaesthetic sensibility. Physiol Rev 58:763 – 820
McCloskey DI (1983) Proprioception and control of movement in man. IUPS Proc XV:464
McCloskey DI, Mitchell JH (1972) Reflex cardiovascular and respiratory responses originating in exercising muscle. J Physiol (Lond) 224:173 – 186
McCloskey DI, Streatfeild KA (1975) Muscular reflex stimuli to the cardiovascular system during isometric contractions of muscle groups of different mass. J Physiol (Lond) 250:431 – 441
McCloskey DI, Matthews PBC, Mitchell JH (1972) Absence of appreciable cardiovascular and respiratory responses to muscle vibration. J Appl Physiol 33:623 – 626
McIntyre AK (1974) Central actions of impulses in muscle afferent fibers. In: Hunt CC (ed) Muscle receptors. Springer, Berlin Heidelberg New York (Handbook of sensory physiology, vol 3/2)
McLachlan EM, Jänig W (1983) The cell bodies of origin of sympathetic and sensory axons in some skin and muscle nerves of the cat hindlimb. J Comp Neurol 214:115 – 130
McRitchie RJ, Vatner SF, Boettcher D, Heyndrickx GR, Patrick TA, Braunwald E (1976) Role of arterial baroreceptors in mediating cardiovascular response to exercise. Am J Physiol 230:85 – 89

Mehler WR (1969) Some neurological species differences – a posteriori. Ann NY Acad Sci 167:424–468
Menétrey D, Giesler GJ, Besson JM (1977) An analysis of response properties of spinal cord dorsal horn neurones to nonnoxious and noxious stimuli in the spinal rat. Exp Brain Res 27:15–33
Menétrey D, Chaouch A, Besson JM (1980) Location and properties of dorsal horn neurons at origin of spinoreticular tract in lumbar enlargement of the rat. J Neurophysiol 44:862–877
Mense S (1977a) Nervous outflow from skeletal muscle following chemical noxious stimulation. J Physiol (Lond) 267:75–88
Mense S (1977b) Muscular nociceptors. J Physiol (Paris) 73:233–240
Mense S (1978) Muskelreceptoren mit dünnen markhaltigen und marklosen afferenten Fasern: Receptive Eigenschaften und mögliche Funktion. Habilitation thesis, Kiel University
Mense S (1982) Reduction of the bradykinin-induced activation of feline group III and IV muscle receptors by acetylsalicylic acid. J Physiol (Lond) 326:269–283
Mense S, Meyer H (1985) Different types of slowly conducting afferent units in cat skeletal muscle and tendon. J Physiol (Lond) 363:403–417
Mense S, Schmidt RF (1974) Activation of group IV afferent units from muscle by algesic agents. Brain Res 72:305–310
Mense S, Stahnke M (1983) Responses in muscle afferent fibers of slow conduction velocity to contractions and ischemia in the cat. J Physiol (Lond) 342:383–397
Mense S, Light AR, Perl ER (1981) Spinal terminations of subcutaneous high-threshold mechanoreceptors. In: Brown AG, Réthelyi M (eds) Spinal cord sensation. Scottish Academic, Edinburgh, pp 79–86
Mense S, Meyer H, Prabakhar NR (1983) Response properties of nociceptors in deep tissues. Neurosci Lett [Suppl] 14:S246
Mense S, Craig AD, Lehmann-Willenbrock E, Meyer H (1985) Neurobiology of small-diameter afferent fibers from deep tissues. In: Rowe M, Willis WD (eds) Development, organization and processing in somatosensory pathways. Liss, New York, pp 299–308
Mesulam M-M (1978) Tetramethyl benzidine for horseradish peroxidase neurohistochemistry: a non-carcinogenic blue reaction-product with superior sensitivity for visualizing neural afferents and efferents. J Histochem Cytochem 26:106–117
Mesulam M-M, Brushart TM (1979) Transganglionic and anterograde transport of horseradish peroxidase across dorsal root ganglia: a tetramethylbenzidine method for tracing central sensory connections of muscles and peripheral nerves. Neuroscience 4:1107–1117
Mesulam M-M, Rosene DL (1979) Sensitivity in horseradish peroxidase neurohistochemistry: a comparative and quantitative study of nine methods. J Histochem Cytochem 27:763–773
Millar J (1975) Flexion-extension sensitivity of elbow joint afferents in cat. Exp Brain Res 24:209–214
Mitchell D, Hellon RF (1977) Neuronal and behavioural responses in rats during noxious stimulation of the tail. Proc R Soc Lond B 197:169–194
Mitchell JH, Schmidt RF (1983) Cardiovascular reflex control by afferent fibers from skeletal muscle receptors. In: Shepherd JT, Abboud FM (eds) Handbook of physiology, Section 2: The cardiovascular system, vol III: Peripheral circulation and organ blood flow, part 2. American Physiological Society, Bethesda, pp 623–658
Mitchell JH, Kaufman MP, Iwamoto GA (1983) The exercise pressor reflex: its cardiovascular effects, afferent mechanisms, and central pathways. Ann Rev Physiol 45:229–242
Mizumura K, Kumazawa T (1976) Reflex respiratory response induced by chemical stimulation of muscle afferents. Brain Res 109:402–406

Moberg E (1983) The role of cutaneous afferents in position sense, kinaesthesia, and motor function of the hand. Brain 106:1 – 19
Moore RM, Moore RE (1933) Studies on the pain-sensibility of arteries. I. Some observations on the pain-sensibility of arteries. Am J Physiol 104:259 – 266
Moore RM, Singleton AO (1933) Studies on the pain-sensibility of arteries. II. Peripheral paths of afferent neurones from the arteries of the extremities and of the abdominal viscera. Am J Physiol 104:267 – 275
Moore RM, Moore RE, Singleton AO (1934) Experiments on the chemical stimulation of pain-endings associated with small blood-vessels. Am J Physiol 107:594 – 602
Morgan DP, Kao F, Lim TPK, Grodins FS (1955) Temperature and respiratory responses in exercise. Am J Physiol 183:454 – 458
Morgan-Hughes JA (1979) Painful disorders of muscle. Br J Hosp Med 22:360 – 365
Mountcastle VB (1957) Modality and topographic properties of single neurons of cat's somatic sensory cortex. J Neurophysiol 20:408 – 434
Mountcastle VB, Powell TPS (1959) Central nervous mechanisms subserving position sense and kinesthesis. Bull Johns Hopkins Hosp 105:173 – 200
Mountcastle VB, Covian MR, Harrison CR (1950) The central representation of some forms of deep sensibility. Res Publ Assoc Res Nerv Ment Dis 30:339 – 370
Mountcastle VB, Poggio GF, Werner G (1963) The relation of thalamic cell response to peripheral stimuli varied over an intensive continuum. J Neurophysiol 26:807 – 834
Munson JB, Sypert GW, Zengel JE, Lofton SA, Fleshman JW (1982) Monosynaptic projections of individual spindle group II afferents to type-identified medial gastrocnemius motoneurons in the cat. J Neurophysiol 48:1164 – 1174
Nahin RL, Madsen AM, Giesler GJ (1983) Anatomical and physiological studies of the gray matter surrounding the spinal cord central canal. J Comp Neurol 220:321 – 335
Nathan PW (1980) Involvement of the sympathetic nervous system in pain. In: Kosterlitz HW, Terenius LY (eds) Pain and society (Dahlem Konferenzen). Chemie, Weinheim, pp 311 – 324
Norrsell V, Wolpow ER (1966) An evoked potential study of different pathways from the hind-limb to somatosensory areas in the cat. Acta Physiol Scand 66:19 – 33
Noth J, Thilmann A (1980) Autogenetic excitation of extensor γ-motoneurones by group II muscle afferents in the cat. Neurosci Lett 17:23 – 26
O'Connor BL (1984) The mechanoreceptor innervation of the posterior attachments of the lateral meniscus of the dog knee joint. J Anat 138:15 – 26
O'Connor BL, Gonzales J (1979) Mechanoreceptors of the medial collateral ligament of the cat knee joint. J Anat 129:719 – 729
O'Connor BL, Kunz B, Peterson RG (1982) The composition of the medial articular nerve to the knee in the dog. J Anat 135:139 – 145
O'Leary J, Heinbecker P, Bishop GH (1935) Analysis of function of a nerve to muscle. Am J Physiol 110:636 – 658
Oscarsson O, Rosén I (1963) Projection to cerebral cortex of large muscle spindle afferents in forelimb nerves of the cat. J Physiol (Lond) 169:924 – 945
Paintal AS (1954) The response of gastric stretch receptors and certain other abdominal and thoracic vagal receptors to some drugs. J Physiol (Lond) 126:271 – 285
Paintal AS (1959) Facilitation and depression of muscle stretch receptors by repetitive antidromic stimulation, adrenaline and asphyxia. J Physiol (Lond) 148:252 – 266
Paintal AS (1960) Functional analysis of group III afferent fibers of mammalian muscles. J Physiol (Lond) 152:250 – 270
Paintal AS (1961) Participation by pressure-pain receptors of mammalian muscles in the flexion reflex. J Physiol (Lond) 156:498 – 514
Paintal AS (1967) A comparison of the nerve impulses of mammalian non-medullated nerve fibers with those of the smallest diameter medullated fibers. J Physiol (Lond) 193:523 – 533

Parida B, Senapati JM, Kalia M (1969) Role of carotid body in hypernea due to stimulation of muscle receptors in the dog. J Appl Physiol 27:519–522
Passatore M, Filippi GM (1982) A dual effect of sympathetic nerve stimulation on jaw muscle spindles. J Auton Nerv Syst 6:347–361
Passatore M, Filippi GM (1983) Sympathetic modulation of periodontal mechanoreceptors. Arch Ital Biol 121:55–65
Paterson WD (1928) Circulatory and respiratory changes in response to muscular exercise in man. J Physiol (Lond) 66:323–345
Paul DH (1961) The effects of anoxia on the isolated rat phrenic nerve-diaphragm preparation. J Physiol (Lond) 155:358–371
Paulev P-E (1973) Cardiac rate and ventilatory volume rate reactions to a muscle contraction in man. J Appl Physiol 34:578–583
Penfield W, Boldrey E (1937) Somatic motor and sensory representation in the cerebral cortex of man as studied by electrical stimulation. Brain 60:389–443
Pérez-González JF, Coote JH (1972) Activity of muscle afferents and reflex circulatory responses to exercise. Am J Physiol 223:138–143
Perl ER (1984) Pain and nociception. In: Darian-Smith (ed) Handbook of physiology, Section 1, The nervous system, vol III, part 2. American Physiological Society, Bethesda, pp 915–975
Perl ER, Whitlock DG (1961) Somatic stimuli exciting spinothalamic projections to thalamic neurons in cat and monkey. Exp Neurol 3:256–296
Perl ER, Kumazawa T, Lynn B, Kenins P (1976) Sensitization of high threshold receptors with unmyelinated (C) afferent fibers. Prog Brain Res 43:263–276
Peschanski M, Lee CL, Ralston HJ (1984) The structural organization of the ventrobasal complex of the rat as revealed by the analysis of physiologically characterized neurons injected intracellularly with horseradish peroxidase. Brain Res 297:63–74
Phillips CG, Powell TPS, Wiesendanger M (1971) Projection from low-threshold muscle afferents of hand and forearm to area 3a of baboon's cortex. J Physiol (Lond) 217:419–446
Pickering GW, Wayne EJ (1933–34) Observations on angina pectoris and intermittent claudication in anaemia. Clin Sci 1:305–325
Piper PJ (1984) Formation and actions of leukotrienes. Physiol Rev 64:744–761
Poggio GF, Mountcastle VB (1963) The functional properties of ventrobasal thalamic neurons studied in unanesthetized monkeys. J Neurophysiol 26:775–806
Pollock LJ, Davis L (1935) Visceral and referred pain. Arch Neurol Psychiatry 24:1041–1054
Pomeranz B, Wall PD, Weber WV (1968) Cord cells responding to fine myelinated afferents from viscera, muscle and skin. J Physiol (Lond) 199:511–532
Proshansky E, Egger MD (1977) Staining of the dorsal root projection to the cat's dorsal horn by anterograde movement of horseradish peroxidase. Neurosci Lett 5:103–110
Ralston HJ, Ralston DD (1979) The distribution of dorsal root axons in laminae I, II and III of the macaque spinal cord: A quantitative electron microscope study. J Comp Neurol 184:643–684
Ranson SW, Davenport HK (1931) Sensory unmyelinated fibers in the spinal nerves. Am J Anat 48:331–353
Réthelyi M, Szentágothai J (1973) Distribution and connections of afferent fibres in the spinal cord. In: Iggo A (ed) Somatosensory system. Springer, Berlin Heidelberg New York (Handbook of sensory physiology, vol 2)
Réthelyi M, Trevino DL, Perl ER (1979) Distribution of primary afferent fibers within the sacrococcygeal dorsal horn: an autoradiographic study. J Comp Neurol 185:603–622
Réthelyi M, Light AR, Perl ER (1982) Synaptic complexes formed by functionally defined primary afferent units with fine myelinated fibers. J Comp Neurol 207:381–393

Rexed B (1952) The cytoarchitectonic organization of the spinal cord in the cat. J Comp Neurol 96:415–496
Rexed B, Therman P-O (1948) Calibre spectra of motor and sensory nerve fibers to flexor and extensor muscles. J Neurophysiol 11:133–139
Richmond FJR, Anstee GCB, Sherwin EA, Abrahams VC (1976) Motor and sensory fibers of neck muscle nerves in the cat. Can J Physiol Pharmacol 54:294–304
Rodbard S (1975) Pain associated with muscular activity. Am Heart J 90:84–92
Roland PE, Ladegaard-Pedersen H (1977) A quantitative analysis of sensations of tension and of kinaesthesia in man. Evidence for a peripherally originating muscular sense and for a sense of effort. Brain 100:671–692
Rowell LB (1974) Human cardiovascular adjustments to exercise and thermal stress. Physiol Rev 54:75–159
Rowell LB, Hermansen L, Blackmon JR (1976) Human cardiovascular and respiratory responses to graded muscle ischemia. J Appl Physiol 41:693–701
Sahlin K, Harris RC, Nylind B, Hultman E (1976) Lactate content and pH in muscle samples obtained after dynamic exercise. Pflügers Arch 367:143–149
Sahlin K, Palmskog G, Hultman E (1978) Adenine nucleotide and IMP contents of the quadriceps muscle in man after exercise. Pflügers Arch 374:193–198
Samuel EP (1952) The autonomic and somatic innervation of the articular capsule. Anat Rec 113:53–70
Sato A, Schmidt RF (1966) Muscle and cutaneous afferents evoking sympathetic reflexes. Brain Res 2:399–401
Sato A, Schmidt RF (1973) Somatosympathetic reflexes: afferent fibers, central pathways, discharge characteristics. Physiol Rev 53:916–947
Sato A, Kaufman A, Koizumi K, McC Brooks C (1969) Afferent nerve groups and sympathetic reflex pathways. Brain Res 14:575–587
Sato A, Sato Y, Schmidt RF (1981) Heart rate changes reflecting modifications of efferent cardiac sympathetic outflow by cutaneous and muscle afferent volleys. J Auton Nerv Syst 4:231–247
Sato A, Sato Y, Schmidt RF (1982) Changes in heart rate and blood pressure upon injection of algesic substances into skeletal muscle. Pflügers Arch 393:31–36
Sato Y, Schaible H-G, Schmidt RF (1983) Types of afferents from the knee joint evoking sympathetic reflexes in cat inferior cardiac nerves. Neurosci Lett 39:71–75
Scadding JW (1981) Development of ongoing activity, mechanosensitivity, and adrenaline sensitivity in severed peripheral nerve axons. Exp Neurol 73:345–364
Schaible H-G, Schmidt RF (1983a) Activation of groups III and IV sensory units in medial articular nerve by local mechanical stimulation of knee joint. J Neurophysiol 49:35–44
Schaible H-G, Schmidt RF (1983b) Responses of fine medial articular nerve afferents to passive movements of knee joint. J Neurophysiol 49:1118–1126
Schibye B, Mitchell JH, Payne FC, Saltin B (1981) Blood pressure and heart rate response to static exercise in relation to electromyographic activity and force development. Acta Physiol Scand 113:61–66
Schmidt A (1916) Zur Pathologie und Therapie des Muskelrheumatismus (Myalgie). Münch Med Wochenschr 63:593–595
Schmidt RF, Schönfuss K (1970) An analysis of the reflex activity in the cervical sympathetic trunk induced by myelinated somatic afferents. Pflügers Arch 314:175–198
Schmidt RF, Kniffki K-D, Schomburg ED (1981) Der Einfluß kleinkalibriger Muskelafferenzen auf den Muskeltonus. In: Bauer HJ, Koella WP, Struppler A (eds) Therapie der Spastik. Verlag für angewandte Wissenschaften, Munich, pp 71–86
Schumann H-J (1972) Überlastungsnekrosen der Skelettmuskulatur nach experimentellem Laufzwang. Zentralbl Allg Pathol 116:181–190
Sell R, Erdelyi A, Schäfer H (1958) Untersuchungen über den Einfluß peripherer Nervenreizung auf die sympathische Aktivität. Pflügers Arch 267:566–581

Seltzer Z, Devor M (1979) Ephaptic transmission in chronically damaged peripheral nerves. Neurology 29:1061 – 1064
Senapati JM (1966) Effect of stimulation of muscle afferents on ventilation of dogs. J Appl Physiol 21:242 – 246
Sherrington CS (1894) On the anatomical constitution of nerves of skeletal muscles; with remarks on recurrent fibers in the ventral spinal nerve-root. J Physiol (Lond) 17:211 – 258
Sherrington CS (1910) Flexion-reflex of the limb, crossed extension-reflex, and reflex stepping and standing. J Physiol (Lond) 40:28 – 121
Shyu BC, Andersson SA, Thorén P (1982) Endorphin mediated increase in pain threshold induced by long-lasting exercise in rats. Life Sci 10:833 – 840
Shyu BC, Andersson SA, Thorén P (1985) Spontaneous running in wheels. A computer assisted method for measuring physiological parameters during exercise in rodents. Acta Physiol Scand: in press
Sicuteri F, Franchi G, Fanciullacci M (1964) Bradichinina e dolore da ischemia. Settim Med 52:127 – 139
Simons DG (1975) Muscle pain syndromes. Part I. Am J Physical Med 54:289 – 311
Simons DG (1976) Muscle pain syndromes. Part II. Am J Physical Med 55:15 – 42
Simons DG, Travell J (1981) Myofascial trigger points, a possible explanation. Pain 10:106 – 109
Sjölund B, Terenius L, Eriksson M (1977) Increased cerebrospinal fluid levels of endorphins after electro-acupuncture. Acta Physiol Scand 100:382 – 384
Skoglund S (1956) Anatomical and physiological studies of knee joint innervation in the cat. Acta Physiol Scand [Suppl 124] 36:1 – 101
Skoglund S (1973) Joint receptors and kinaesthesis. In: Iggo A (ed) Somatosensory system. Springer, Berlin Heidelberg New York (Handbook of sensory physiology, vol 2)
Smith CL (1983) The development and postnatal organization of primary afferent projections to the rat thoracic spinal cord. J Comp Neurol 220:29 – 43
Snyder RL (1982) Light and electron microscopic autoradiographic study of the dorsal root projections to the cat dorsal horn. Neuroscience 7:1417 – 1437
Sprague JM, Ha H (1964) The terminal fields of dorsal root fibers in the lumbosacral spinal cord of the cat, and the dendritic organization of the motor nuclei. Prog Brain Res 11:120 – 154
Stacey MJ (1969) Free nerve endings in skeletal muscle of the cat. J Anat 105:231 – 254
Staton WM (1951) New approach to muscle soreness. Athletic J (Chicago) 31:24 – 61
Stauffer EK, Watt DGD, Taylor A, Reinking RM, Stuart DG (1976) Analysis of muscle receptor connections by spike-triggered averaging. 2. Spindle Group II afferents. J Neurophysiol 39:1393 – 1402
Staunton HP, Taylor SH, Donald KW (1964) The effect of vascular occlusion on the pressor response to static muscular work. Clin Sci 27:283 – 291
Steffen RP, McKenzie JE, Haddy FJ (1982) The possible role of acetate in exercise hyperemia in dog skeletal muscle. Pflügers Arch 392:315 – 321
Stegemann J (1963) Zum Mechanismus der Pulsfrequenzeinstellung durch den Stoffwechsel. I – IV. Pflügers Arch 276:481 – 524
Stegemann J, Kenner T (1971) A theory on heart rate control by muscular metabolic receptors. Arch Kreislaufforsch 64:185 – 214
Stilwell DL (1957a) The innervation of tendons and aponeuroses. Am J Anat 100:289 – 317
Stilwell DL (1957b) The innervation of deep structures of the foot. Am J Anat 101:59 – 74
Stilwell DL (1957c) The innervation of deep structures of the hand. Am J Anat 101:75 – 100

Taira N, Nakayama K, Hashimoto K (1968) Vocalization response of puppies to intra-arterial administration of bradykinin and other algesic agents, and mode of actions of blocking agents. Tohoku J Exp Med 96:365 – 377

Tallarida G, Baldoni F, Peruzzi G, Brindisi F, Raimondi G, Sangiorgi M (1979) Cardiovascular and respiratory chemoreflexes from the hindlimb sensory receptors evoked by intra-arterial injection of bradykinin and other chemical agents in the rabbit. J Pharmacol Exp Ther 208:319 – 329

Tallarida G, Baldoni F, Peruzzi G, Raimondi G, Massaro M, Abate A, Sangiorgi M (1983) Different patterns of respiratory reflexes originating in exercising muscle. J Appl Physiol 55:84 – 91

Tan Ü (1983) Excitatory and inhibitory effects of repetitive stimulation of group I and II extensor afferents on homonymous motoneurones. Arch Ital Biol 121:167 – 186

Thompson FJ, Barnes CD (1969) Evidence for thermosensitive receptors in the femoral vein. Fed Proc 28/1:722

Thompson FJ, Barnes CD, Wald JR (1982) Interactions between femoral venous afferents and lumbar spinal reflex pathways. J Auton Nerv Syst 6:113 – 126

Tibes U (1977) Reflex inputs to the cardiovascular and respiratory centers from dynamically working canine muscles: some evidence for involvement of group III or IV nerve fibers. Circ Res 41:332 – 341

Tiegs OW (1953) Innervation of voluntary muscle. Physiol Rev 33:90 – 144

Tominaga S, Curnish RR, Belardinelli L, Rubio R, Berne RM (1980) Adenosine release during early and sustained exercise of canine skeletal muscle. Am J Physiol 238: H156 – H163

Torebjörk HE, Ochoa JL, Schady W (1984a) Referred pain from intraneural stimulation of muscle fascicles in the median nerve. Pain 18:145 – 156

Torebjörk HE, Schady W, Ochoa J (1984b) Sensory correlates of somatic afferent fiber activation. Hum Neurobiol 3:15 – 20

Tracey DJ (1979) Characteristics of wrist joint receptors in the cat. Exp Brain Res 34:165 – 176

Travell JG, Simons DG (1983) Myofascial pain and dysfunction. The trigger point manual. Williams and Wilkins, Baltimore

Trevino DL, Maunz RA, Bryan RN, Willis WD (1972) Location of cells of origin of the spinothalamic tract in the lumbar enlargement of cat. Exp Neurol 34:64 – 77

Turner DF, Marfurt CF (1983) Electron microscopic demonstration of horseradish peroxidase-tetramethylbenzidine reaction product as a method for identifying sensory nerve fibers in the rat tooth pulp. Neurosci Lett 41:213 – 218

Wagman IH, Price DD (1969) Responses of dorsal horn cells of M. mulatta to cutaneous and sural nerve A and C fiber stimuli. J Neurophysiol 32:803 – 817

Wallin G, Torebjörk E, Hallin R (1976) Preliminary observations on the pathophysiology of hyperalgesia in the causalgic pain syndrome. In: Zotterman Y (ed) Sensory functions of the skin in primates. Pergamon, Oxford, pp 489 – 502

Webb-Peploe MM, Brender D, Shepherd JT (1972) Vascular responses to stimulation of receptors in muscle by capsaicin. Am J Physiol 222:189 – 195

Webster ME, Skinner NS, Powell WJ (1967) Role of the kinins in vasodilatation of skeletal muscle of the dog. Am J Physiol 212:553 – 558

Weiner IH, Weiner HL (1980) Nocturnal leg muscle cramps. JAMA 244:2332 – 2333

Westrum LE, Canfield RC, Black RG (1976) Transganglionic degeneration in the spinal trigeminal nucleus following removal of tooth pulps in adult cats. Brain Res 101:137 – 140

Whipp BJ (1983) Ventilatory control during exercise in humans. Ann Rev Physiol 45:393 – 413

Wiesenfeld-Hallin Z, Hallin RG (1984) The influence of the sympathetic system on mechanoreception and nociception. A review. Hum Neurobiol 3:41 – 46

Wietoska B, Böning D (1979) Was ist eigentlich Muskelkater? – Gesichertes und Ungesichertes in der medizinischen Literatur. Dtsch Z Sportmed 30:395 – 401
Wildenthal K, Mierzwiak DS, Skinner NS, Mitchell JH (1968) Potassium-induced cardiovascular and ventilatory reflexes from the dog hindlimb. Am J Physiol 215:542 – 548
Willis WD, Trevino DL, Coulter JD, Maunz RA (1974) Responses of primate spinothalamic tract neurons to natural stimulation of hindlimb. J Neurophysiol 37:358 – 372
Willis WD, Leonard RB, Kenshalo DR (1978) Spinothalamic tract neurons in the substantia gelatinosa. Science 202:986 – 988
Willis WD, Kenshalo DR, Leonard RB (1979) The cells of origin of the primate spinothalamic tract. J Comp Neurol 188:543 – 574
Woodworth RS, Sherrington CS (1904) A pseudaffective reflex and its spinal path. J Physiol (Lond) 31:234 – 243
Woollard HH, Roberts JEH, Carmichael EA (1932) An inquiry into referred pain. Lancet 2:337 – 338
Yao T, Andersson S, Thorén P (1982) Long-lasting cardiovascular depressor response following sciatic stimulation in spontaneously hypertensive rats. Evidence for the involvement of central endorphin and serotonin systems. Brain Res 244:295 – 303
Young EW, Sparks HV (1980) Prostaglandins and exercise hyperemia of dog skeletal muscle. Am J Physiol 238:H191 – 195
Zotterman Y (1936) Specific action potentials in the lingual nerve of cat. Skand Arch Physiol 75:105 – 119

Subject Index

MIX
Papier aus verantwortungsvollen Quellen
Paper from responsible sources
FSC® C105338

If you have any concerns about our products,
you can contact us on
ProductSafety@springernature.com

In case Publisher is established outside the EU,
the EU authorized representative is:
Springer Nature Customer Service Center GmbH
Europaplatz 3, 69115 Heidelberg, Germany

Printed by Libri Plureos GmbH
in Hamburg, Germany